ECOLOGICAL CHANGE
IN THE UPLANDS

ECOLOGICAL CHANGE IN THE UPLANDS

SPECIAL PUBLICATION NUMBER 7 OF THE
BRITISH ECOLOGICAL SOCIETY

EDITED BY

M. B. USHER
Department of Biology
University of York
York YO1 5DD

D. B. A. THOMPSON
Chief Scientist Directorate
Nature Conservancy Council
12 Hope Terrace
Edinburgh EH9 2AS

BLACKWELL SCIENTIFIC PUBLICATIONS

OXFORD LONDON EDINBURGH

BOSTON PALO ALTO MELBOURNE

1988

© 1988 by the British Ecological Society
and published for them by
Blackwell Scientific Publications
Editorial offices:
Osney Mead, Oxford OX2 0EL
  (*Orders*: Tel. 0865 240201)
8 John Street, London WC1N 2ES
23 Ainslie Place, Edinburgh EH3 6AJ
3 Cambridge Center, Suite 208
  Cambridge, Massachusetts 02142, USA
667 Lytton Avenue, Palo Alto
  California 94301, USA
107 Barry Street, Carlton
  Victoria 3053, Australia

First published 1988

Set by Times Graphics, Singapore
Printed in Great Britain at the University Press, Cambridge

DISTRIBUTORS

USA and Canada
  Blackwell Scientific Publications Inc
  PO Box 50009, Palo Alto
  California 94303
  (*Orders*: Tel. (415) 965-4081)

Canada
  Oxford University Press
  70 Wynford Drive
  Don Mills
  Ontario M3C 1J9
  (*Orders*: Tel. (416) 441-2941)

Australia
  Blackwell Scientific Publications
  (Australia) Pty Ltd
  107 Barry Street
  Carlton, Victoria 3053
  (*Orders*: Tel. (03) 347 0300)

British Library
Cataloguing in Publication Data

Ecological change in the uplands.
  – (Special publication number 7 of the
  British Ecological Society)
  1. England. Upland regions. Vegetation.
  Change. Ecological aspects
  I. Usher, M.B. II. Thompson, D.B.A.
  III. Series
  581.5′264

  ISBN 0-632-02293-0

Library of Congress
Cataloguing-in-Publication Data

Ecological change in the uplands.

  (Special publication number 7 of the British
Ecological Society)
  Includes indexes.
  1. Ecology—Great Britain. I. Usher, Michael B.,
1941–   . II. Thompson, D. B. A., 1958–   .
III. Series: Special publication of the British
Ecological Society; no. 7.
  QH137.E36 1988   574.5′264′0941   88-5072
  ISBN 0-632-02293-0

Contents

Contents

Land-use changes and their impacts

Management of ecological change

Preface

This volume contains 32 of the papers presented at a joint meeting of the Ecological Affairs Committee of the British Ecological Society and the Nature Conservancy Council. The meeting, held in Edinburgh from 22 to 24 September 1987, attracted nearly 250 delegates, virtually all from the UK due to the meeting's focus on the British uplands. We welcomed as our guest H. E. Professor Olav Gjaerevoll, Norway's first Minister of the Environment. His paper in this volume sets out the Norwegian approach to conservation, contrasting a country with relatively little agricultural development (not more than 3 per cent of the area growing agricultural crops) with the UK, where over 70 per cent of the land area is in agricultural usage.

Although there has been a long-standing interest in the uplands of Britain, the Edinburgh meeting was planned during a time of mounting interest in the British upland scene. To some extent this interest was stimulated in 1984 by the Countryside Commission's uplands debate in England and Wales (*A Better Future for the Uplands*. Countryside Commission, Cheltenham). It has, however, continued to be highlighted by various environmental issues of widespread public concern. There have been the Nature Conservancy Council's 1986 report (*Nature Conservation and Afforestation in Britain*. Nature Conservancy Council, Peterborough), the House of Lords report on the EEC forestry industry, and the National Audit Office's recent report on economic aspects of tree planting in the UK. Agriculture has also been at the forefront of discussion. The EEC Common Agricultural Policy and the Agricultural and Island Development Programmes in the Scottish Highlands and Islands have major influences on the uplands. Up to one million hectares of agricultural ground may go out of production over the next decade, with the alternatives being forestry, small-scale industries, many forms of tourism and recreation, and less intensive forms of traditional farming. The Agriculture Bill received the royal assent on 25 July 1986; section 18 of the Act enables Ministers to designate areas of high landscape or nature conservation value as Environmentally Sensitive Areas. The House of Lords' report on research at the agriculture/environment interface highlighted the needs for further integrated research.

The efforts to conserve common land in England and Wales have increased as semi-natural ground elsewhere has diminished. The Common Land Forum, initiated by the Countryside Commission in 1983 and involving many other interested parties, has passed four measures so as to maintain the attractiveness of the commons. With impending de-commonization, there is particular interest in the likely consequences for wildlife of increased recreational use. In Scotland ski developments continue to dominate the recreational scene (see paper by Bayfield, Watson & Miller).

The widespread environmental effects of acidification probably receive more publicity abroad than in this country, but this is likely to change with the impending reports from the Nature Conservancy Council and the Forestry Commission on the impacts of acid rain on trees. Several studies still in progress are revealing some very dramatic impacts of acidic deposition on both terrestrial (see papers by Woodin and by Lee, Tallis & Woodin) and aquatic (see paper by Gee & Stoner) ecosystems in the uplands.

Against this background our aim in planning the meeting about the uplands was to link together as many facets of pure and applied ecological research as was possible within the space of two-and-a-half days. Thus, as well as looking at the upland environment and the changes which have taken, and are taking, place, we wanted to incorporate aspects of forestry, agriculture, recreation, acid pollution and wildlife conservation. During the meeting 27 papers were read, including the opening address by the President of the British Ecological Society (Professor C. H. Gimingham) and a 'summing up' address by the Chairman of the Nature Conservancy Council's Committee for Scotland (Mr A. R. Trotter). In addition to the verbal presentations, there were 36 poster papers displayed for all or part of the meeting; although many more were offered, space availability precluded more being shown. This book contains most of the papers that were presented verbally, together with a few papers that are based on the poster presentations.

The need to integrate the various ecological interests in the uplands has been one of our considerations in designing this volume. Although the meeting itself was divided into themes, starting with 'the upland environment', then 'ecological impacts: agriculture and land use', 'ecological impacts: forestry', 'other ecological impacts' and finishing with 'management and conservation', it was apparent that many of the papers transcended such rigid categories. This publication is, therefore, only loosely divided into sections. The first dozen papers are primarily concerned with describing upland environments and investigating the processes that cause long-term changes. These are followed by a collection of papers that broadly address land-use change and the various impacts associated with these changes. The volume is concluded with nine papers that are management orientated; with the interests of the Nature Conservancy Council and the British Ecological Society, these mostly focus attention on conservation management.

In his introductory paper, Gimingham addresses the question 'What is going on in the uplands?' The answer is clearly 'a lot', with the papers in this publication indicating both the diversity of subjects studied and the variety of approaches adopted. As well as discussing what is happening, Ratcliffe & Thompson highlight what they consider to be the significant features of the British uplands. These include the unique plant communities, the extensive development of blanket bogs, phytogeographical novelties and the breeding bird assemblages, together with the oceanic and widespread anthropogenic influences.

Acidic deposition, as a major perturbation, is only now coming to light. Lee, Tallis & Woodin document the increasing acidification of the Pennines since the industrial revolution, and Woodin discusses the implications of this for wildlife conservation. Grace & Unsworth, in their discussion of upland climate and microclimate, describe the mechanisms for acidic deposition on vegetated surfaces. However, to understand these changes, and the time scales over which they operate, they must be set into the whole of the upland scene. The soils, vegetation and fauna of the uplands are described by G. Hudson, Miles and Coulson, respectively. As well as acidic deposition, which is a relatively unperceived perturbation of the uplands that has been operating gradually over the last two centuries or so, public recreation (Bayfield, Watson & Miller) is a more directly observable perturbation of the last two or three decades. These two very different perturbations indicate how important it is to think of the time scales of ecological processes in the uplands.

A variety of factors, operating at distinctly different time scales, have shaped the British uplands. These scales may vary from thousands of years on a geological scale (Birks) to one or two hundred years in successional studies (Hester; Miles) or to just a decade or two in terms of management activities (Usher & Gardner). At whatever time scale a process is studied, the upland communities are dynamic, responding to changes in climate, successional age or management activity. It is the degree of predictability of these changes, or, viewed the other way, the amount of stochasticity in them, that indicates how much is understood about the ecology of the uplands.

Change is often perceived but much less often recorded in detail. It is this detailed recording of an individual farm that makes McAdam & Chance's study so valuable; indeed, the amount of change during a period that has seen a virtual agricultural revolution is quite remarkable. The effects of agricultural change on bird populations (Bibby) highlight what ecologists have known for a long time; the effects of land-use change are felt far outside the territorial boundaries of that change. Planning for agricultural change is the theme of Smith's paper, which looks at the new concept of an Environmentally Sensitive Area, concentrating particularly on the Pennine Dales ESA. The policy issues behind these changes are discussed by Mowle & Bell.

Afforestation is arguably the most controversial form of land-use change in Britain in the 1980s, though many of the arguments tend to be more emotional or aesthetic (the 'seried ranks of conifers' type of statement) than scientific. Two papers in particular attempt to assess the effects of afforestation in a cool, scientific manner. Gee & Stoner are mainly concerned with acidic deposition deriving from forests (some of the mechanisms for this are discussed by Grace & Unsworth) and its effects on water quality. Thompson, Stroud & Pienkowski are specifically interested in the effects of new forests on bird populations. The loss of ground for breeding, or the loss of feeding habitat, is a clearly demonstrable effect of afforestation; what is much less clear are the interactions across the forest boundary between the forest animal community and the animal community in unplanted areas. Once forests are established a variety of ecological influences change. Watt & Leather investigate a pest species, the pine beauty moth, in young lodgepole pine plantations, and Stevens *et al.* investigate the effects of management activities on water quality and site nutrient status. However, Price reminds us not to take ourselves too seriously; economists are really no better than ecologists either at agreeing amongst themselves or at building predictive models of forests!

In relating ecological research to management, the important factors to determine are the management objectives. Definition of these is difficult, especially as there is neither guidance nor agreement on whether the objectives should be formulated 'on high' (i.e. within the EEC or by national governments) or whether they should be discussed and agreed at a more regional or local level. The paper describing the conservation framework in Norway (Gjaerevoll) is interesting in that the conservation objectives have been clearly formed by the national government, though the processes are regionalized or localized in the lengthy discussions and consultations on the national plans. Norway, of course, has a rich natural environment with the National Parks planned to cover 12 per cent of the country. This harmonious duality of the Norwegian approach certainly seems to be lacking in both Wales and New Zealand (Dixon).

For management purposes, one must be able to answer questions such as 'How much of a resource is there?' or 'How has the amount of that resource changed?' The paper by Bunce & Barr demonstrates a clear methodology for looking at amounts of different land uses in the uplands. More technical studies, utilizing satellite imagery but tying it in with ground surveys, are those of Jones & Wyatt and of Williams. On a less universal scale, Hobbs focuses attention on one rare type of upland heath, that in which the leafy liverworts are abundant, and describes a survey to ascertain both the extent and the distribution of such heaths, and how these have changed.

There are innumerable interrelationships between plants and animals in the uplands. If one considers the larger grazing animals, such as deer (Albon & Clutton-Brock) or sheep (Grant & Maxwell), their grazing has a clearly defined impact on the upland vegetation. The smaller animals, such as grouse (P. Hudson) or invertebrates (Usher & Gardner; Coulson), usually have less spectacular impacts on the vegetation, though for the former the *Calluna*-dominated moors are burnt in regular rotations and the latter may occasionally defoliate the heathland (e.g. the heather beetle). If one considers the plants, then their management depends greatly on the influence of herbivores (Sydes & Miller), and indeed the eco-physiological reactions of plants may be regulated, at least to some extent, by herbivore grazing (Torvell, Common & Grant).

This publication contains, either in the papers or in the references, a lot of information about the British uplands. One of the purposes of a symposium meeting should be to synthesize the data and information which currently exist. With rapid changes in agricultural structures, there is no time for further research relating to decisions that have to be made now or within the next few months. The synthesis of our present knowledge must be used to answer questions which are being asked now. Monitoring or research has a three- to five-year lead in period; this can be used to check that the decisions made today are appropriate, or to provide data which could be used subsequently to modify these decisions, or indeed even to provide new information in areas which may become more important in the future. If only ecologists and land managers had a crystal ball to show them what decisions would need to be taken in four or five years' time, the planning of research now would be so much easier!

The meeting did identify some research priorities, however. As first foreshadowed by Gimingham, but repeated throughout, there is still a compelling need for two types of monitoring. First, in the long term, monitoring would establish climatic, hydrological, physiographic and other trends, as well as the prevalence of some of the more contemporary pressures, such as pollution. Second, in the short term, we need a better understanding of ecological processes on which to base predictions about further change. Miles, Grant & Maxwell and Thompson *et al.* have indicated the need for such research on soils, vegetation, and birds, respectively. As well as fundamental research contributing to the science base, we also need more synthetic approaches. At least two of these are important. First, a better dialogue between the many research institutes and sponsors would hopefully lead to a more goal-orientated form of research. Second, desk studies are needed to review the literature, synthesize the present state of scientific knowledge and indicate the way ahead. It was a feature of the Edinburgh meeting that participants started to think laterally, becoming aware that other interest groups had already addressed similar problems in upland research or management. Conservationists face a mammoth task, for there are urgent problems in identifying priorities for management when objectives differ between individual land managers, to say nothing of differences between the major land users. International cooperation can certainly help, and a broader perspective is clearly needed of both the land-use changes and the wildlife or landscape features that are affected.

A good scientific meeting should, slightly like the crystal ball for predicting research needs, be doing at least two things. First, it should raise new questions in the minds of the participants and, second, it should stimulate new research. It is still far too soon to know if any new research or investigations have been stimulated, but we do know that participants left Edinburgh with a keener and wider interest in the British upland scene. As editors, we hope that this volume will also increase general interest in the uplands, as well as raising questions in the minds of its readers. All papers published here, whether they were presented verbally or as posters at the meeting, have been refereed. They were available before the meeting finished, and indeed the majority had been refereed by at least two anonymous referees prior to the meeting. Authors were allowed a few weeks after the meeting to modify their papers in the light of the referees' and editors' comments, and in the light of discussion and comments that were received from other participants during the meeting. We should like to thank the large number of referees for the speed and efficiency with which they tackled their task; their help has made the editorial task much easier.

We should also like to thank other people who helped us in a variety of ways with the organization of the meeting, especially the officers of the British Ecological Society and the Nature Conservancy Council for their financial support. Edinburgh University provided excellent conference facilities and accommodation; we are grateful to Bill Campbell and Linda Crease for their efforts on the University's behalf. We should particularly like to thank our secretarial staff, Susan Whyte, Marianne Lean and Sheena Gardner (in Edinburgh) and Mags Wetherill and Andrea Johnson (in York), who have handled the bookings and correspondence with both participants and authors. We are especially indebted to Susan Whyte for help in many ways before, during and after the meeting. David Horsfield, Alison Hobbs, Alan Brown, John Halliday and Hector Galbraith were indispensable throughout the conference.

DESMOND B.A. THOMPSON
MICHAEL B. USHER
November 1987

Introduction: Ecological change in the uplands

C. H. GIMINGHAM

*Department of Plant Science, University of Aberdeen,
St Machar Drive, Aberdeen AB9 2UD*

WHAT IS GOING ON IN THE UPLANDS?

It gives me great pleasure to introduce this Symposium on Ecological Change in the Uplands. The importance of the theme and its relevance to current problems are such as to make its development a matter of urgency. This is something of which the British Ecological Society is very much aware, and I therefore welcome the opportunity to express the Society's appreciation of the initiative taken by the organizers and of the partnership between the Nature Conservancy Council and the Society's Ecological Affairs Committee in setting up this joint meeting. It is also particularly appropriate that this meeting falls within the European Year of the Environment and forms part of the British Ecological Society's contribution to the EYE programme. The facts that several organizations have devoted recent meetings and publications to this subject, and that funds are being allocated to various research efforts bearing on 'The Uplands', give some indication of the importance attached to the theme in both Government and scientific circles. It is especially timely to review current thinking in this field through the medium of a symposium.

The uplands, substantial in area though sparsely populated, are an important and complex component of the British countryside. They have for many generations provided a livelihood for rural communities which, although much depleted today, are still significant in terms of both culture and the national economy. In addition, they attract visitors in numbers far greater than the resident populations, although the tourists tend to congregate in relatively few, highly popular locations. Over the greater part of their area, vegetation and fauna have been strongly influenced and modified by mankind, yet that influence is generally much less than in the lowlands. Larger tracts of country survive in at least a semi-natural state in the uplands than elsewhere, and they provide important refuges for wildlife. This has been the situation in the uplands for a long time: profound changes have taken place in the past, but in the main they have been gradual. In places it is already evident that the rate of change is increasing, and the signs are that this trend will continue and become even more dramatic in the near future, as new factors come into play. It is in this context that we have to be sure that we know the answers to the question 'What is going on in the uplands?' Without these answers there will be no way of controlling the new and rapid changes, and much of value in the uplands, in terms both of economics and of conservation, may be lost.

The question is far from simple and, in introducing the Symposium, I suggest that it is necessary to consider it on several levels. First, it is important that we should be aware of long-term ecological processes which give rise to continuing change and create a dynamic backdrop against which other influences must be viewed. Second, there are the many changes which have been initiated by mankind in the past, some of them setting in motion long chains of consequences, others apparently leading to a degree of stability. Third, it is necessary to recognize new human impacts which may cause change in directions quite different from those of the past. Finally, we need also to relate the question 'What is going on?' to action which should be taken, through research and monitoring, to detect and understand these various types of change.

LONG-TERM PROCESSES

To say that no ecological system is entirely stable, and that all are undergoing some form of change, is a truism. However, there is a tendency to regard upland systems, particularly those of the higher altitudes, as more stable than most. While there may be some truth in this, it would be a mistake to ignore the various continuing, long-term processes of change. Unless these are recognized, the exact nature of anthropogenic changes cannot be properly assessed. Quaternary climatic trends and fluctuations are now relatively well understood, and as far as the uplands are concerned their effects, for example on the natural altitudinal tree limit, have been documented for certain localities. Less is known about the past fluctuations in the altitudinal ranges of individual species or communities. What is even more important is to be aware of any such trends taking place, even if almost imperceptibly, at the present time. It may be that the disappearance of a species from a particular locality, or an increase in the population of another, is in some instances attributable to this kind of cause rather than to human influence.

Another long-term aspect of change is that of weathering and erosion. Here it may be particularly difficult to disentangle the effects of continuing natural processes from those due directly or indirectly to human activity, including, for example, fire and grazing by domestic animals or deer (the population levels of which are much influenced by man). In the same way, continued interactions between climate, soil and vegetation, leading to processes such as acidification and podzolization, must be thoroughly understood if we are to gauge the effects of recent developments such as atmospheric pollution (including the acidification of rain-water resulting from industrial emissions).

ANTHROPOGENIC CHANGES DATING FROM THE PAST

The far-reaching effects of human occupation over the centuries have to be examined within this context of long-term 'natural' trends. The greatest cause of change affecting all but the highest altitudes has probably been deforestation. Apart from small, much modified, surviving fragments of native forest, the British uplands have been almost completely denuded of woodland. Beginning probably in late Neolithic or Bronze Age times, and continuing with varying intensity until the nineteenth century, this process was the outcome of needs not only for timber for house building, boat building and fuel, but also for open grazing lands. Furthermore, as numbers of livestock and deer increased, they inhibited tree regeneration and accelerated the change. The more recent advent of game shooting did nothing to reverse the loss of native forest.

Loss of the forests undoubtedly led to shifts in local climate and in edaphic conditions, superimposed on the long-term processes already mentioned. In some areas this may have facilitated peat formation and in others led to erosion. Throughout the potential forest zone we therefore now have semi-natural systems, determined largely by the soil parent material. Where this is acid and oligotrophic, heath, moor and wet, acid grasslands result; where the nutrient status is higher, better quality grasslands occur. The composition of these communities depends very largely on the grazing regime, and in many cases they are far from stable. The sensitive balance between them and their reactions to changes in management have been the subject of numerous studies, and will be further considered in this Symposium. However, the possibility that some of the changes may be irreversible, or partly so, should not be overlooked. Some of the semi-natural grasslands, once they have become almost pure stands of *Molinia caerulea* or *Nardus stricta*, show signs of relative stability over rather long periods. It may be important to discover the extent to which they will or will not undergo change if management alters or ceases, and what kind of manipulations will be needed if change is desired.

NEW OR RECENT IMPACTS

This is the kind of knowledge which is required if we are to account for present-day communities, which are the resultant not only of past influences but also of contemporary factors. It is the essential basis for recognition of what is new in the picture, whether the novelty lies in altered rates of change or entirely fresh influences. Some of these are already upon us, others can only be guessed at. In either case, their consequences may be difficult to predict, but this is what both ecologists and conservationists need to be able to do, if only within certain levels of probability.

Among the new developments which are already in evidence is the massive reintroduction of trees to the uplands, to which we are devoting a whole section of the Symposium. It is not possible to predict ecological changes arising from recent afforestation by treating it as a return to the type of ecosystem in existence prior to the impact of man. Most of the new forests are and will be different in terms of species composition, structure, nutrient cycling, hydrology and many other features. They are creating new and distinct local environmental conditions as regards the light climate and atmosphere within the canopy, as well as soil factors. Some of these effects spread out from the margins of forests to an unknown extent. Silvicultural practices prior to planting may affect the drainage patterns and nutrient status of quite extensive areas. The cycles of planting, thinning, felling and second rotation have major effects upon the populations of native plant and animal species in the vicinity of the forest. Also, the size and pattern of the plantation blocks introduce ecological as well as landscape effects on a broader scale.

The various aspects of land use in the uplands, including forestry, agriculture and other activities, are closely interrelated. The future of sheep farming, for instance, will depend to a large extent on the level of European Community subsidies and the perceived viability of other concerns, notably forestry. Cattle have declined in numbers in the uplands for some time now, and there is evidence that where sheep are the only domestic grazer the botanical composition of the vegetation is affected. Agricultural surpluses in Europe have not as yet involved sheep products but, if there should be an increase in lowland sheep farming to compensate for reductions in arable land, then the uplands may experience a corresponding decline. Some of the consequences of this may be predictable, but if certain areas go out of sheep production altogether it is not easy, on the basis of existing knowledge, to forecast either the nature or the rate of the resulting vegetation change. Other changes may follow from the implementation of a 'set-aside' policy in respect of arable land. If upland farms are affected, the abandonment of fields may lead locally to significant changes in vegetation. Similarly, the suggested de-commonization in England and Wales may initiate change.

Other aspects of management may also alter, among them moor burning. While grouse moor management and sheep farming in heather areas still demand regular burning, this would be discontinued should interest in either of these activities wane. As it is, the expansion of forestry has already led to a reduction in systematic burning in the neighbourhood of plantations because of the risks involved. Where heather burning is relaxed, the stands become degenerate and open to invasion by grasses and bracken or, if seed parents are present, shrubs and trees.

Another possibility is the introduction of entirely new grazing animals. Already there is growing interest in goats for cashmere production. Goat grazing could have major effects on the vegetation: if carefully controlled these may be beneficial to farmers, for example by eliminating rushes (*Juncus* spp.) from damp pastures, but uncontrolled goat grazing in our uplands would be damaging.

While the changes introduced by afforestation and altered grazing patterns may be widespread, other influences new to the uplands are more local, though intensive. The popularity of upland areas for various kinds of recreation has had profound effects where pressures are greatest. These have been the subject of a considerable volume of research, but it is still difficult to achieve any generalized assessment of the magnitude of the changes induced, or of the best ways to control or reverse them. Even more difficult, though very necessary, is the task of quantifying changes caused by atmospheric pollution. The methods of historical ecology are beginning to point to changes in community composition, such as the decline in amounts of *Sphagnum* in the uplands of northern England, which are contemporaneous with industrial development. The Scottish uplands have seemed more remote from such influences, but recent monitoring shows unequivocally that they are not exempt. Yet we still do not know much about what is going on as a result.

THE NEED FOR ACTION

The Symposium will be taking up many of these questions, and offering at least partial answers to some of them. But one question remains: what action is being, or should be, taken? Certain practical steps are already well in hand, such as the designation of areas for protection: National Parks, National Scenic Areas, Nature Reserves, Forest Parks, Sites of Special Scientific Interest, Environmentally Sensitive Areas and others. Valuable as these are, they cannot resist changes the causes of which originate beyond their boundaries, sometimes at a considerable distance. Furthermore, protection of these special areas involves decisions about whether or not they should be preserved indefinitely in their present condition, and the amount of change to be permitted. If they are managed systems, management has to be continued and even then, unless the ecological processes are fully understood, unknown changes may be occurring.

What else is needed? In the first place, monitoring of two main kinds is required. On the one hand, monitoring of the environment is necessary to establish long-term climatic, physiographic and other trends, as well as to detect contemporary events such as pollution. On the other hand, biological monitoring is needed, on several scales, to provide the facts about change on which predictions can be based. Considerable effort is already going into both, and some of it will be reported in the Symposium. On a broad scale, monitoring of countryside features and major vegetation categories is being carried out on a sampling basis by the Countryside Commissions, the Institute of Terrestrial Ecology, the Nature Conservancy Council and others, but it is perhaps on the smaller scale of detailed changes in community composition that the monitoring effort is still inadequate.

In the second place, there is scope for increased support of strategic and applied research, especially at the systems level. Agricultural and forestry research organizations have a long tradition of work in this area, which is currently being extended into new fields such as agro-forestry and new types of grazing management. What may be profitable in the future is experimental research on an estate or regional scale, when the changes that follow from new types of land use (or new ways of integrating existing uses) may be examined and models constructed from which new approaches to management may arise.

Finally, there may soon be improved opportunities for taking the initiative in promoting ecological change in the uplands, rather than merely reacting to changes dictated by economic and social pressures. This will be highlighted towards the end of the Symposium, but I think it is fair to say that thinking along these lines is still only in its early stages. Ecological change of a successional nature is inevitable if management is withdrawn, and in some instances this may be regarded as desirable from the viewpoint of nature conservation. However, we need to know how this will affect habitats and populations of plants and animals in the countryside generally, and whether there may be other sorts of system which it is desirable to create in the interests of environmental or wildlife conservation.

This introduction has served, I hope, to underline the complexity of the subject. Ecological change in the uplands is a compound of changes taking place at several levels. These need to be distinguished and assigned to their various causes before judgements can be made about specific impacts or the ecological implications of particular kinds of land use. The papers making up the Symposium will contribute significantly to this analysis, and eventually to a synthesis in which ecological change in the uplands can be recognized for what it is, what has caused it, and how it can be controlled and channelled into desired pathways, whatever these may be.

UPLAND ENVIRONMENTS
AND LONG-TERM CHANGES

The British uplands: their ecological character and international significance

D. A. RATCLIFFE AND D. B. A. THOMPSON*

*Chief Scientist Directorate, Nature Conservancy Council,
Northminster House, Peterborough PE1 1UA
and *Chief Scientist Directorate, Nature Conservancy Council,
12 Hope Terrace, Edinburgh EH9 2AS*

SUMMARY

1 The uplands cover almost a third of Britain. The vegetation is mainly sub-montane, anthropogenic dwarf-shrub heath and grassland. Remaining natural, climatic 'climax' vegetation is predominantly montane heath or extensive ombrogenous blanket bog.

2 Regional differences are related principally to increasing oceanicity from east to west, and decreasing temperature from south to north. Local variations reflect influences of topography, geology and both past and present land use.

3 The oceanic climate at the Atlantic edge of the European continent, mountainous terrain and widespread anthropogenic impacts (massive forest clearance, burning, grazing and recent acidic deposition) have produced a unique landscape with an exceptional range of podzolic soils and blanket bogs.

4 Many of the plant communities do not occur anywhere else. The ericaceous dwarf shrubland, *Ulex* scrub and anthropogenic grasslands are extensive; the large bryophyte-rich and pteridophyte-abundant communities have localized world distributions. The phytogeographical interest includes (i) blending of many climatic elements, (ii) many ecotypes (or unique species mixtures) and (iii) several species with disjunct world distributions.

5 The breeding bird assemblage is an unusual mixture of species (e.g. boreal-arctic peatland and montane communities). At least five species have main strongholds or breed at relatively high densities in upland Britain. The zoogeographical interest lies in (i) the extreme southern or western distribution of at least 14 species, (ii) disjunct species ranges and (iii) racial differentiation in at least four species.

6 The relative depletion of some predatory vertebrates (especially mammals) and the scarcity of 'alien' species (particularly invertebrates) are significant.

INTRODUCTION

The uplands, embracing the hills, moors and mountains, form the largest extent of undeveloped wildlife habitat remaining in Britain. They are typically above the limits of enclosed farmland and are composed predominantly of dwarf shrub heaths, grasslands and peat bogs (Pearsall 1950; Ratcliffe 1977). Their area, *c.* 6·5 million ha, is almost 30 per cent of Britain (Table 1).

TABLE 1. Extent of upland ground in Britain. Natural or semi-natural 'rough grazing' amounts to approximately 66,000 km² (29·1 per cent of Britain's land surface). Figures differ quite considerably between different sources (e.g. Ball, Radford & Williams 1983; Forestry Commission 1987). Out of the total below, between 60,000 and 70,000 km² are composed of open grasslands, heaths and bogs, making up Britain's typical hill, moor and mountain resource. The tabulated figures have been adapted from Ball *et al.* (1983)

Altitude (m)	Main land type	% Britain's surface	Number of km² Britain	Number of km² Scotland	Number of km² England	Number of km² Wales
123–244	Marginal agricultural ground	23·9	54,324	19,944	28,869	5,511
245–610	Hill pasture and moorland	20·8	47,315	27,030	12,363	7,922
611–914	Mountain range	2·3	5,263	4,645	394	224
>915	High mountains	0·2	402	394	2	6

Although relatively small in area and altitudinal range compared with the great alpine ranges of the world, the British uplands contain a great variety of landscapes and dependent biotic communities. These arise from marked differences in climate, geology, topography, soils and past land use, and complex interactions between these factors are such that no two upland districts are alike. The first half of this paper reviews the relevant aspects of the underlying pattern of variation; the second half highlights the upland features which are rare or unusual on a global scale, or especially well represented in Britain, being those with special international significance.

PRIMARY SOURCES OF VARIATION

Post-glacial climatic change

During the height of the last glaciation (50,000–15,000 BP) the present North Sea was ice, and Britain and the rest of Europe formed a continuous land mass. Although Britain and Ireland, and perhaps even the Western Isles and mainland Scotland, were not joined there may have been narrow connections (Godwin 1975). Unglaciated south-east Britain held a transition from permanent ice through tundra and steppe to open scrub. Land bridges allowed subsequent migrations by plants and animals, but sea-level rises ended further immigrations, thus causing a limitation to species diversity. Some native races of species began to diverge as a consequence (Yalden 1982; Ratcliffe 1989), but isolation has been far too short for a significant degree of endemism to develop in the British biota. There is a relatively large variety of lower plants, perhaps because these are dispersed by wind-borne spores (Birks 1976) whereas plants with large fruits/ seeds are often less readily dispersed. The immensely richer flora of the Alps re- flects not only freedom of post-glacial migration across a continuous land mass, but also a far greater extent of suitable habitat persisting in recent times. The invertebrates are also influenced fundamentally by climate, with Britain's complement of butterflies (Rhopalocera) and dragonflies (Odonata) being relatively poor, especially in the north where there are few sunshine hours.

An ameliorating climate since the last glaciation combined with widespread human influence has further impoverished or severely modified the upland flora and fauna. The climate has become milder since deglaciation around 10000 years BP, and in the uplands was at its warmest in the period 8000–9000 BP (Birks, this symposium) and not the early Atlantic period, 7000–6000 BP, as widely believed (Godwin 1975). From around 4000 BP the climate became more oceanic (Birks 1986, 1988). The influence of man on vegetation dates at least from 5000 BP (Turner 1965) and became significant from around 3900–3000 BP (Birks 1988). Ensuing deforestation, burning and agricultural intensification in the uplands have produced mainly sheepwalk, deer forest and grouse moor with notable anthropogenic vegetation (Tansley 1939; Pearsall 1950; Ratcliffe 1977). 'Natural' vegetation is found only on the least productive ground in certain areas of the montane zone (the lower boundary corresponding with the climatic tree line — one of the lowest in highland Europe), in the wettest bogs and in inaccessible situations such as cliffs or lake islands. Many areas of the remaining uplands have 'semi-natural' vegetation (*sensu* Tansley 1939), being dominated by native plants which, although much changed in abundance, have been present since early Holocene times.

The British upland climate and major gradients

Britain is subjected to prevailing westerly winds from the Atlantic. The resulting oceanic climate of the British mountains is unusual elsewhere in the world. Its main features are small seasonal variation in cloudiness, precipitation and temperature; high atmospheric humidity and precipitation (mostly falling as rain); low insolation; and high winds (Manley 1952). While there is a gradient towards more continental conditions in an easterly direction, annual variation in monthly mean temperature remains small; local maxima of 12 °C occur in the Scottish Highlands and Southern Uplands, far smaller than those found abroad in more continental countries (Barry 1981; Grace & Unsworth 1988). The 'growing season', taken by agronomists as the period over which average daily mean temperature is above a given threshold for growth (6 °C for grass), decreases northwards and with altitude. The altitudinal lapse rate (the average temperature decrease with height) is relatively steep, the standard for mean temperature being 1 °C per 150 m.

Altitude, local climate and effects on vegetation pattern

Altitudinal deterioration in climate is matched by parallel change in vegetation and dependent animals, reflecting a response to the fall in temperature and increase in windspeed, rainfall and cloud cover. The resulting sequence of life zones has been greatly modified in Britain by the widespread impact of human activity over the uplands. From the extensive occurrence of some similar vegetation types and the fragmentary presence of others, it seems a fair inference that the altitudinal zonation in the British mountains was closely comparable to that observable today in the mountains of extreme south-west Norway, e.g. Sogn, the continental region of greatest climatic and biogeographical similarity to the Scottish mountains. It is noteworthy that such altitudinal zonation is a small-scale and local representation of the massive latitudinal zonation of vegetation formations which occurs across northern Eurasia from the temperate to the polar regions.

Rainfall increases with altitude but tends to be heaviest towards the leeward of high peaks. Cloud cover increases and sunlight diminishes at greater heights; north-west to east facing slopes receive least direct sun. There can be considerable variations in microclimate, depending principally on aspect and shelter. Windiness can rise dramatically in terms of average speed and gustiness, although topography rather than altitude *per se* is a major influence on velocity (Barry 1981). The combination of wind and rain produces a chill factor important in the biology of some upland vertebrates (Thompson, Thompson & Nethersole-Thompson 1986). The fraction of annual precipitation falling as snow, number of days with snowfall and duration of snow cover also increase with altitude (Manley 1945, 1952; Barry 1981). Although snow lie limits the length of growing season for plants, it protects underlying vegetation from extreme frosts, and even provides shelter for birds such as ptarmigan (*Lagopus mutus*) and red grouse (*L. lagopus scoticus*) and for some small mammals. Again, topography exerts a substantial influence on snow lie and melt, and on unprotected plateaux much snow may be blown clear and the ground subjected to cryoturbation and wind action, producing many unusual surface and geomorphological features such as solifluction hummocks, vegetation stripes and ridges, terraces, sorted polygons, summit tors and a range of boulder fields and scree slopes (Ratcliffe 1977; Thompson, Galbraith & Horsfield 1987). Other factors being similar, the extent of snow gathering grounds is important in determining the degree of late snow bed development in sheltered hollows below; large plateaux produce much more late snow than sharp peaks and narrow ridges (McVean & Ratcliffe 1962).

Vegetation zones under natural conditions

In Britain, the lower mountain slopes were covered naturally by forest, except where the ground was too rocky or wet. In the south this was usually of oak (*Quercus petraea*), alone or mixed with other broadleaves. More typically farther north, there was Scots pine (*Pinus sylvestris*) with a good deal of birch (*Betula pubescens* and *B. pendula*) on more fertile soils. The difference corresponds to the latitudinal change from cool temperate broadleaf (nemoral) to boreal coniferous and birch forest (taiga). With increasing altitude under natural conditions there is a decline in stature of the trees until they occur only as tall shrubs and then disappear. Above this in Norway is a zone of medium shrubs, mostly willows (*Salix* spp.), low juniper (*Juniperus communis*) and taller forms of dwarf birch (*Betula nana*). In Scandinavia, the upper zone of natural birchwoods and medium shrubs is distinguished as the sub-alpine zone. On top of this are successive zones of dwarf shrubs (low-alpine), high level grasslands, moss and lichen heaths, and fellfields (middle-alpine). On the highest mountains there is a final zone of stone deserts and permanent ice and snow (high-alpine).

This climatic sequence is complicated at all altitudes by the topographically determined influence of varying wetness of ground and length of snow cover. Wet ground supports a range of vegetation types from ombrogenous and soligenous bogs to flushes, springs and rills. Depending on the shelter/exposure balance, there is a range of plant communities showing the effects of increasing snow cover, to the point where snow fields are permanent and vegetation non-existent.

British sub-montane and montane zones

The sub-alpine zone of birchwood and medium shrubs has very largely been eradicated in Britain. Patches of birchwood and juniper scrub survive but are mostly not the climax type, and willow scrub exists only on cliff ledges. The natural tree-line can be seen only as a few fragments, mainly in the Cairngorms (Nethersole-Thompson & Watson 1981), and the upper limits of woodland are mostly well below the natural potential limit. Whilst in Britain the potential upper limit of tree growth forms a useful boundary between sub-alpine and alpine types of vegetation, we prefer to use the terms 'sub-montane' and 'montane', the former including all vegetation derived from forest above the limits of enclosed farmland, and the latter applying to everything lying above the potential tree-line. Anthropogenic effects can extend well into the montane zone, however, disturbing the original pattern of vegetation there (Thompson *et al.* 1987).

Geographical gradients

There is a gradual descent of the life zones towards the north-west and north of Scotland; this results from the markedly decreasing mean temperature northwards and increasing windspeed and oceanicity westwards and towards the coast (Fig. 1a). The potential tree line occurs at around 650–700 m in the Cairngorms and 300 m in north-west Sutherland and is almost non-existent in Orkney and Shetland. The lower limit of prostrate dwarf *Calluna*, marking the bottom of the montane zone, occurs at between 700 and 800 m in the Cairngorms and central Grampians, 550 m in Sutherland, 350 m in farthest north-west Sutherland and only 200–300 m in Orkney and Shetland (McVean & Ratcliffe 1962; Thompson *et al.* 1987). Higher up, barren fellfield areas typically occur above 900 m in the central Highlands but at only 450 m on Ronas Hill, Shetland. Average snow lie increases northwards, but higher winter temperature gives rather limited or more intermittent snow cover in the west.

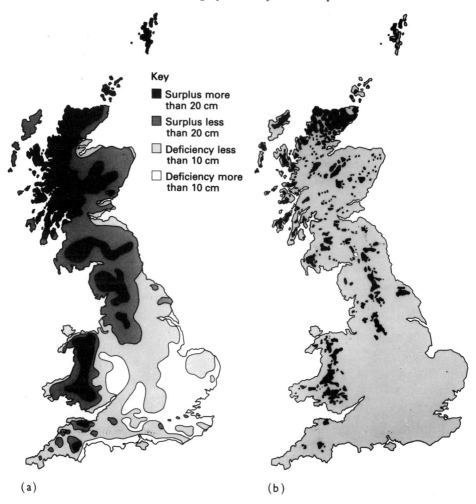

FIG. 1. (a) Oceanicity gradient across Britain giving difference between precipitation and evapo-transpiration during April–September. (b) Distribution of substantial blanket bogs (coloured dark) in Britain (both from Stroud *et al.* 1987). There is a surplus if precipitation exceeds evapo-transpiration and a deficiency if evapo-transpiration exceeds precipitation.

The extent of blanket bog increases in a westerly and northerly direction as climate becomes wetter and cooler (Fig. 1b). Bryophyte-rich communities also increase towards the west and north (Fig. 2a, c). Lichen heaths are favoured by relatively continental conditions in the east (Fig. 2b, d), but have some anomalous coastal outliers.

Fig. 3 gives an example of how relative oceanicity (and topography) can influence vegetation. Across the six sample sites combined there is an increase towards the east in the percentage cover of *Calluna–Eriophorum vaginatum* bog and sub-montane *Calluna* heath, but an increase towards the west in the cover of

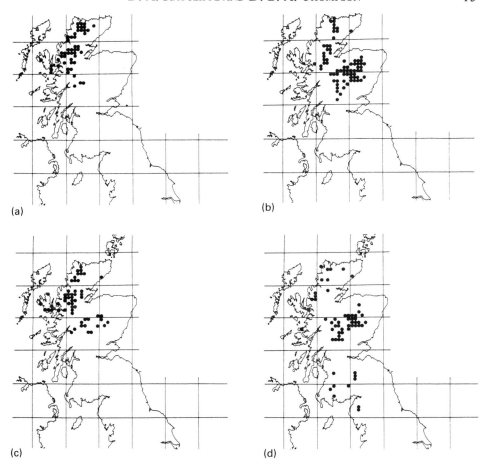

FIG. 2. British distribution of four upland vegetation types (from Rodwell 1987; McVean & Ratcliffe 1962; Ratcliffe 1977). (a) *Calluna vulgaris–Racomitrium lanuginosum* (from 150 to 800 m); (b) *C. vulgaris–Cladonia arbuscula* (from 600 to 1000 m); (c) *Vaccinium myrtillus–R. lanuginosum* (from 400 to 1000 m); (d) *V. myrtillus–C. arbuscula* (from 600 to 1000 m). All occur on rankers and podzols, but (a) and (c) receive over 2000 mm rain or more than 220 wet days year^{-1}, whereas (b) and (d) receive 1000–1200 mm rain or 160 wet days, and less than 1600 mm rain or 180–200 wet days, respectively. Notice more southerly extent of *Vaccinium*, due possibly to greater tolerance of higher temperatures and to overgrazing effects resulting in the loss of *C. vulgaris*.

Scirpus cespitosus–Calluna, Molinia caerulea–Calluna and *Scirpus–Myrica gale* vegetation. On Ben Wyvis there is less *Calluna* heath than expected because some of the sub-montane ground has been afforested, and the topography is more suited to *Calluna–Eriophorum* bog development. Some exemplary clines in oceanic to relatively continental vegetation types are found across Beinn Eighe–Fionn Bheinn–Fannichs–Wyvis (Ross); Reay Forest/Foinaven–Ben Hope–Ben Loyal–Morven and Scaraben (Sutherland and Caithness); Glenfinnan/

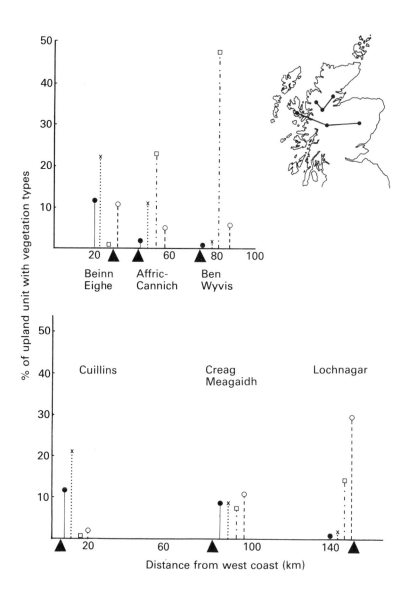

FIG. 3. The proportion of six upland units composed of four vegetation types (Birks & Ratcliffe 1980) in relation to distance eastwards from the west coast. *Scirpus caespitosa–Myrica gale* bog (●); *S. caespitosa–Calluna vulgaris/Molinia caerulea–C. vulgaris* bog (X); *C. vulgaris–Eriophorum vaginatum* bog (□); and sub-montane *C. vulgaris* heaths (○). The altitude ranges are considerable, e.g. for sub-montane *C. vulgaris* 200–730 m (Cuillin), 600–900 m (Creag Meagaidh) and 740–1100 m (Lochnagar). Both oceanicity gradients and topography have partially determined the extent of these typically western or eastern upland vegetation types.

Lochaber range (western Inverness-shire)–Perthshire/Drumochter range–Cairngorms (as far north-east as the Ladder Hills); and western Pennines–North York Moors (northern England).

Variations in geology and soils

Geology, especially hardness of rock and extent of glaciation, mediates these climatic influences through its effect on topography. Towards the west of Britain many mountain massifs are very rugged and steep. Snowdonia, Lakeland and much of the western Highlands have heavily glaciated mountains, whereas the Pennines and Cheviots are lower and more gently contoured with large plateaux and broad valleys. The highest mountains are concentrated in the Scottish Highlands, and those of most massive form are clustered towards the centre. The greatest extent of montane and especially late snow bed vegetation occurs on the high mountains of the central Highlands, rather than in the extreme north. Conversely, the topography which favours most blanket bog development occurs extensively in the far north-east of Scotland, and not the west. In addition to effects on vegetation, the species diversity and abundance of birds are greater on more gently contoured areas (e.g. compare the Pennines and Cheviots with Lakeland and Snowdonia for waders).

The cool, wet climate has caused the development of prevailing podzolic soils by leaching of exchangeable base ions and reduced chemical weathering, and has inhibited humus decomposition. Soil types range from skeletal brown semi-podzols on steeper ground, through free-draining podzols, to ombrogenous peat on permanently waterlogged, flatter sites, or various types of gleys where water-logging is seasonal or shows lateral movement (Pearsall 1950; Curtis, Courtney & Trudgill 1976). Soil erosion is often marked and varies from truncated profiles to complete denudation. The rather shallow and raw soils of the montane zone are distinctly unstable, and solifluction, gravitational movement and erosion are fairly widespread, because of both climatic severity and human influence (Miles 1985, 1988; Birks 1986).

A depauperate flora and a predominance of calcifuge species are characteristic of the British uplands, where calcium is the key base ion. Calcareous rocks are extremely localized. However, flushing produces noticeable local nutrient enhancement effects, and, where there is base-richness (especially deriving from calcareous parent materials), there is a considerable increase in diversity and productivity of the flora and fauna. Good examples are found on Ordovician pumice-tuff in Snowdonia, the Carboniferous limestone in the Pennines, Dalradian limestones and calcareous schists in Perthshire, Argyll and Angus, and Durness dolomitic limestone in the north-western Highlands. The fairly basic Old Red Sandstone of Scotland and Wales, the Moine Series of the central Highlands, and the Ordovician and Silurian greywackes and shales in the Scottish Southern Uplands are also locally quite productive for variety and biomass of plants and animals. Hard, siliceous rocks such as quartzite, millstone grit,

Torridonian sandstone, granite, rhyolite and granophyre produce the most acidic and some of our least productive soils, and in the unimproved state these have a low carrying capacity for herbivores.

SECONDARY LAND-USE VARIATION AND INFLUENCE ON THE UPLANDS

Land use has exerted a profound but varying, comparatively recent and secondary effect on the upland biota. The climatic and geological patterns have contributed to this by influencing upland carrying capacity and therefore the timing, location and intensity of change. Other factors affecting the pattern of human colonization of the uplands have nevertheless been involved. According to Birks (this symposium) the principal impact, extensive deforestation, has occurred in four spatio-temporal phases, the first dating from around 3900 BP (north-western Highlands and eastern Skye) while the fourth is as recent as 300–400 BP (Grampians and Cairngorms).

Britain's total forest cover is amongst the smallest in Europe, although the present rate of afforestation is the greatest. Taken as a whole, Britain now has more of its surface devoted to agriculture than is found in any other major country for which statistics are available (Table 2).

Predominantly grazing and burning impacts on vegetation

Vegetation changes

Upland areas cleared of forest became dominated by dwarf-shrub heaths and small areas of grassland, with dwarf-shrub heaths then losing ground to acidic grassland where they were grazed and/or burned intensively (McVean & Ratcliffe 1962; Table 3). Grass-dominated sheepwalk has become widespread (e.g. Wales, Lakeland, Pennines) and in some areas tends to prevail on the richer soils (e.g. Southern Uplands, Breadalbane Hills). Some uplands, typically lying towards the east, have remained heather-dominated and are managed as grouse moor (e.g. north-east and south-east Scotland, north-east England). Dating from *c.* 1840, and reaching its peak *c.* 1900, grouse management has involved rotational burning, control of sheep or deer numbers, and predator destruction. Many of the higher and more rugged upland ranges of the Highlands and Islands are managed as deer forest, and consist of grassland and dwarf-shrub heath mixtures.

Soil fertility is important in the conversion of dwarf-shrub heath to grassland because richer areas have been grazed preferentially and for longer periods, and have higher stocking rates. Areas of basic soils thus show the greatest modification of previous vegetation; tall herbaceous vegetation is confined to cliff ledges or other ungrazed situations, calcicolous or herb-rich dwarf-shrub heaths are relatively scarce, and willow (*Salix* spp.) scrub is vestigial (see Ratcliffe 1977). Many montane plateaux south of the Highlands appear to have been modified

TABLE 2. International comparisons of land use and extent of change in forest cover

Country	Area (million hectares)[1]	% land devoted to Agriculture	Forestry	Net change in woodland cover 1950–1970 (thousand hectares)[2]	Change as % of country area[2]
USSR	2,227·5	27	42	–	–
Canada	916·7	7	48	–	–
USA	912·5	60	33	–	–
France*	54·3	60	28	+454	0·83
Sweden	41·1	8	68	–	–
Japan	37·1	15	68	–	–
Norway	30·8	3	28	–	–
Finland	30·5	10	76	–	–
Italy*	30·1	63	27	+453	1·50
W. Germany*	24·3	58	30	+56	0·23
Great Britain*	22·7	77	10	+590	2·60
Spain*	19·9	41	25	–	–
Greece*	12·9	30	45	–	–
Portugal*	8·6	56	35	–	–
Ireland*	6·9	68	6	–	–
Denmark*	4·2	69	11	+40	0·95
Netherlands*	3·4	72	10	+29	0·85
Belgium/Luxembourg*	3·3	49	23	+13/0	0·39/0

*Denotes member country of European Economic Community.
[1]Data from Forestry Commission (1984), who give original sources.
[2]Data calculated from Peterken (1981), who gives original sources. Net change refers to the difference between areas afforested and cleared.

from largely moss to grass-dominated heaths and this appears to be due to prolonged and intensive sheep grazing (Thompson *et al.* 1987). The montane and sub-montane complement of vascular plants in the southern mountains has also been depleted by sheep grazing (Ratcliffe 1977). It is possible, however, that occult deposition of pollutants, particularly nitrates, has accounted for damage and contraction of some montane heaths (Lee, Tallis & Woodin 1988; see Table 3).

Soil and peat erosion

Extensive screes are often associated with the loss of dwarf-shrub heaths on steep slopes, though some were natural and fairly stable under tree cover (Innes 1983; Ballantyne 1986). Grazing and, particularly, burning account for widespread gullying and sheet erosion of blanket bogs (Tallis 1964, 1985; Lindsay *et al.* 1988) and may accelerate podzolization (Dearing, Elner & Happey-Wood 1981; Innes 1983). However, atmospheric pollution has contributed to erosion through the loss of peat-forming *Sphagnum* spp. in the southern Pennines (Lee *et al.* 1988;

Bowler & Bradshaw 1985), and possibly to dominance of *Vaccinium myrtillus, V. vitis-idaea* and *Empetrum nigrum* on eroding Peak District blanket bog.

Consequences for animal communities

Several accounts of the upland animal communities are available, either being general (Pearsall 1950; Ratcliffe 1977; Usher & Gardner, this symposium) or concentrating on invertebrates (Coulson & Whittaker 1978; Coulson, this symposium), birds (Fuller 1982; Ratcliffe 1989) or mammals (Yalden 1982; Nethersole-Thompson & Watson 1981). Here we mention three points relevant in part to land-use influences.

Soil-determined differences

Although it is mainly low temperatures that account for the lower species diversity of invertebrates in the uplands compared with lowland regions, there are also the major influences of rainfall, snow lie, peat formation and soil type, as well as vegetation. Mineral soils contain a significantly greater biomass of invertebrates than peat, and a different species composition — although not a greater species richness (Coulson 1988). This reflects the presence in peat of more organic matter, a lower rate of litter decomposition, less nitrogen and phosphorus, a greater degree of wetness and fewer nutritious living plants. Base-saturated (especially calcareous) soils with pH greater than 6·0 consistently have a higher animal biomass production than either acidic mineral soils or peats. The soil-induced differences in vegetation influence the densities of herbivorous birds, e.g. red grouse (Watson, Moss & Parr 1984) and ptarmigan (Watson 1979), and mammals, e.g. red deer (*Cervus elaphus*) (Staines, Crisp & Parish 1982; Clutton-Brock & Albon 1988), mountain hare (*Lepus timidus*) (Moss & Hewson 1985) and sheep (Welch 1981; Grant & Maxwell 1988). Even some invertebrate feeders, and especially those depending on earthworms (e.g. golden plover, *Pluvialis apricaria*, and mole, *Talpa europaea*), achieve higher densities in base-rich areas.

Losses through persecution, pollution and vegetation change

Forest clearance obviously resulted at first in a substantial increase in upland animals (especially birds) which need or exploit treeless ground, at the expense of those which need trees and woodland. But there were subsequently further large readjustments. Management for grouse, and in some areas for sheep, has resulted in a marked lack of top predators. Of the mammals, wild cat (*Felis sylvestris*), pine marten (*Martes martes*) and stoat (*Mustela erminea*) have been reduced in distribution or abundance through trapping (Pearsall 1950; Langley & Yalden 1977). McVean & Lockie (1969) argue that in the western Highlands and Islands predominantly extractive land-use, with continued grazing and burning, has run down soil and vegetation fertility to lower levels than before the turn of this century. On much of this land, grouse moor has been abandoned since *c*. 1920 to sheep and deer, and there is evidence of a permanent long-term decline in numbers of many moorland mammals and birds (MacKenzie 1924).

The fate of the birds is better documented. Information provided by Harvie-Brown & Buckley (1887), MacKenzie (1924), Pearsall (1950), Sharrock (1976), Cramp & Simmons (1977–1988), Ratcliffe (1986, 1989) and Thompson, Stroud & Pienkowski (1988) gives suggestive evidence that, of 71 species associated with the uplands, at least 10 predators and 2 waders have been reduced in number or distribution by persecution. There are at least five other anthropogenic influences on the typical open-ground species including (i) the use of pesticides (5 species appear to have declined); (ii) loss of dwarf-shrub heaths through heavy grazing pressure (at least 8 species appear to have declined; but twite, *Acanthis flavirostris*, and skylark, *Alauda arvensis*, have increased locally); (iii) afforestation (at least 14 species have declined); (iv) agricultural drainage (at least 4 species have declined) and reseeding/reclamation (at least 5 species have declined, but at least another 8 have increased); and (v) atmospheric pollution (3 species appear to have declined). These are quite separate from apparent effects of climatic change over the past 200 years (apparently affecting at least 42 species), interspecific competition (involving 5 species) and novel migration/settlement patterns (involving 3 species). Lack of adequate control data and a preponderance of circumstantial information, however, mean that these figures should be regarded as approximations (notably for climatic effects).

In parts of Britain persecution continues, yet in spite of this numbers of predators seem to remain static in a few areas (Hudson, this symposium). This may arise because persecution brings populations down below levels at which density-dependent mortality (particularly in winter) operates, or because young birds are restocking depleted territories or ranges from more productive, comparatively unpersecuted areas. Both situations are likely to be relatively unstable, however, because chance events such as severe weather or habitat loss will limit the reservoir of potential recruits.

Increases through beneficial management

Densities of red grouse, and more locally numbers of red deer and mountain hare, have obviously increased because of habitat management. Rotational burning produces more nutritious young heather for grouse near to older, rank stands which provide nesting cover for grouse as well as for merlin (*Falco columbarius*), hen harrier (*Circus cyaneus*) and several small passerines. Scavenging birds have also benefited through increased carrion food supplies related to density-dependent mortality in high-density populations of sheep and deer (Dare 1986; Watson, Langslow & Rae 1987). Species typically found on marginal or even arable agricultural ground (e.g. lapwing, *Vanellus vanellus*, and oystercatcher,

Haematopus ostralegus) have extended their range into the uplands as a result of land improvement. Perhaps local populations of waders (e.g. golden plover and greenshank, *Tringa nebularia*) have thrived because burning and grazing have reduced vegetation height so that foraging efficiency and chick mobility are improved.

These changes appear to be most marked where there is improvement, particularly soil fertilization, of the base-poor ground towards the north-west of Britain (Hobbs & Gimingham 1987). More details of these secondary land-use impacts on the upland biota are given in Table 3.

TABLE 3. Six examples of how past and recent man-induced changes may have influenced upland vegetation and wildlife. The order is roughly chronological. Data for 1, 2 in McVean & Ratcliffe (1962), Ratcliffe (1977, 1988), Thompson *et al.* (1987); for 5 see Lee *et al.* (1988), Battarbee *et al.* (1985), Ormerod, Tyler & Lewis (1985), and Sharrock (1976); for 3, 4 and 6 see text

Form of anthropogenic change	Examples of resulting features
1. Extensive forest clearance (3900–300 years BP)	Replacement by dwarf-shrub heaths (*Calluna vulgaris, Erica cinerea, E. tetralix, Vaccinium myrtillus, V. vitis-idaea, Empetrum nigrum*) and some rather local grasslands.
	Loss of many woodland bird populations (e.g. goshawk, capercaillie, woodcock, crested tit, siskin).
2. Extensive use as grazing range for sheep (mainly from 1750–1830 onwards, but locally up to 500 BP)	Conversion of sub-montane heath to *Festuca ovina, Deschampsia flexuosa, Nardus stricta, Juncus squarrosus, Molinia caerulea; Pteridium aquilinum* on well-drained soils; *Scirpus cespitosus, Eriophorum vaginatum, E. angustifolium, M. caerulea* on wetter ground, especially in W. Scotland. Changes are due to grazing and burning.
	Conversion of montane *Racomitrium lanuginosum–Carex bigelowii* continuous heath to *F. ovina–D. flexuosa–R. lanuginosum* heath, and then possibly to *F. ovina–D. flexuosa–Argostis canina* grassland.
	Depletion of sub-montane and montane vascular plant complement, especially south of Scottish Highlands. Most montane *Salix, Dryas* heath and tall forb communities converted to species-rich grassland.
	Increase in densities of carrion-eating birds.
3. Land 'improvement' in form of grouse moor management (1800 onwards in England, 1840 onwards in Scotland to peak around 1900) and ploughing, reseeding and fertilization (1820–40 onwards) to enhance carrying capacity for sheep and deer	Marked increase in distribution and abundance of red grouse and more locally increases in mountain hare and red deer.
	Increases in density and biomass of soil invetebrates (e.g. Lumbricida, Nematoda, Collembola) in response to fertilization and gradual agricultural intensification.
	Expansion of breeding range of typically agricultural birds (e.g. buzzard, lapwing, skylark) or birds benefiting from more prey or nesting sites (e.g. red kite, wheatear, whinchat).

4. Persecution related to sheep and game management (1760–1914; local occurrences prevail)	Marked reductions in population of predatory birds, e.g. golden eagle (from late 18th C.), buzzard (from 1850), red kite (1880–90 almost extinct), hen harrier (lowest numbers during 19th C.–1900), raven (very low between 19th C. and 1914). Noticeable gap in predator component in NE England and SE Scotland (primary grouse moor). Dotterel and red-necked phalaropes were shot, or eggs were collected by trophy hunters.
5. Industrial acidification (from 200 BP in S. Pennines and 150 BP in Galloway to present)	Loss of *Sphagnum* spp. and increase in peat erosion: spread of *Molinia caerulea*; contraction of Lobarion lichen community. Acidification of water bodies; reduction in fresh-water invertebrate abundance and species diversity; contraction of dipper's breeding range and fall in breeding density. Possible local declines of nuthatch and spotted flycatcher in upland woods.
6. Extensive conifer afforestation (mainly 1919 onwards)	On poor soils greater podzolization and accumulation of surface organic matter; increase in water run-off rate; increase in water sediment load and acidity (exacerbates acid pollution effects). Depletion of at least 63 vascular plant species (e.g. *Equisetum variegatum*, *Trollius europaeus*, *Viola lutea*, *Arctostaphylos uva-ursi*, *Listera cordata*, *Carex pauciflora*). Substantial reduction in numbers of at least 14 bird species (e.g. raven, dunlin, golden plover, red grouse, wheatear) through loss of nesting/feeding habitat.

INTERNATIONAL SIGNIFICANCE

Alongside the great mountain systems of the world, the British uplands are insignificant in extent and stature. However, the hyper-oceanic climate at the most insular, western Atlantic edge of the European continent, combined with historically widespread anthropogenic influences, has produced a distinctive landscape and range of ecosystems which are not duplicated anywhere in the World. We now consider some of the internationally significant features within the range of variation in uplands already discussed.

Wide range of podzolic soils

Elsewhere in the world, podzols and weakly podzolized soils are found mainly in Finland, the USSR central belt across to Kamchatka, and North America (from Newfoundland to the western part of Northwest Territories; Curtis *et al.* 1976). None of these countries, however, has a comparable range and mixture of podzols and related peat types over such a small area as that found in Britain. Miles (1985, 1988) comprehensively reviewed soil and vegetation successional changes and dynamics in the British uplands; we believe that many of these are unique to this country and Ireland.

Large extent of blanket bog

Ombrotrophic bog, in which the living surface to the peat body receives its nutrients largely through precipitation, is a localized global type. The extreme form, blanket bog, is still rarer on the world scale, but is especially widely and extensively developed in the British uplands, which support at least 7 per cent and perhaps up to 13 per cent of the total known amount, 10×10^6 ha (Gore 1983; Lindsay *et al.* 1988). The underlying peat of blanket bog is perhaps the most distinctive and unusual of all British upland soil features, compared with other countries, except Ireland.

Caithness and Sutherland have probably the largest single expanse of blanket bog in the world: 401,000 ha before afforestation began (Stroud *et al.* 1987; Lindsay *et al.* 1988). According to Lindsay *et al.* (1988) other major occurrences are, in Europe, in Ireland (771,000 ha), western Norway, the Pyrenees and possibly western Iceland; in the New World, in eastern Canada (especially southwest Newfoundland), the Pacific coast of North America, Tierra del Fuego and the Falkland Islands; in Asia, in west Kamchatka, USSR, and east Hokkaido, Japan; in South Island of New Zealand and other islands of the southern ocean; and possibly in Ruwenzori, Uganda. With increasing distance from Britain, these other blanket bog areas diverge increasingly from our types in floristics and, hence, in peat composition. In some of these other areas, the occurrences are highly dissected and sometimes fragmentary.

Many British blanket bogs also show highly developed surface patterns of pools with intervening hummocks and ridges. These patterned bogs are especially well represented in western and northern Scotland, where they show a great deal of variation in size, shape and configuration of the pools, evidently in relation to a complex of climatic, topographic and hydrological factors. The pool systems are the recognizable analogues of patterned surfaces widely distributed on raised bogs and other mires within the Boreal forest zone of northern Europe and Canada. Their occurrence on blanket bogs appears to be confined to Britain and Ireland (Lindsay *et al.* 1988).

Blanket bogs have additional importance in regard to Quaternary ecology, since their peat contains a record of vegetational, climatic and land-use events from the time that they began to form. Their extremely variable age is itself a feature of interest, and the tree remains so often preserved within them give direct information on forest history and its pattern of change (McVean & Ratcliffe 1962; Birks 1988). The northern Scottish blanket bogs support a noteworthy southern outlier of the northern Eurasian tundra fauna. Afforestation has accounted for the loss of over 1×10^6 ha of open upland since 1920, and much new planting is now concentrated on lower level blanket bogs, posing the single most serious threat to Britain's upland wildlife and its habitat (Stroud *et al.* 1987; Thompson *et al.*, 1988).

Extensive development of acidophilous dwarf-shrub heaths and grasslands

Ericaceae

A major influence on soil acidification accompanying the wide and intensive impact of man in the uplands has been the expansion of vegetation dominated by acidophilous Ericaceae. This dwarf-shrub heath, with *Calluna vulgaris, Erica cinerea, E. tetralix, Vaccinium myrtillus* and *Empetrum nigrum*, is extensive in Britain. With the exception of *Erica tetralix*, this heath occurs in a fragmentary form in the rest of Europe and is regarded as an oceanic plant formation (Good 1974; Gimingham 1975). *Calluna* heath is the best developed type, occurring in a community which is dominant in an ecologically optimal habitat mainly in the eastern uplands of Britain. In addition to representing a unique ecosystem managed principally for red grouse, it has been the subject of some important studies of interactions between vegetation dynamics, herbivores and fire (reviewed by Hobbs & Gimingham 1987) and the dynamics of successional change in vegetation (reviewed by Miles 1985, 1988). It contains a diverse invertebrate and vertebrate fauna but a relatively depauperate flora. *Vaccinium vitis-idaea* (which grades into *V. myrtillus* heath) and *E. nigrum* heaths are well developed in the Peak District and North Wales, and locally in South Wales. These probably represent a marked transitional phase from formerly extensive *Calluna* heath and dried *Eriophorum vaginatum* bog subjected to much grazing and atmospheric pollution. *Arctostaphylos uva-ursi* is an important component locally in the Highlands.

Ulex *scrub*

No continental European mountains have a comparable extent of vegetation dominated by *Ulex gallii*, although it is widespread in the Irish uplands (Fitter 1978). In Britain this is widespread in the uplands of south-west England, Wales and north Midlands where, typically with *Calluna* and *E. cinerea*, it thrives on base-poor soils in a mild oceanic climate (Ratcliffe 1977; Rodwell 1987). *U. gallii* with *Agrostis curtisii* is confined mainly to south and south-west England and is sometimes regarded as a lowland heath type. *Ulex europaeus* scrub is extensive but also has a fairly restricted world distribution (east Netherlands, Belgium, north-east France). It provides an important nesting and feeding habitat for heathland and upland passerines.

Grasslands

The sub-montane grassland communities mostly reflect a greater degree of anthropogenic impact than the ericoid heaths. *Juncus squarrosus* associated with *Nardus stricta* grassland is widespread on the moist, well-aerated acid to peaty mineral soils of the wet and cool oceanic lowlands and upland fringes of western Britain; yet elsewhere in the European mountains it is fairly localized. The characteristic dominant species of anthropogenic upland grasslands (and bogs) include *Festuca ovina, Agrostis* spp., *Nardus stricta, Molinia caerulea, Scirpus cespitosus, Eriophorum* spp. and *Narthecium ossifragum* (see Table 3 and Ratcliffe 1977). These vegetation types are locally associated with high densities of breeding snipe (*Gallinago gallinago*), redshank (*Tringa totanus*) and, in places, lapwing and curlew (*Numenius arquata*); and they have a limited but specialized invertebrate fauna (Coulson 1988). These anthropogenic plant communities are an excellent example of succession occurring under man-induced changes. Their component species are mostly widely distributed in continental Europe, but in different community relationships.

Montane plant communities

These represent southern and oceanic outliers of arctic-alpine fellfield and mountain tundra. Although they cover only a small area compared with the main continental occurrences, the British examples show considerable diversity and include several types which are either highly local or apparently absent elsewhere. They reach their greatest extent in the Highlands, and especially in the Cairngorms, which have the largest area of land above 900 m, and a relatively continental climate, giving the most arctic conditions in Britain. The international importance of the high Cairngorm plateau has been asserted by Curry-Lindahl (1974), Nethersole-Thompson & Watson (1981) and the World Wilderness Congress (1983).

The most characteristic montane community found first above the tree-line is some form of prostrate *Calluna* heath, a type apparently rare outside this country. Sometimes *Calluna* is dominant, with few other species, but there are facies varying according to abundance of lichens, mosses (including *Racomitrium lanuginosum*) or dwarf shrubs (*Empetrum* spp., *Loiseleuria procumbens, A. uva-ursi, Actous alpina, J. communis* subsp. *nana*). Where snow lie increases there is change to dominance of *V. myrtillus*, sometimes with abundance of *Empetrum hermaphroditium*. Above these montane dwarf-shrub heaths, there is typically a higher zone of *Racomitrium lanuginosum*-dominated heath occupying flat or gently sloping ground of summit plateaux and upper spurs. This may be species-poor, although usually containing an abundance of *Carex bigelowii* and *Salix herbacea*, or, especially in the northern Highlands, it may have great quantities of cushion herbs, *Silene acaulis, Cherleria sedoides* and *Armeria maritima*. It grades into stony ablation surfaces and solifluction forms, where the open ground has a still larger variety of species. *Racomitrium* heath is an oceanic community absent from the more continental European mountains; other notable areas, west Norway, Iceland and New Zealand (Dickson 1973), are very local.

These types have largely been replaced through heavy grazing and possibly atmospheric acid deposition on mountains south of the Highlands. Their derivatives are either a *V. myrtillus–E. nigrum* heath (often with much lichen) or a species-poor *Festuca–Agrostis* grassland. Montane blanket bog occurs widely in the Highlands; *Calluna* tends to be replaced by *Vaccinium* spp., and *Empetrum* spp., *Rubus chamaemorus* and *Sphagnum fuscum* are usually present. In the Cairngorms, the highest vegetation zone is one of *Juncus trifidus* heath, with varying amounts of lichen. Similar communities are widespread in Norway but absent from the Alps. On mountains where snow lies longest, there is a sequence of communities, similar to those found extensively in western Norway, in which the sequence with increasing snow cover is (i) *Vaccinium* (with *Empetrum*) to (ii) *Nardus* (with *Scirpus cespitosus*) and then to (iii) dwarf herbs, sparse grasses, and moss and liverwort communities. These also are most extensively and fully developed in the Cairngorms. Most of these communities are described by McVean & Ratcliffe (1962).

Extensive development of fern and bryophyte-rich vegetation

The humid climate and varied topography of Britain are favourable for bryophytes and pteridophytes (Ratcliffe 1968; Conolly & Dahl 1970; Birks 1976). Britain has a richer Atlantic bryophyte flora than anywhere else in Europe; North Wales, Lakeland and the north-west Highlands and Islands are especially species-rich (Ratcliffe 1968).

Bryophytes

Racomitrium lanuginosum occurs in abundance over a wide range of altitudes. Apart from the predominantly western montane heaths, it occurs widely on sub-montane blanket bog and wet heath where *S. cespitosus, E. vaginatum* and/or *Calluna* are dominant; it is often considered to indicate bog or wet heath dried through burning (McVean & Ratcliffe 1962).

Three other community types should be mentioned here. First, the *Rhytidia-delphus loreus* heath (also containing other pleurocarpous mosses and dwarf *Deschampsia cespitosa*) occurs in areas of late snow lie in the north-west Highlands; species-rich facies on basic soils occur extremely locally (McVean & Ratcliffe 1962). Second, there are very localized hepatic carpets of disjunct oceanic species, best developed as an understorey of dwarf-shrub heath or grassland mainly in damp shady mountain areas in parts of north-west Scotland and Ireland. Liverwort-rich *C. vulgaris* heath is a good example (community H21 in Rodwell 1987; Hobbs 1988); it has an understorey of liverworts, many of which have spectacular world disjunct distributions, including *Herbertus aduncus, Anastrepta orcadensis* and *Scapania nimbosa*. The liverwort understorey of *J. communis* subsp. *nana* is also found only in the north-western Highlands; and on Beinn Eighe, the dominant liverwort is *Herbertus borealis*, found nowhere else in Britain, and abroad only in west Norway. Third, there are mossy facies of dwarf-shrub heaths and grasslands, instead of the continental European lichen facies.

Pteridophytes

Birks's (1976) analysis of the European Pteridophyta revealed a unique flora in Britain, characterized by a large element of Atlantic, Mediterranean and arctic-alpine elements, many of the species being typical of upland habitats.

There is an abundance of ferns in some sub-montane vegetation. The cosmopolitan but genetically variable bracken (*Pteridium aquilinum*) is locally dominant on sheep-grazed hillsides, especially in Wales and northern England, on ground often originally occupied by woodland (Page 1976). It frequently advances as heather moor recedes under heavy grazing. Although generally regarded as a major pest plant in Britain, it is an important habitat for passerine birds, notably whinchat (*Saxicola rubetra*). *Oreopteris limbosperma* locally invades *Festuca–Agrostis* grassland, and other species are characteristic of ungrazed situations on both acidic and basic soils, e.g. the widespread *Dryopteris filix-mas* and *D. pseudomas*. Open growth of *Cryptogramma crispa* appears to represent an arrested successional stage on scree, and is widespread in North Wales, northern England and parts of the Southern Uplands.

Of montane fern communities, *Athyrium distentifolium–C. crispa* is commonly characteristic of high Highland corries where snow lies late (McVean and Ratcliffe 1962), and *Dryopteris oreades* and *D. expansa* are also sometimes present. *D. oreades* appears to have European strongholds only in Britain (not Ireland) and Iceland (Jalas & Suominen 1972). The possibly endemic *Athyrium distentifolium* var. *flexile* occurs in several Highland corries. On the other hand, arctic-alpine ferns such as *Cystopteris montana, Woodsia ilvensis, W. alpina* and *Polystichum lonchitis* have a much more limited ecological and geographical range than in the mountains of continental Europe. The abundance of *Hymenophyllum wilsonii* in Britain is distinctive and unusual, while the occurrence of *H. tunbrigense* up to 400 m reflects the mildness of the upland climate in south-western Britain.

Phytogeographical interest

It is becoming increasingly apparent that most of Britain's upland plant communities are peculiar to this country in terms of species composition. For example, of the 82 British upland communities recognized by the National Vegetation Classification (Rodwell 1987) only 23 occur in Norway (H. J. B. Birks, pers. comm.). The northern and montane vascular flora of Britain is drawn from at least seven phytogeographical elements (Matthews 1955). The mixture of alpine (e.g. *Cherleria sedoides, Gentiana verna*), arctic-alpine (e.g. *Dryas octopetala, Silene acaulis, Lloydia serotina*) and arctic (e.g. *Alopecurus alpinus, Koenigia islandica, Artemisia norvegica*) species is unique. Globally disjunct species include *Alopecurus alpinus, Arenaria norvegica, Artemisia norvegica, Saxifraga hirculus, Oxytropis halleri* and *O. campestris*.

Climatic blending also results in some characteristically southern and northern species overlapping. The Burren, in Ireland, provides a spectacular example with occurrences of the arctic-alpine *D. octopetala* and *A. uva-ursi* with the Mediterranean *Adiantum capillus-veneris* and *Neotinea intacta*. Less extreme examples are found in western Britain, where several southern Atlantic ferns and bryophytes (e.g. *Trichomanes speciosum, Dryopteris aemula, Jubula hutchinsiae*) occur within a close linear and altitudinal distance of montane vascular plants (e.g. *Saxifraga oppositifolia, S. aizoides, W. ilvensis*).

Some characteristic upland species show such different ecological relationships in Britain, compared with continental Europe, that different ecotypes are evidently involved. Examples include (i) *Betula nana*, widespread dwarf to medium shrub in Fenno-Scandia but limited in the Highlands to high ridges/ hummocks in undamaged bogs or more widely on damaged bogs and wet heaths, and occurring only as a depauperate, usually spreading form; (ii) *A. uva-ursi*, mainly a woodland species in Fenno-Scandia but typical of heaths and mires in Britain; (iii) *Pleurozia purpurea*, common in western bogs of Scotland and Ireland, but a woodland plant in Norway; (iv) *A. alpina*, which is a strict calcicole in the Alps and occupies dry areas on continental mountains, but is a strict calcifuge species and grows in bogs as well as dry heaths in Britain; (v) *Trollius europaeus*, which is locally common in west Scotland but does not occur in western areas of Norway; (vi) *Saxifraga nivalis*, exclusive to basic rocks in Britain, but considerably less selective in Scandinavia; and (vii) *Myosotis alpestris* and *Draba incana* in Upper Teesdale, and *Artemisia norvegica* on Ingleborough (Yorkshire) and in the Highlands, which appear to occur as endemic ecotypes within Britain. *C. sedoides* and *S. acaulis* are apparently represented by different ecotypes for they are confined to basic rocks in the south of their range, but in the north occur on most soil and rock types, including highly leached substrates on high moss heaths and fellfields.

Animal communities and assemblages

International comparisons of bird populations are especially far advanced compared with those for the flora or other animal groups. Detailed reviews are provided by Hoffmann (1974), Cramp & Simmons (1977–1988), Piersma (1986) and Ratcliffe (1989).

Breeding bird assemblages

There is probably a greater *mixture* of boreal, low-, mid- and high-arctic, temperate and continental species breeding in the British uplands than in any other comparably sized part of Europe. The species composition of at least two habitats is particularly distinctive. First, the boreal-arctic peatland combination (with its often relatively large continental element) has no counterpart elsewhere. Within regions such as the north, north-west, eastern and central Highlands, Shetland, Inner and Outer Hebrides, Cheviots, Southern Uplands, Pennines, Dartmoor and Wales there are unique assemblages of Anatidae, Falconidae,

Charadrii, Laridae and passerines. Although most of the species are not scarce (some are fairly abundant abroad), the assemblage *per se* certainly is scarce. The Caithness and Sutherland peatlands, for example, contain 15 of Britain's 17 upland breeding waders, 6 of the 7 typical upland birds of prey, and at least 11 small passerines. Each has somewhat different distributions varying from boreal-high arctic (e.g. red-throated diver, *Gavia stellata*; arctic skua, *Stercorarius parasiticus*) to typically continental (e.g. lapwing).

There are also some unusual adaptations. Greenshanks typically nest in forest marshes and open forests just south of the tundra (Knystautas 1987), but in Britain they nest mainly in north-western open, treeless blanket bogs (Nethersole-Thompson & Nethersole-Thompson 1986). The curlew appears to substitute for the whimbrel (*Numenius phaeopus*) and possibly the black-tailed godwit (*Limosa limosa*) as the typical long-billed upland wader, and in Britain is more numerous than elsewhere in Europe (Britain has 18–20 per cent of European population; Piersma 1986). Subspecies *tundrae* of ringed plover, *Charadrius hiaticula*, does not breed in Britain, yet subspecies *hiaticula*, which does and also occupies stony or mossy ground in Iceland, only occupies some upland river and lake margins. Combinations of migration tendencies, population pressure, predation risk and interspecific competition influence such divergences producing geographically complex patterns of distribution and subspeciation (Hale 1980; Pienkowski & Evans 1985).

The second important element belongs to the montane plateaux and corries; it contains obvious outliers of Arctic Eurasia including dotterel (*Charadrius morinellus*), snow bunting (*Plectrophenax nivalis*) and the circumpolar boreal to arctic-alpine ptarmigan, which is resident. Several other species, such as dunlin (*Calidris alpina schinzii*), golden plover and wheatear (*Oenanthe oenanthe*) nest there, but also at much lower altitudes. At least three other 'fringe' species have summered or bred in Britain only on the high plateaux (Nethersole-Thompson & Nethersole-Thompson 1986; Ratcliffe 1989). Some of these northern birds appear to be interesting indicators of climatic change, appearing or disappearing from Britain as conditions become colder or warmer.

Large population and high density bird assemblages

On an international scale five species are significant in having main strongholds or very high population densities in Britain: peregrine (*Falco peregrinus*) (Ratcliffe 1980), golden eagle (*Aquila chrysaetos*) (Watson *et al.* 1987), red grouse (Cramp & Simmons 1977–1988); ptarmigan (Galbraith *et al.* 1988) and raven (*Corvus corax*) (Ratcliffe 1962, 1988; Dare 1986). Others, reaching almost greater numbers, at least locally, in upland Britain than elsewhere in Europe, include hen harrier (*Circus cyaneus*), merlin (*Falco columbarius*), golden plover, dunlin, curlew, redshank (*Tringa totanus*) and lapwing; the last three reach especially high densities mainly when associated with marginal agricultural ground in the uplands. Several other species are important because locally large populations seem to have been more comprehensively studied in Britain than elsewhere in Europe: red kite (*Milvus milvus*), buzzard (*Buteo buteo*), dotterel, greenshank, common sandpiper (*Actitis hypoleucos*) and dipper (*Cinclus cinclus*).

The high densities of grouse relate to the unique system of habitat management to maintain heather dominance in varying age classes. Associated birds of heather moor benefit from this, and the continuous loss of grouse moor will reduce the international standing of Britain's upland bird communities. Extensive range management for sheep and deer (in many areas associated with large numbers of rabbits on marginal agricultural ground) largely accounts for the remarkably high densities of scavenging/predatory birds, notably raven and buzzard, especially in the Welsh uplands (Newton, Davis & Davis 1982; Dare 1986), Lake District and Scottish Highlands and Islands; the southern Uplands also had good populations of both species before afforestation (see Marquiss, Newton & Ratcliffe 1978; Thompson *et al.*, this symposium).

Zoogeographical interest in upland birds

Southern and/or western fringe populations may be important for the evolution of novel adaptations to changing environmental conditions, and in any case fascinate ecologists seeking to explain patterns in breeding distributions. The following upland species are relevant here: snowy owl (*Nyctea scandiaca*), white-tailed eagle (*Haliaeetus albicilla*), dotterel, ruff (*Philomachus pugnax*), Temminck's stint (*Calidris temmincki*), purple sandpiper (*C. maritima*), whimbrel, green sandpiper (*Tringa ochropus*), wood sandpiper (*T. glareola*), turnstone (*Arenaria interpres*), red-necked phalarope (*Phalaropus lobatus*), bluethroat (*Luscinia suecica*), brambling (*Fringilla montifringilla*), shorelark (*Eremophila alpestris*), snow bunting, Lapland bunting (*Calcarius laponicus*), and even North American spotted sandpiper (*Actitis macularia*) and pectoral sandpiper (*Tringa melanotos*). All of these breed in the British uplands, or may do so soon; the reasons for some but not others settling to breed are unclear. The racial differentiation of some species (red grouse, golden plover, twite and snow bunting) and disjunct species distributions (ptarmigan, great skua (*Stercorarius skua*) and twite) also merit further zoogeographical research.

Other animal communities

Other animal communities have been reviewed by Usher & Gardner (this symposium), who consider natural or semi-natural origins. Ratcliffe (1977), Nethersole-Thompson & Watson (1981) and D. Horsfield (pers. comm.) have provided fairly detailed lists of montane invertebrates. There are, for example, 13 strictly montane Lepidoptera, 17 montane spiders, 22 montane beetles (Coleoptera), at least 13 montane Diptera and at least 14 species of strictly montane sawflies (Symphata). Virtually nothing is known about montane invertebrate community structure or functioning. Although the locally anthropogenic vegetation provides potentially novel food plants for invertebrates in such an extreme environment, particularly south of the Scottish Highlands such as in North Wales, it appears that more widespread species rather than montane specialists utilize these plants. Usher & Gardner's (1988) point about the relative depletion of predatory mammals and the scarcity of 'alien' or invasive species is certainly apposite; mink (*Mustela vison*) is one noticeable exception, however.

CONSERVATION OBLIGATIONS

The existing series of National Nature Reserves, established by the Nature Conservancy Council, and other nature reserves (mainly of the Royal Society for Protection of Birds) should safeguard some of the most important areas of upland Britain. It remains to be seen whether the further series of upland Sites of Special Scientific Interest (SSSIs) will give sufficient protection to a larger number of important mountain and moorland areas. Currently there are 206 upland SSSIs covering just over 82×10^4 ha (12.4% of upland Britain). Even if these designations are successful conservation measures, a much larger area of the British uplands will still be at risk to damaging land-use change.

If the natural and internationally significant upland features are to be maintained adequately, further urgent action is needed. Obvious priorities involve halting the further expansion of damaging blanket conifer afforestation; limiting the amount of ecologically damaging grazing, burning and agricultural reclamation; reducing industrial pollution; and minimizing undesirable accompaniments of recreation such as excessive erosion and construction of hill roads or downhill ski facilities. With impending changes in the European Community's rules governing support for sheep farmers, there are excellent opportunities to negotiate a reduction in sheep stocking rates (Mowle & Bell 1988) and to restore the cover of *Calluna* heath in many southern districts. This should improve the quality and international standing of our upland habitat resource, and for the predatory and even scavenging birds the provision of more live prey is likely to improve population viability.

Some of the National Parks of England and Wales provide for some regulation of damaging land uses and adverse changes, but in Scotland the absence of National Parks is a cause for concern. The Scottish Highlands were identified in the World Conservation Strategy (Anon. 1983) as a priority area for the establishment of protected areas, on the grounds of intrinsic quality and importance but also because of present inadequacy of safeguard measures applied on the ground in this region. Conservationists are presently urging the British Government to find ways to protect the unafforested parts of the Sutherland and Caithness blanket bogs, as an international as well as national obligation, and there is renewed pressure for National Parks and further nature reserves in Scotland. There are also clear pointers to needs for monitoring and for fundamental and applied biological research and socio-economic studies. These should be pursued as soon as possible if ecologists are to be able to predict how land-use changes will affect the upland biotic resource. Such work would also indicate how social and economic interests can be met in ways compatible with the conservation of both wildlife and scenic beauty.

ACKNOWLEDGMENTS

We thank the following for discussions or the extensive use of their published or unpublished work: John and Hilary Birks, Desmond Nethersole-Thompson, Adam Watson, Charles Gimingham, John Miles, Richard Lindsay, David Stroud, Chris Sydes, Alan Brown, David Horsfield, Alison Hobbs, Hector Galbraith, Michael Usher, Keith Kirby, Gordon Miller, Paul Haworth, Derek Langslow, Peter Hudson and Terry Burke. Sandra Lackie, Jean Short, Norma Mayer and Pam Piggott very kindly typed the manuscript.

REFERENCES

Anon. (1983). *The Conservation and Development Programme for the UK: a Response to the World Conservation Strategy.* Kogan Page, London.

Ball, D.F., Radford, G.L. & Williams, W.M. (1983). *A Land Characteristic Databank for Great Britain.* Occasional Paper No. 13, Institute of Terrestrial Ecology, Bangor.

Ballantyne, C.K. (1986). Landslides and slope failures in Scotland: a review. *Scottish Geographical Magazine*, 102, 134–150.

Barry, R.C. (1981). *Mountain Weather and Climate.* Methuen, London.

Battarbee, R.W., Flower, R.J., Stevenson, A.C. & Rippey, B. (1985). Lake acidification in Galloway: a palaeoecological test of competing hypotheses. *Nature*, 314, 350–353.

Birks, H.J.B. (1976). The distribution of European pteridophytes: a numerical analysis. *New Phytologist*, 77, 257–287.

Birks, H.J.B. (1986). Late-Quaternary biotic changes in terrestrial and lacustrine environments, with particular reference to north-west Europe. *Handbook of Holocene Palaeoecology and Palaeohydrology* (Ed. by B.F. Berglund), pp. 3–65. Wiley, Chichester.

Birks, H.J.B. (1988). Long-term ecological change in the British uplands. *This volume.*

Birks, H.J.B. & Ratcliffe, D.A. (1980). Upland Vegetation Types: a list of *National Vegetation Classification Plant Communities.* Nature Conservancy Council, Peterborough.

Bowler, M. & Bradshaw, R.H.W. (1985). Recent accumulation and erosion of blanket peat in the Wicklow Mountain, Ireland. *New Phytologist*, 101, 543–550.

Clutton-Brock, T.H. & Albon, S.D. (1988). *Red Deer in the Highlands: the Ecology of a Marginal Population.* Blackwell Scientific Publications, Oxford.

Conolly, A.P. & Dahl, E. (1970). Maximum summer temperature in relation to the modern and Quaternary distributions of certain arctic-montane species in the British Isles. *Studies in the Vegetational History of the British Isles* (Ed. by D. Walker & R.G. West), pp. 159–223. Cambridge University Press, Cambridge.

Coulson, J.C. (1988). The structure and importance of invertebrate communities in peatlands and moorland, and effects of environmental and management changes. *This volume.*

Coulson, J.C. & Whittaker, J.B. (1978). Ecology of moorland animals. *Production Ecology of Moors and Montane Grasslands.* (Ed. by O.W. Heal & D.F. Perkins), pp. 52–93. Springer-Verlag, Berlin.

Cramp, S. & Simmons, K.E.L. (Eds) (1977–1988). *The Birds of the Western Palearctic, Vols. 1–5.* Oxford University Press, Oxford.

Curry-Lindahl, K. (1974). *IUCN Survey of Northern and Western European National Parks and Equivalent Reserves. Report on Great Britain.* International Union of Conservation of Nature, London.

Curtis, L.F., Courtney, F.M. & Trudgill, S.T. (1976). *Soils in the British Isles.* Longmans, London.

Dare, P.S. (1986). Raven, *Corvus corvus* populations in two upland regions of North Wales. *Bird Study,* **33**, 179–189.

Dearing, J.A., Elner, J.K. & Happey-Wood, C.M. (1981). Recent sediment flux and erosion processes in a Welsh upland lake-catchment based on magnetic susceptibility measurements. *Quaternary Research,* **16**, 356–372.

Dickson, J.H. (1973). *Bryophytes of the Pleistocene: the British Record and its Chronological and Ecological Implications.* Cambridge University Press, Cambridge.

Fitter, A. (1978). *An Atlas of the Wild Flowers of Britain and Northern Europe.* Collins, London.

Forestry Commission (1984). *Census of Woodlands and Trees 1979–82: Great Britain.* Forestry Commission, Edinburgh.

Forestry Commission (1987). *Forestry Facts and Figures.* Forestry Commission, Edinburgh.

Fuller, R.J. (1982). *Bird Habitats in Britain.* Poyser, Calton.

Galbraith, H.G., Kinnes, L. Watson, A. & Thompson, D.B.A. (1988). Pressures on ptarmigan. *Game Conservancy Annual Review,* **19**.

Gimingham, C.H. (1975). *An Introduction to Heathland Ecology.* Oliver & Boyd, Edinburgh.

Godwin, H. (1975). *The History of the British Flora,* 2nd edn. Cambridge University Press, Cambridge.

Good, R. (1974). *The Geography of the Flowering Plants.* Longman, London.

Gore, A.J.P. (Ed.) (1983). *Mires: Swamp, Bog, Fen and Moor; Regional Studies.* Elsevier, Amsterdam.

Grace, J. & Unsworth, M.H. (1988). Climate and microclimate of the uplands. *This volume.*

Grant, S.A. & Maxwell, T.J. (1988). Hill vegetation and grazing by domesticated herbivores: the biology and definition of management options. *This volume.*

Hale, W.G. (1980). *Waders.* Collins, London.

Harvie-Brown, J.A. & Buckley, T.E. (1887). *A Vertebrate Fauna of Sutherland, Caithness and West Cromarty.* Douglas, Edinburgh.

Hobbs, A.M. (1988). Conservation of leafy liverwort-rich *Calluna vulgaris* heath in Scotland. *This volume.*

Hobbs, R.J. & Gimingham, C.H. (1987). Vegetation, fire and herbivore interactions in heathland. *Advances in Ecological Research,* **16**, 87–173.

Hoffman, R.S. (1974). Terrestrial vertebrates. *Arctic and Alpine Environments* (Ed. by J.D. Ives & R.G. Barry), pp. 475–567. Methuen, London.

Hudson, P.J. (1988). Spatial variations, patterns and management options in upland bird communities. *This volume.*

Innes, J.L. (1983). Lichenometric dating of debris — floor deposits in the Scottish Highlands. *Earth Surface Processes and Landforms,* **8**, 579–588.

Jalas, J. & Suominen, J. (1972). *Atlas Florae Europaeae, I. Pteridophyta.* Akateeminen Kirjakauppa, Helsinki.

Knystautas, A. (1987). *The Natural History of the USSR.* Century Hutchinson, London.

Langley, P.J.W. & Yalden, D.W. (1977). The decline of the rarer carnivores in Great Britain during the nineteenth century. *Mammal Review,* 7, 95–116.

Lee, J.A., Tallis, J.H. & Woodin, S.J. (1988). Acidic deposition and British upland vegetation. *This volume.*

Lindsay, R.A., Charman, D.T., Everingham, F., O'Reilly, R.M., Palmer, M.A., Rowell, T.A. & Stroud, D.A. (1988). *The Flow Country: the Peatlands of Caithness and Sutherland.* Nature Conservancy Council, Peterborough.

MacKenzie, O.H. (1924). *A Hundred Years in the Highlands.* Edward Arnold, London.

McVean, D.N. & Lockie, J.D. (1969). *Ecology and Land Use in Upland Scotland.* Edinburgh University Press, Edinburgh.

McVean, D. & Ratcliffe, D.A. (1962). *Plant Communities of the Scottish Highlands.* HMSO, London.

Manley, G. (1945). The effective rate of altitudinal change in temperate Atlantic climates. *Geographical Review,* **35**, 408–417.

Manley, G. (1952). *Climate and the British Scene.* Collins, London.

Marquiss, M., Newton, I. & Ratcliffe, D.A. (1978). The decline of the raven, *Corvus corax,* in relation to afforestation in southern Scotland and northern England. *Journal of Applied Ecology,* **15**, 129–144.

Matthews, J.R. (1985). *Origin and Distribution of the British Flora*. Hutchinson, London.

Miles, J. (1985). The pedogenic effects of different species and vegetation types and the implications of succession. *Journal of Soil Science*, 36, 571–584.

Miles, J. (1988). Vegetation and soil change in the uplands. *This volume*.

Moss, R. & Hewson, R. (1985). Effects on heather of heavy grazing by mountain hares. *Holarctic Ecology*, 8, 280–284.

Mowle, A. & Bell, M. (1988). Rural policy factors in land-use change. *This volume*.

Nethersole-Thompson, D. & Nethersole-Thompson, M. (1986). *Waders: Their Breeding, Haunts and Watchers*. Poyser, Calton.

Nethersole-Thompson, D. & Watson, A. (1981). *The Cairngorms: Their Natural History and Scenery*. Melven, Inverness.

Newton, I., Davis, P.E. & Davis, J.E. (1982). Ravens and buzzards in relation to sheep-farming and forestry in Wales. *Journal of Applied Ecology*, 19, 681–706.

Ormerod, S.J., Tyler, S.J. & Lewis, J.M. (1985). Is the breeding distribution of dippers influenced by stream acidity? *Bird Study*, 32, 33–40.

Page, C.N. (1976). The taxonomy and phytogeography of bracken — a review. *Botanical Journal of the Linnaean Society*, 73, 1–34.

Pearsall, W.H. (1950). *Mountains and Moorlands*. Collins, London.

Peterken, G.F. (1981). *Woodland Conservation and Management*. Chapman & Hall, London.

Pienkowski, M.W. & Evans, P.R. (1985). The role of migration in the population dynamics of birds. *Behavioural Ecology: Ecological Consequences of Adaptive Behaviour* (Ed. by R.M. Sibly & R.H. Smith), pp. 331–352. Blackwell Scientific Publications, Oxford.

Piersma, T. (Ed.) (1986). Breeding waders in Europe: a review of population size estimates and a bibliography of information sources. *Wader Study Group Bulletin*, 48, Supplement.

Ratcliffe, D.A. (1962). Breeding density in the peregrine, *Falco peregrinus* and raven, *Corvus corax*. *Ibis*, 104, 13–39.

Ratcliffe, D.A. (1968). An ecological account of Atlantic bryophytes in the British Isles. *New Phytologist*, 67, 365–439.

Ratcliffe, D.A. (Ed.) (1977). *A Nature Conservation Review. Vol 1*. Cambridge University Press, Cambridge.

Ratcliffe, D.A. (1980). *The Peregrine Falcon*. Poyser, Calton.

Ratcliffe, D.A. (1986). The effects of afforestation on the wildlife of open habitats. *Trees and Wildlife in the Scottish Uplands* (Ed. by D. Jenkins), pp. 46–54. Symposium No. 17, Institute of Terrestrial Ecology, Banchory.

Ratcliffe, D.A. (1988). The British upland scene. *Upland Seminar* (Ed. by D.B.A. Thompson & S. Whyte), pp. 4–26. Nature Conservancy Council, Peterborough.

Ratcliffe, D.A. (1989). *Upland Birds*. Cambridge University Press, Cambridge.

Rodwell, J.A. (1987). *National Vegetation Classification: upland heaths and grasslands*. Duplicated. Nature Conservancy Council, Peterborough.

Sharrock, J.T.R. (1976). *Atlas of Breeding Birds in Britain and Ireland*. Poyser, Berkhamsted.

Staines, B.W., Crisp, J.M. & Parish, T. (1982). Differences in the quality of food eaten by red deer (*Cervus elaphus*) stags and hinds in winter. *Journal of Applied Ecology*, 19, 65–77.

Stroud, D.A., Reed, T.M., Pienkowski, M.W. & Lindsay, R.A. (1987). *Birds, Bogs and Forestry: the Peatlands of Caithness and Sutherland*. Nature Conservancy Council, Peterborough.

Tallis, J.H. (1964). Studies on Southern Pennine peats. III. The behaviour of *Sphagnum*. *Journal of Ecology*, 52, 345–353.

Tallis, J.H. (1985). Mass movement and erosion of a Southern Pennine blanket peat. *Journal of Ecology*, 73, 283–315.

Tansley, A.G. (1939). *The British Islands and their Vegetation*. Cambridge University Press, Cambridge.

Thompson, D.B.A., Thompson. P.S. & Nethersole-Thompson, D. (1986). Timing of breeding and breeding performance in a population of greenshanks (*Tringa nebularia*). *Journal of Animal Ecology*, 55, 181–199.

Thompson, D.B.A., Galbraith, H. & Horsfield, D.H. (1987). Ecology and resources of Britain's mountain plateaux: conflicts and land use issues. *Agriculture and Conservation in the Hills and*

Uplands (Ed. by M. Bell & R.G.H. Bunce), pp. 22–31. Institute of Terrestrial Ecology, Merlewood.

Thompson, D.B.A., Stroud, D.A. & Pienkowski, M.W. (1988). Afforestation and upland birds: consequences for population ecology. *This volume.*

Turner, J. (1965). A contribution to the history of forest clearance. *Proceedings of the Royal Society, B*, **161**, 343–354.

Usher, M.B. & Gardner, S.M. (1988). Animal communities in the uplands: how is naturalness influenced by management? *This volume.*

Watson, A. (1979). Bird and mammal numbers in relation to human impacts at ski lifts on Scottish hills. *Journal of Applied Ecology*, **16**, 753–764.

Watson, A., Moss, R. & Parr, R. (1984). Effects of food enrichment on numbers and spacing behaviour of red grouse. *Journal of Animal Ecology*, **53**, 663–678.

Watson, J., Langslow, D.R. & Rae, S.R. (1987). *The Impact of Land-use Changes on Golden Eagles* (Aquila chrysaetos) *in the Scottish Highlands*. CSD Report 720, Nature Conservancy Council, Peterborough.

Welch, D. (1981). Diurnal movements of Scottish Blackface sheep between improved grassland and heather hill in north-east Scotland. *Journal of Zoology*, **194**, 267–271.

World Wilderness Congress (1983). Congress resolutions. *World Wilderness Congress*, 3.

Yalden, D.W. (1982). When did the mammal fauna of the British Isles arrive? *Mammal Review*, **12**, 1–57.

Long-term ecological change in the British uplands

H. J. B. BIRKS

Botanical Institute, University of Bergen, Allégaten 41, N-5007 Bergen, Norway

SUMMARY

1 Ecological change over periods of 100–1000 years in the British uplands is reconstructed from palaeoecological data, mainly pollen and macrofossils preserved in organic sediments.

2 Woodland covered much of the English uplands, and up to 50 per cent of the altitudinal range in mainland Scotland. Shetland, Orkney, Lewis, some of the Inner Hebrides and parts of Caithness supported open scrub.

3 Woodland composition varied in the uplands, with birch, pine and hazel being the major upland trees.

4 Extensive deforestation occurred, for different reasons, at 3700–3900, 2100–2600, 1400–1700 and 300–400 BP. Soils quickly became acid and blanket mire developed at different times in different areas.

5 Maximum summer warmth probably occurred *c.* 8000–9000 BP.

6 'Little Ice Age' climatic effects may have been ecologically important in extreme, marginal habitats.

7 Atmospheric pollution has been a major cause of recent vegetational change in the uplands.

8 A generalized model of ecological change in the uplands highlights the importance of edaphic and climatic factors in influencing the upland–lowland contrast within Britain.

INTRODUCTION

The complex of environmental factors that influence vegetational composition, structure and performance varies in time and space. The result is continuous vegetational and hence ecological change over a variety of temporal and spatial scales. Ecological change through time can be monitored by direct observation over periods of 10–100 years. However, important ecological changes in the British uplands have been operating since deglaciation *c.* 10,000 years ago. Such changes over periods of 100–10,000 years cannot be observed directly but can be reconstructed indirectly by detailed analyses of pollen and macrofossils (e.g. seeds, wood) preserved in peats and lake sediments. Such analyses provide a means of reconstructing the past flora, vegetation and, by inference, environment (e.g. soils, climate and biotic factors including human influence). An independent chronology for these reconstructions is provided by radiocarbon dating.

In this paper, uplands are defined as areas above the 'upper limit of enclosed

37

land' (Ratcliffe 1977) and thus comprise the sub-montane and montane zones. The upper limit of the sub-montane zone is drawn at the potential tree-limit (Ratcliffe 1977). The British uplands, as used here, thus include extensive areas of Scotland, large parts of northern England and Wales, and small areas of south-west England.

Despite the abundance of bogs and lakes in the uplands, comparatively little is known about their detailed ecological history. This contrasts with the enormous amount of information available from parts of lowland Britain (e.g. East Anglia, Somerset) and results from the historical development of Quaternary palynology in Britain, with its original geographical distribution centred on southern England. An advantage of this early neglect of the uplands by palaeoecologists is that nearly all upland studies are comparatively recent and hence have been implemented with modern techniques, such as ^{14}C-dating and identification of herb pollen types. Thus what may be lacking in quantity in upland palaeoecology is hopefully compensated for by its quality!

This paper draws on the results of recent studies to reconstruct (i) the horizontal and vertical extent of woodland and hence the extent of the montane zone during the Holocene (≡ post-glacial, Flandrian), (ii) the past patterns of woodland distribution in northern and western Britain, (iii) the timing of extensive deforestation, (iv) patterns of soil and bog development, (v) the role of climatic changes in influencing vegetational patterns through time, and (vi) recent ecological changes in the uplands.

Plant nomenclature follows Clapham, Tutin & Warburg (1962).

HOLOCENE EXTENT OF THE WOODLAND ZONE

Recent pollen-analytical studies demonstrate that Shetland, Orkney, the Outer Hebrides, parts of Caithness and some of the smaller Inner Hebridean islands never supported extensive woodland, even at low elevations, at any time during the Holocene (Birks 1977; Peglar 1979; Keatinge & Dickson 1979; Birks & Williams 1983; Walker 1984a; Johansen 1985). This lack of extensive woodland cover presumably resulted from exposure to sea spray and westerly gales, and a lack of summer warmth and hence a short growing season (McVean & Ratcliffe 1962). These areas have thus always been within the upland zone, either sub-montane or montane, during the last 10,000 years.

In Shetland, open herbaceous vegetation persisted until *c.* 9500 BP, after which *Betula pubescens* and *Corylus avellana* scrub developed locally in sheltered areas with *Salix* and *Juniperus communis* (Johansen 1985). Scrub persisted locally to *c.* 4500 BP (Birnie 1984). Tall-herb vegetation with *Filipendula ulmaria* and ferns was widespread, but was progressively replaced by grassland, heath and bog from *c.* 7500 BP (Johansen 1985). On Orkney Mainland and north-east Caithness (Fig.1), *Betula*, *Corylus* and *Salix* scrub developed locally, but the predominant vegetation prior to human interference was grassland and tall-herb stands with *Rumex acetosa*, *Filipendula* and abundant ferns (Keatinge & Dickson 1979; Peglar 1979).

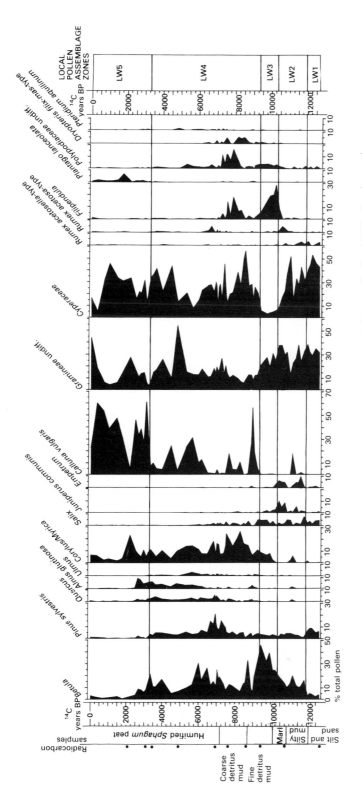

FIG. 1. Pollen diagram from Loch of Winless, north-east Caithness (Peglar 1979), showing major pollen and spore types plotted against sample age. All pollen and spore types are calculated as percentages of total pollen.

On Lewis the regional vegetation was similarly a mosaic of grassland, heath, bog and tall-herb stands with some *Salix* and ferns (Fig. 2). Occasional *Betula*, *Corylus* and *Pinus sylvestris* copses were present in locally favourable sites (Birks & Madsen 1979; Wilkins 1984; Bohncke 1988). Small areas of *Betula*, *Corylus* and possibly *Alnus glutinosa* scrub may have occurred locally on Canna and Tiree, even though the predominant regional vegetation was grassland and heath (Birks & Williams 1983). Trees were totally absent from St Kilda, where a mosaic of heath, fern-rich stands and maritime grassland prevailed (Walker 1984b).

This reconstructed absence of extensive woodland cover in Shetland, Orkney, Lewis and parts of Caithness based on pollen analysis contrasts with the local occurrence of wood remains preserved in peats in these areas (Birnie 1984), including pine stumps in Caithness and Lewis (Bennett 1984; Wilkins 1984). These wood remains do not, however, contradict the palynological reconstructions (Wilkins 1984). They simply indicate that in some areas small populations of trees occurred locally at certain times between 8000 and 4000 BP whereas the regional vegetation, as recorded by pollen deposited in basins of 10–100 ha and with pollen-source areas of 100–1000 km^2 (Birks 1986), was predominantly but not entirely treeless (Birks 1988). The reality of this reconstructed mosaic has been elegantly demonstrated by Bohncke (1988), who analysed peats from small (<5 ha) basins at Callanish, Lewis, and showed local tree growth in some areas and not in others. A possible modern analogue for early or mid-Holocene tree extent in northernmost and westernmost Scotland is the forest–tundra transition in Scandinavia, with widespread treeless vegetation and occasional small copses of trees in locally favourable sites.

The vertical extent of woodland and the position of the altitudinal tree-line are notoriously difficult to establish pollen-analytically (Maher 1964, 1972; Maguire & Caseldine 1985). A combination of pollen analysis, including estimates of annual pollen-accumulation rates, and macrofossils (Davis, Spear & Shane 1980) is required for reliable tree-line reconstructions. In the absence of such studies in the British uplands, I have assembled crude estimates of the maximum elevation of the Holocene tree-line in different areas based on my interpretation of the available pollen and macrofossil data (Table 1). My criteria for indicating woodland presence are the occurrence of tree macrofossils or tree pollen values in excess of 50 per cent total pollen. These tree-line estimates are approximate and will undoubtedly be modified in light of further, more detailed studies. As the elevation of the highest point and hence the maximum altitude to which trees could, in theory, grow, is different in the different areas, the estimates are standardized as percentages of the maximum altitude in the area (Fig. 3). These percentages decrease northwards and westwards and indicate that even at the time of maximum woodland cover, *c.* 8000–7000 BP, there were extensive areas above tree-line in North Wales, the Lake District and Scotland. In contrast, the Pennines, the North York Moors and Bodmin Moor were probably tree-covered and much of Dartmoor was similarly forested. Within the Pennines, however, there were treeless sites such as exposed summits, cliffs, screes, areas of

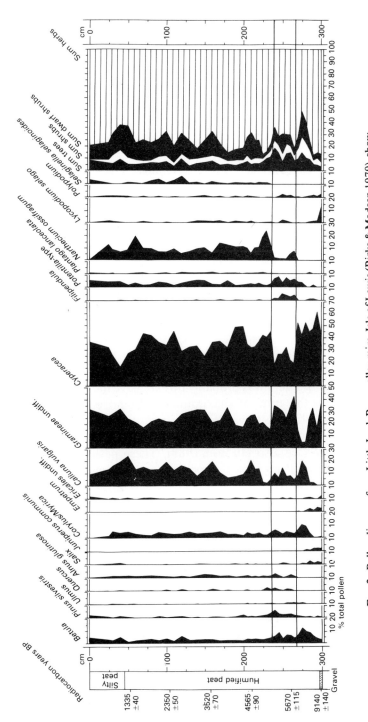

FIG. 2. Pollen diagram from Little Loch Roag valley mire, Isle of Lewis (Birks & Madsen 1979), showing major pollen and spore types plotted against sample depth. All pollen and spore types are
calculated as percentages of total pollen.

TABLE 1. Estimates of the maximum Holocene tree-line in different upland areas, the elevation of the highest point in these areas, and the tree-line elevation expressed as percentage of the elevation of the highest point

	Tree-line estimate (m)	Highest point (m)	$\frac{\text{Tree-line}}{\text{Highest point}}$ %	Reference
Bodmin Moor	419	419	100	Brown (1977)
Dartmoor	551	621	89	Maguire & Caseldine (1985)
S. Wales	>715	886	>81	Chambers (1982)
N. Wales	>635	1085	>59	Walker (1978)
North York Moors	454	454	100	Simmons (1969)
Southern Pennines	595	634	94	Tallis & Switsur (1983)
Northern Pennines	893	893	100	Turner (1984)
Lake District	760	978	78	Pennington (1970)
Galloway	>457	843	>54	Birks (1972a)
Western Grampians	716	1214	59	Donner (1962)
Eastern Grampians	700	1068	65	Huntley (1981)
Cairngorms	793	1311	60	Pears (1968)
North-west Highlands	520	1081	48	H. J. B. Birks & H. H. Birks (unpublished)
Isle of Skye	457	1008	45	H. J. B. Birks (unpublished)

shallow soil, mires, springs and flushes (Turner *et al.* 1973) in which shade-intolerant plants survived (e.g. *Polemonium caeruleum, Dryas octopetala, Helianthemum* spp.). There was probably no extensive montane vegetation south of North Wales at the time of maximum woodland extent. The gradient of lowered relative tree-line northwards and westwards (Fig. 3) is paralleled today in the altitudinal limit of enclosed land and hence, by definition, in the lower limit of the upland zone (see also Hudson 1988). These patterns presumably result from the gradients of increased exposure and decreased summer warmth northwards and westwards.

Floristically it is interesting that several shade-intolerant taxa of open habitats formerly grew on Bodmin Moor, the North York Moors, or the Southern Pennines and became locally extinct in the early Holocene (e.g. *Astragalus alpinus, Betula nana, Epilobium alsinifolium, Hippophae rhamnoides, Luzula arcuata, Lycopodium annotinum, Salix herbaceae* and *Saxifraga stellaris*). Extinctions also occurred within Scotland, for example the local extinction of *S. aizoides* and *L. annotinum* in Galloway and of *Koenigia islandica* on the Scottish mainland. There are several reasons for these extinctions, depending on the species concerned.

HOLOCENE WOODLAND PATTERNS

Comparison of the estimated maximum tree-line in the Holocene (Table 1 and Fig. 3) and the actual limit of woodland today indicates that many, but not all,

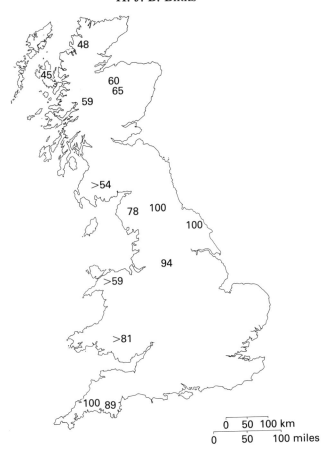

FIG. 3. Maximum Holocene tree-line estimates for different upland areas; estimates are expressed as percentages of the maximum altitude in each area.

upland areas, although treeless today, once supported trees. In this section I consider the geographical variation in the composition of this tree cover at *c.* 7000–5000 BP, just prior to extensive human interference (Birks, Deacon & Peglar 1975). The reconstruction for Scotland (Fig. 4) draws on many palynological sources, referenced in Birks (1977, 1980) and Walker (1984a), and largely confirms the reconstruction of potential woodland composition by McVean & Ratcliffe (1962) based on existing woodland fragments and ecological knowlege.

Betula pubescens–Sorbus aucuparia–Corylus avellana-dominated woodland, often with *Populus tremula* and *Prunus padus*, occurred widely in north-west Scotland and extended south to Loch Assynt. It was also common on northern and western Skye. On the mainland, *Pinus sylvestris* was an important but not dominant component in the Loch Assynt, Inverpolly and Ullapool areas. *Pinus* was the dominant tree, with *B. pubescens*, *S. aucuparia*, and *P. tremula* in the

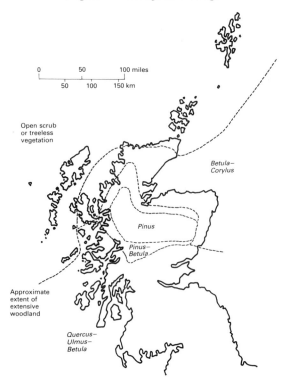

Fig. 4. Reconstruction of major woodland patterns for the mid-Holocene (7000–5000 BP) for Scotland.

Loch Maree–Torridon areas, and extended south-east to the Cairngorms. The relative importance of *Pinus* increased eastwards, reaching a maximum in Upper Deeside and Upper Speyside. Pine dominance did not, however, extend east of Dinnet. In the Aberdeenshire lowlands, *Betula* and *Corylus* were prominent, with some *Quercus* and *Ulmus*. Mixed *Pinus–Betula* stands occurred locally south of the *Pinus*-dominance zone, for example on eastern Skye and inland as far south as Rannoch Moor and Glen Falloch. In Southern Scotland and extending up the west coast through Knapdale, Mull and Ardnamurchan to southern Skye, *Quercus–Ulmus glabra–Betula* woods were widespread, with *Corylus*, *S. aucuparia* and *Ilex aquifolium*. *Alnus glutinosa* and *Salix* spp. occurred widely but locally throughout Scotland on wet soils.

Contrary to popular belief, Caledonian pine woodland was not the widespread 'primeval' woodland type of Scotland. Pine woods were centred on the acid, well-drained soils of the Cairngorms, Upper Deeside and Upper Speyside (Birks *et al.* 1975). *Pinus* extended westwards through, for example, Glen Affric, Glen Strathfarrar and Strath Bran to the Loch Maree–Torridon areas where it formed a major woodland component with *Betula*, *S. aucuparia* and *Ilex*. Between 5000 and 4000 BP *Pinus* extended north, west and south of its presumed native range today to form mixed pine–birch woods (Birks 1980). Remains of pine stumps occur widely in peats (Birks 1975; Bennett 1984). However, these reflect the local growth of pine on dried bog surfaces rather than the former extent of the Caledonian woodland, and indicate that pine bog-woodlands developed widely but locally from Rannoch Moor to A'Mhoine between 5000 and 4000 BP.

Many pine stumps preserved in peats in eastern Sutherland and western Caithness are only c. 20 cm diameter and yet represent 200+ years growth, indicating an annual growth increment of 1 mm or less. This extremely poor growth coupled with the absence of any extensive natural woodland cover in the last 10,000 years in this area are perhaps potential ecological indicators of the area's suitability for productive forestry. Despite these indicators, planting of non-native conifers is currently widespread in the area. It is striking that in western Scotland there are virtually no wood remains younger than c. 4000 BP (Birks 1975) and that pine became extinct over large areas of western Scotland at c. 4000 BP, including exposed coastal areas such as Cape Wrath, Point of Stoer, Rubha Coigeach and Lewis. Possible reasons for this are discussed later.

The woodland composition at or near the tree-line in Scotland is poorly known, but it is likely that in western and southern Scotland *Betula*, *P. tremula* and *S. aucuparia* formed an upper zone above the mixed deciduous woodlands. In the Cairngorms, *Pinus*, possibly with some *Betula*, may have formed the upper tree-limit, as it does today on Creag Fhiaclach (Miller & Cummins 1982) and in eastern Norway, with a fringing belt of juniper scrub. In the extreme north and west, *Betula* and *S. aucuparia* probably formed the tree-line, perhaps with an upper fringe of juniper (McVean 1961) and *Salix* scrub.

Further south, the composition of the uppermost woodland zone is better known. *Betula* and *Corylus*, often with *U. glabra* on richer soils and *Pinus* on poorer sites, appear to have formed the upper woodlands in the Lake District. In North Wales the patterns were similar except that *Quercus* was commoner and *Pinus* less frequent than in Lakeland. In the Southern Pennines both *Quercus* and *Pinus* were important at high elevations, along with *Betula* and *Corylus*. Further south (Dartmoor, Bodmin Moor and South Wales), *Corylus* scrub with some *Betula* and *Quercus* was abundant, whereas on the North York Moors *Betula*–*Corylus* scrub with some *Quercus* and, more locally, *Pinus* was widespread. In the Northern Pennines detailed analyses by Turner & Hodgson (1979, 1981, 1983) indicate considerable fine-scale differentiation related to soils, altitude and local climate. For example, *Pinus* was prominent in Upper Teesdale and the Derwent area; *Corylus* and *Ulmus* were important at high elevations in the west on a range of soils, but elm was least frequent on the Millstone Grit; and *Betula* was most abundant above 600 m and in areas lacking *Pinus*.

DEFORESTATION IN THE UPLANDS

Calculations based on Milankovitch's theory of global climatic change predict an 8 per cent increase in summer solar radiation and a corresponding decrease in winter insolation between 10000 and 8000 BP in the northern hemisphere (Kutzbach & Guetter 1986). This resulted from the precession cycle of the earth's orbital axis. The earth was nearest to the sun in July *c.* 10,000–8000 years ago, leading to increased summer solstice radiation. This, in turn, may have resulted in a maximum altitudinal extent of trees in the early Holocene in the uplands, similar to the maximum latitudinal tree-extent in north-west Canada *c.* 9000–8000 BP. (Ritchie, Cwynar & Spear 1983). There are indications from the uplands that maximum tree-lines were attained by 8000–7000 BP, e.g. Dartmoor (Caseldine & Maguire 1986), Cairngorms (Dubois & Ferguson 1985) and Southern Pennines (Tallis & Switsur 1983), suggesting that the woodland zone was at its Holocene maximum at about these times, and thus that the montane zone had its minimal extent then.

The first minor deforestation of the uplands may have occurred soon after this tree-line maximum. There is increasing evidence from charcoal and pollen stratigraphy in upland peats for a tree-line depression of 100–200 m in parts of England and Wales from *c.* 7500 BP, perhaps as a result of 'regular and recurrent burning by Mesolithic hunting populations' leading to 'a permanent suppression of a closed tree cover above *c.* 350 m altitude' (Jacobi, Tallis & Mellars 1976). This may be the case in Dartmoor, South Wales, the North York Moors, the Southern Pennines and, possibly, the Northern Pennines.

In considering when the uplands were first deforested, it is impossible to document all the local temporary clearances (*sensu* Turner 1965) made from *c.* 5000 BP by Neolithic and Bronze Age people. From an ecological viewpoint it is more relevant to establish when extensive, permanent deforestation occurred. In Table 2, I have assembled for different areas the known ages for the first extensive deforestation phase, defined as when tree pollen values drop to 50 per cent of their Holocene maximum percentages. Four major phases of extensive deforestation can be recognized: (i) 3700–3900 BP, confined to the north-west Highlands and eastern Skye; (ii) 2100–2600 BP (pre-Roman Iron Age), during which deforestation occurred in all upland areas of Wales, England (except the Lake District), northern Sutherland and northern Skye; (iii) 1400–1700 BP (post-Roman), with extensive deforestation in the Lake District, Galloway, Knapdale–Ardnamurchan and southern Skye; (iv) 300–400 BP, restricted to the Grampians and Cairngorms.

Before discussing these patterns, it should be emphasized that local clearances, burning and agriculture occurred in nearly all areas prior to extensive deforestation but these phases were often short-lived and temporary (Walker 1984a). The small areas of scrub on Shetland, Orkney, Lewis and north-east Caithness were largely destroyed *c.* 5000–4000 BP as a result of Neolithic clearance, burning and grazing. There were extensive Neolithic and Bronze Age settlements in these areas with, for example, field systems dating to 4800 BP now buried under peat on Shetland (Whittington 1978) and cereal cultivation on Orkney (Walker 1984a).

Some interesting patterns emerge from Table 2. The three later phases of

TABLE 2. Age in ^{14}C-years BP for the earliest extensive deforestation phase in different upland areas

	Earliest extensive clearance BP	Reference
Bodmin Moor	2400	Brown (1977)
Dartmoor and Exmoor	2200	Merryfield & Moore (1974)
S. Wales	2200	Chambers (1982)
N. Wales	2100	Walker (1978)
North York Moors	2300	Atherden (1976)
Southern Pennines	2400	Hicks (1971)
Northern Pennines	2500	Turner et al. (1973)
Lake District	1700	Pennington (1970)
Galloway	1500	Birks (1972a)
Knapdale–Ardnamurchan	1400	Birks (1980)
Western Grampians	400	H. J. B. Birks (unpublished)
Eastern Grampians	300	Huntley (1981)
Cairngorms	300	O'Sullivan (1974)
North-west Highlands	3700	Birks (1980)
Northern Sutherland	2500	Birks (1980)
Isle of Skye (S.)	1700	Birks & Williams (1983)
Isle of Skye (N.)	2600	Birks & Williams (1983)
Isle of Skye (E.)	3900	Birks & Williams (1983)

extensive deforestation undoubtedly result from human interference in the uplands. The pre-Roman Iron Age clearances were confined in Scotland to areas of dominant birch–hazel woodland in the far north and west, whereas the post-Roman clearances affected the mixed-deciduous woodland areas of western Scotland. Extensive deforestation was comparatively recent in the central and eastern Highlands. The causes of the early deforestation phase in the north-west Highlands and eastern Skye are less clear. This phase follows the widespread pine decline at c. 4000 BP and is associated with the expansion of blanket mires. The reasons for the spectacular pine decline over such a wide area (Lamb 1964; Birks 1975; Bennett 1984) are not fully known. An interaction between human activity, including burning, and a change to a more oceanic climate with increased precipitation and strong winds resulting from a shift in the position of Atlantic storm tracks would have caused waterlogging, thereby encouraging bog spread, and would have inhibited pine regeneration on mineral soils by reducing the frequency of good seed-years (Birks 1972b). There is independent evidence from various sources for increased oceanicity and gale frequency at c. 4000 BP, e.g. machair and dune formation in the Uists and Orkney (Birks 1988). Such a climatic change would have inhibited pine growth in many exposed coastal areas, e.g. Cape Wrath and Stoer.

The available palaeoecological data suggest that extensive deforestation of the uplands occurred, for different reasons, from c. 4000 BP to 300 BP.

SOIL CHANGES AND BOG DEVELOPMENT

Progressive leaching of exchangeable ions, often with acid humus accumulation, is, under the cool, moist climate of the uplands, an inevitable pedogenic process independent of climatic change and human influence (Birks 1986). The end result is a preponderance of acidic, base-deficient soils, often with mor humus (e.g. podzols, semi-podzols, rankers and peats), except in areas of limestone and other base-rich rocks and in locally flushed situations. However, at the time of deglaciation, upland soils freshly deposited by glacial activity and disturbed by periglacial activity would have been skeletal, devoid of humus and richer in bases than today. Chemical investigations of lake sediments in the uplands (such sediments are primarily derived from catchment material and are thus an indirect record of catchment soils (Mackereth 1965)) indicate that acidic soils had already developed by *c.* 8000 BP following rapid leaching of soluble bases in the early Holocene. Reconstructions of lake acidity from stratigraphical diatom studies in the Lake District (Haworth 1969) and Galloway (Jones, Stevenson & Battarbee 1986) show lakes were acid (e.g. pH 5·5–6·0 in Galloway) early in the Holocene, presumably as a result of catchment soil deterioration. Lake acidity was then stable for over 8000 years despite expansion of blanket mires in the catchment during mid-Holocene times, in contrast to the rapid increase in lake acidity (pH <5·0) in the last 100 years (Jones *et al.* 1986).

Extensive mor accumulation and formation of humus-iron pans in the uplands would, above the pan, have inhibited vertical drainage of soil water, decreased soil aeration and resulted in waterlogging and widespread paludification; hence the development of the extensive blanket mires so characteristic of the British uplands today. Some blanket mires developed as a result of natural soil degradation in the absence of human disturbance or climatic change as early as 9000 BP in the Scottish Highlands (Birks 1975). In other areas climatic change at, for example, 4000 BP may have initiated blanket-mire formation or expansion (e.g. Pennington *et al.* 1972). Prehistoric people, through woodland clearance, burning and animal grazing, may also have altered local hydrological balances and accelerated pedogenic processes, particularly in more southerly upland areas such as Exmoor, Dartmoor and Wales (Moore, Merryfield & Price 1984). Outside the Scottish Highlands, blanket-peat formation may have begun at any time between *c.* 7500 and 2000 BP, depending on altitude and local topography (Tallis 1964a; Chambers 1981). Peat generally developed earlier on high-altitude plateaux and in basins than in valley bottoms and on gentler slopes at lower altitudes, whereas accumulation on exposed sites at a range of altitudes started even later (Tallis 1964a; Edwards & Hirons 1982). There is thus no general date or single cause for upland blanket-mire origin.

Soil erosion is a further important pedogenic process in the uplands, resulting from deforestation, burning, intensive grazing and drainage. Spectacular examples of erosion resulting from human interference are recorded by 10 m of gravel deposited in the last 1700 years in Llyn Peris, Snowdonia (Tinsley & Derbyshire 1976) and the widespread formation of truncated podzols and associated erosion at the time of extensive post-Roman clearances in the Lake District uplands (Pennington 1970). Modern (1965–75) erosion rates in Snowdonia are estimated to be c. 0·42 t ha^{-1} year^{-1} compared with 0·05 t ha^{-1} year^{-1} in 1800, probably resulting from excessive grazing by sheep (Dearing, Elner & Happey-Wood 1981). The majority of hillslope debris flows in the Scottish Highlands have formed in the last 250 years (Innes 1983). The formation of such flows is most commonly initiated by intense rainstorms (Ballantyne 1986a). However, debris-flow formation is much more frequent and more destructive today than in the past (Ballantyne 1986a), probably as a result of recent land-use changes, particularly burning and intensive grazing (Innes 1983).

Very little is known about the extent or history of periglacial processes in the montane zone during the Holocene. There is evidence for renewed high-altitude amorphous solifluction (*sensu* Dahl 1956) in the Cairngorms and the north-west Highlands at c. 4000 and 2500 BP and during the 'Little Ice Age' of AD 1550–1750 (Sugden 1971; Mottershead 1978; Ballantyne 1986b). Structured or patterned (stripes, gliding blocks) and amorphous solifluction features in the Rhinogs may similarly date from the 'Little Ice Age' (Goodier & Ball 1969). Modern rates of down-slope movement of solifluction lobes in the Scottish Highlands are c. 6–35 mm year^{-1}, whereas Holocene rates probably did not exceed 6 mm year^{-1} (Ballantyne 1986b).

Other mechanisms of soil erosion in the montane zone include wind erosion, particularly in the north-west Highlands where strong winds with gusts exceeding 100 km h^{-1} and even 200 km h^{-1} occur (Taylor 1976). Deflation surfaces, fellfield vegetation and deposits of windblown sands formed by niveo-aeolian activity occur locally on, for example, An Teallach and Ben More Coigach (Ballantyne & Whittington 1987) and Ben Arkle (Pye & Paine 1984). Sand accumulation began on An Teallach c. 8000 years ago but was greatly reduced by mid-Holocene vegetation development. Massive recent erosion and redeposition have occurred in the last 200 years as a result of excessive grazing by sheep (Ballantyne & Whittington 1987). In contrast, snow-patch erosion appears to be relatively unimportant under present conditions (Ballantyne 1985), as are rockfall accumulation and talus formation (Ballantyne & Eckford 1984).

CLIMATIC CHANGE

Ideas about the nature of climatic change in the Holocene have been revolutionized in the last decade by the convincing demonstration that the Milankovitch astronomical theory of orbitally induced global climatic change is the principal determinant of Quaternary glacial–interglacial cycles (Hays, Imbrie & Shackleton 1976; Berger 1981; Berger *et al.* 1984). It is now possible to predict, using climatic and atmospheric dynamic models, past climatic patterns for different regions from the Milankovitch theory (Berger 1978; Kutzbach & Guetter 1986). These predictions show how variations in, for example, summer radiation resulting from the obliquity, precession and eccentricity cycles of the earth's orbit produce geographically distinct atmospheric, oceanic, continental and glacial responses (Imbrie 1985).

Although these developments are in their infancy, enough is now known to refute assumptions such as synchronous climatic change over large geographical areas during the Holocene (Kutzbach & Guetter 1986), and the long-held idea of alternating cool or warm and wet or dry phases in the Blytt–Sernander scheme (e.g. Boreal, Atlantic, etc.). There is no unambiguous evidence for the validity of this scheme in Britain or Scandinavia (Birks 1982); moreover the Milankovitch model predicts continuous rather than discontinuous climatic change at differing rates in time and space.

There are numerous problems in reconstructing past climate from palaeo-ecological data (Birks 1981). A conservative interpretation of the available data, using current (but limited) knowledge of climatic tolerances of the taxa concerned, is that maximum summer warmth in the uplands occurred *c.* 8000–9000 BP and not in the so-called mid-Holocene 'thermal optimum'. Many climatic changes presumably occurred in the last 8000 years but these changes would only have been important in influencing vegetational change in particular ecologically critical situations. The inferred increase in oceanicity *c.* 4000 BP illustrates this well. There are independent reconstructions from stable-isotope ratios (D/H) preserved in pine stumps from the Cairngorms of major increases in precipitation at 4200–3900 BP (and also at *c.* 7300, 6200–5800 and 3300 BP; Dubois & Ferguson 1985). However, no major vegetational change occurred in the area at this time, presumably because no critical ecological threshold was crossed. In contrast, the contemporaneous increase in oceanicity had major ecological effects along the western seaboard of Scotland and Ireland and may have resulted in the extinction of pine in much of the extreme west.

Turning to historically documented climatic change, particularly the 'Little Ice Age' of AD 1550–1750, there is some tentative evidence from lichen measurements inside and outside boulder moraines to suggest that corrie glaciers formed in some of the Cairngorm corries in the seventeenth to nineteenth centuries (Sugden 1977). The expansion of maritime grassland at the expense of heath and grassland on St Kilda at *c.* 500 BP (Walker 1984b) probably resulted from increased westerly winds and storminess in the North Atlantic, features that accompanied the cooler and wetter conditions of the 'Little Ice Age'.

It is likely that important ecological changes in habitats at high elevations or in extreme situations in the upland zone may have been initiated by climatic change associated with the 'Little Ice Age'. Much remains to be discovered about the effects of recent climatic change on upland biota, particularly the importance of climatic extremes and marginality in the uplands (Parry & Carter 1985).

RECENT ECOLOGICAL CHANGES IN THE UPLANDS

Ecological changes over the last 100–200 years are poorly documented, partly because palaeoecologists are primarily interested in longer time periods and broad-scale changes. There is, however, evidence from various palaeoecological sources to indicate major changes in the uplands that occurred largely as a result of recent land-use change or atmospheric pollution.

Complete woodland clearance (*sensu* Turner 1965) of much of Scotland occurred in the last two to three centuries, e.g. in Skye (Birks & Williams 1983) and north-west Sutherland (Birks 1980). The first extensive deforestation of the central and eastern Highland glens was similarly very recent (Table 2). In addition, increased sheep and deer stocking in the last two centuries, for example in Caenlochan Glen and Coire Fee, resulted in grassland expansion and restriction of tall herbs and ferns to cliff ledges and other inaccessible sites and in the impaired regeneration and even degeneration of high-elevation birchwoods (Huntley 1981). These areas appear to have been relatively undisturbed by humans until the recent past.

The dominance of *Eriophorum vaginatum* in the Southern Pennines appears to have developed as a result of intense grazing and burning of these uplands *c.* 500 years ago (Tallis 1964b). Other ecological changes in the area are almost certainly responses to the atmospheric pollution of the industrial revolution; these changes include the virtual disappearance of *Sphagnum* from blanket-mires (Tallis 1964b; Lee, Tallis & Woodin, 1988). Chambers, Dresser & Smith (1979) have shown in South Wales that the change from heather moor to the dominance of *Molinia caerulea*, so characteristic of upland Wales and Galloway today, occurred after the start of the industrial revolution and suggested that the high deposition of soot ($2 \cdot 5$–4 g m^{-2} year^{-1}) may have had important ecological effects.

The recent increase in the extent and rate of blanket-mire erosion in the Southern Pennines may, in part, be a response to intense land use and atmospheric pollution (Tallis 1985; cf. Bowler & Bradshaw 1985). Recent erosion and inwashing of peat into lakes had important limnological effects in Galloway, with a change from clear, oligotrophic conditions with submerged aquatics such as *Isoetes lacustris* to dystrophic conditions lacking isoetids (Birks 1972a). The most important recent ecological change in several lakes in the uplands is the onset of major lake acidification from about AD 1840, e.g. in Galloway (Battarbee *et al.* 1985) and Wales (Fritz *et al.* 1986). These changes, reconstructed from detailed stratigraphical diatom analyses of recent lake sediments, occur in the absence of afforestation or significant land-use changes and greatly exceed long-term changes in lake acidification (Jones *et al.* 1986). They result from acid atmospheric deposition in areas of acidic bedrock and have had major ecological impacts on, for example, fish populations in some upland areas.

CONCLUSIONS

Palaeoecological data demonstrate that very considerable ecological changes have occurred in the uplands over time scales of 100–10,000 years. They also show important geographical variation in the patterns of change, for example

TABLE 3. Major phases in the Holocene (post-glacial) ecological history of the lowlands and uplands of Britain

Phase	General characteristics	Lowlands	Uplands
Cryocratic	Cool, dry climate. Base-rich skeletal mineral soils	>13500 BP	>10000 BP (last glaciation)
Protocratic	Increasingly temperate climate. Unleached base-rich soils	13500–9500 BP Grassland and open woodland	13500–9000 BP Grassland, scrub, and some woodland
Mesocratic	Temperate climate. Fertile brown earths	9500–5000 BP Mixed deciduous forest	9000–8000 BP Tall-herb and fern-rich vegetation. Some woodland
Oligocratic	Temperate climate. Leached soils, podzols, peats	? Absent in Holocene (Coniferous woodland, heath, bog in previous interglacials)	8000–2500 BP Heaths, blanket mires
Homo sapiens	Temperate climate. Eroded or fertilized soils	5000–0 BP Grassland, agriculture	2500–0 BP Heaths, grasslands, blanket mires

deforestation times and onset of paludification. Despite this spatial and temporal variability, some general patterns of ecological change in the uplands can be recognized and integrated into a simple descriptive model. This model is Birks's (1986) five-phase modification of Iversen's (1958) glacial–interglacial cycle. The principal features of the model are summarized in Table 3. The approximate duration and broad characteristics of the various phases in the lowlands and uplands are also shown.

This model emphasizes the major ecological contrasts between the lowland and upland regions of Britain. The mesocratic phase with fertile brown earths supporting luxuriant and productive vegetation was short-lived (1000 years) in the uplands compared with southern and eastern areas. The oligocratic phase with widespread acid soils, podzols and peats began in the uplands as early as 8000 BP, whereas there is little or no evidence for natural oligotrophication during the Holocene in the lowlands. The *Homo sapiens* phase, the period of intensive human impact on vegetation, began *c.* 5000 BP in the lowlands in contrast to the uplands where intensive disturbance did not commence until *c.* 2500 BP.

The principal ecological factor that characterizes the long-term ecological history of the uplands is the widespread development of infertile acidic soils (oligocratic phase), as a result of the acidic nature of much of the local bedrock and till, coupled with intensive leaching under a cool, moist climate. This long-term natural soil acidification over much of the Holocene has been greatly increased by the rapid recent acidification from acid atmospheric deposition in certain upland areas. Fundamental edaphic and climatic distinctions between the lowlands and uplands have always existed and have resulted in the long-standing vegetational, archaeological, land-use and economic differentiation of Britain into a fertile, prosperous and densely populated lowland zone and an infertile, poor and lightly populated upland zone.

ACKNOWLEDGMENTS

In writing this review I have drawn extensively on the published work of many palaeoecologists, too numerous to cite because of space limitations. I gratefully acknowledge their contribution. I am also indebted to Colin Ballantyne, Hilary Birks and Derek Ratcliffe for invaluable discussions, to John Miles for useful comments on an earlier draft, and to Hilary Birks and Annechen Ree for indispensable help.

REFERENCES

Atherden, M.A. (1976). Late Quaternary vegetational history of the North York Moors. III. Fen Bogs. *Journal of Biogeography*, **3**, 115–124.

Ballantyne, C.K. (1985). Nivation landforms and snowpatch erosion on two massifs in the Northern Highlands of Scotland. *Scottish Geographical Magazine*, **101**, 40–49.

Ballantyne, C.K. (1986a). Landslides and slope failures in Scotland: a review. *Scottish Geographical Magazine*, **102**, 134–150.

Ballantyne, C.K. (1986b). Late Flandrian solifluction on the Fannich mountains, Rosshire. *Scottish Journal of Geology*, **22**, 395–406.

Ballantyne, C.K. & Eckford, J.D. (1984). Characteristics and evolution of two relict talus slopes in Scotland. *Scottish Geographical Magazine*, **100**, 20–33.

Ballantyne, C.K. & Whittington, G. (1987). Niveo-aeolian sand deposits on An Teallach, Wester Ross, Scotland. *Transactions of the Royal Society of Edinburgh (Earth Sciences)*, **78**, 51–63.

Battarbee, R.W., Flower, R.J., Stevenson, A.C. & Rippey, B. (1985). Lake acidification in Galloway: a palaeoecological test of competing hypotheses. *Nature*, **314**, 350–352.

Bennett, K.D. (1984). The post-glacial history of *Pinus sylvestris* in the British Isles. *Quaternary Science Reviews*, **3**, 133–155.

Berger, A.L. (1978). Long-term variations of caloric insolation resulting from the Earth's orbital elements. *Quaternary Research*, **9**, 139–167.

Berger, A.L. (1981). The astronomical theory of paleoclimates. *Climatic Variations and Variability: Facts and Theories* (Ed. by A. Berger), pp. 501–525. Reidel, Dordrecht.

Berger, A., Imbrie, J., Hays, J., Kukla G. & Saltzman, B. (1984). *Milankovitch and Climate*. Reidel, Dordrecht.

Birks, H.H. (1972a). Studies in the vegetational history of Scotland II. Two pollen diagrams from the Galloway Hills, Kirkcudbrightshire. *Journal of Ecology*, **60**, 183–217.

Birks, H.H. (1972b). Studies in the vegetational history of Scotland III. A radiocarbon-dated pollen diagram from Loch Maree, Ross and Cromarty. *New Phytologist*, **71**, 731–754.

Birks, H.H. (1975). Studies in the vegetational history of Scotland IV. Pine stumps in Scottish blanket peats. *Philosophical Transactions of the Royal Society of London B*, **270**, 181–226.

Birks, H.J.B. (1977). The Flandrian forest history of Scotland: a preliminary synthesis. *British Quaternary Studies Recent Advances* (Ed. by F.W. Shotton), pp. 119–135. Clarendon Press, Oxford.

Birks, H.J.B. (1980). Quaternary vegetational history of west Scotland. *5th International Palynological Conference, Cambridge. Excursion C8 Guidebook*, pp. 1–70.

Birks, H.J.B. (1981). The use of pollen analysis in the reconstruction of past climates: a review. *Climate and History* (Ed. by T.M.L. Wigley, M.J. Ingram & G. Farmer), pp. 111–138. Cambridge University Press, Cambridge.

Birks, H.J.B. (1982). Holocene (Flandrian) chronostratigraphy of the British Isles: a review. *Striae*, **16**, 99–105.

Birks, H.J.B. (1986). Late-Quaternary biotic changes in terrestrial and lacustrine environments, with particular reference to north-west Europe. *Handbook of Holocene Palaeoecology and Palaeohydrology* (Ed. by B.E. Berglund), pp. 3–65. Wiley, Chichester.

Birks, H.J.B. (1988). Floristic and vegetational history of the Outer Hebrides. *Flora of the Outer Hebrides* (Ed. by R.J. Pankhurst). British Museum (Natural History), London.

Birks, H.J.B. & Madsen, B.J. (1979). Flandrian vegetational history of Little Loch Roag, Isle of Lewis, Scotland. *Journal of Ecology*, **67**, 825–842.

Birks, H.J.B. & Williams, W. (1983). Late-Quaternary vegetational history of the Inner Hebrides. *Proceedings of the Royal Society of Edinburgh*, **838**, 269–292.

Birks, H.J.B., Deacon, J. & Peglar, S. (1975). Pollen maps for the British Isles 5000 years ago. *Proceedings of the Royal Society of London B*, **189**, 87–105.

Birnie, J. (1984). Trees and shrubs in the Shetland Islands: evidence for a postglacial climatic optimum? *Climatic Changes on a Yearly to Millenial Basis* (Ed. by N.-A. Mörner & W. Karlén), pp. 155–161. Reidel, Dordrecht.

Bohncke, S.J.P. (1988). Vegetation and habitation history of the Callanish area, Isle of Lewis, Scotland. *The Cultural Landscape — Past, Present, Future* (Ed. by H.H. Birks, H.J.B. Birks, P.E. Kaland & D. Moe). Cambridge University Press, Cambridge.

Bowler, M. & Bradshaw, R.H.W. (1985). Recent accumulation and erosion of blanket peat in the Wicklow Mountains, Ireland. *New Phytologist*, **101**, 543–550.

Brown, A.P. (1977). Late-Devensian and Flandrian vegetational history of Bodmin Moor, Cornwall. *Philosophical Transactions of the Royal Society of London B*, **276**, 251–320.

Caseldine, C.J. & Maguire, D.J. (1986). Late glacial/early Flandrian vegetation change on northern Dartmoor, south-west England. *Journal of Biogeography*, **13**, 255–264.

Chambers, F.M. (1981). Date of blanket peat initiation in upland South Wales. *Quaternary Newsletter*, **35**, 24–29.

Chambers, F.M. (1982). Two radiocarbon-dated pollen diagrams from high-altitude blanket peats in South Wales. *Journal of Ecology*, **70**, 445–459.

Chambers, F.M., Dresser, P.Q. & Smith, A.G., (1979). Radiocarbon dating evidence on the impact of atmospheric pollution on upland peats. *Nature*, **282**, 829–832.

Clapham, A.R., Tutin, T.G. & Warburg, E.F. (1962). *Flora of the British Isles*, 2nd edn. Cambridge University Press, Cambridge.

Dahl, E. (1956). Rondane — mountain vegetation in south Norway and its relation to the environment. *Skrifter Utgitt av det Norske Videnskaps-Akademi i Oslo I. Mat.-Naturv. Klasse*, **3**, 1–374.

Davis, M.B., Spear, R.W. & Shane, L.C.K. (1980). Holocene climate of New England. *Quaternary Research*, **14**, 240–250.

Dearing, J.A., Elner, J.K. & Happey-Wood, C.M. (1981). Recent sediment flux and erosional processes in a Welsh upland lake-catchment based on magnetic susceptibility measurements. *Quaternary Research*, **16**, 356–372.

Donner, J.J. (1962). On the post-glacial history of the Grampian Highlands of Scotland. *Societas Scientiarum Fennica Commentationes Biologicae*, **24**(6), 1–29.

Dubois, A.D. & Ferguson, D.K. (1985). The climatic history of pine in the Cairngorms based on radiocarbon dates and stable isotope analysis, with an account of the events leading up to its colonization. *Review of Palaeobotany and Palynology*, **46**, 55–80.

Edwards, K.J. & Hirons, R.K. (1982). Date of blanket peat initiation and rates of spread — a problem in research design. *Quaternary Newsletter*, **36**, 32–37.

Fritz, S., Stevenson, A.C., Patrick, S.T., Appleby, P., Oldfield, F., Rippey, B., Darley, J. & Battarbee, R.W. (1986). *Palaeoecological Evaluation of the Recent Acidification of Welsh Lakes 1. Llyn Hir, Dyfed*. Working Paper 16, Palaeoecology Research Unit, Department of Geography, University College, London.

Goodier, R. & Ball, D.F. (1969). Recent ground pattern phenomena in the Rhinog Mountains, North Wales. *Geografiska Annaler*, **51A**, 121–126.

Haworth, E.Y. (1969). The diatoms of a sediment core from Blea Tarn, Langdale. *Journal of Ecology*, **57**, 429–439.

Hays, J.D., Imbrie, J. & Shackleton, N.J. (1976). Variations in the Earth's orbit; pacemaker of the Ice Ages. *Science*, **194**, 1121–1132.

Hicks, S.P. (1971). Pollen-analytical evidence for the effect of prehistoric agriculture on the vegetation of North Derbyshire. *New Phytologist*, **70**, 647–667.

Hudson, G. (1988). Effects of latitude, oceanicity and geology on alpine and sub-alpine soils in Scotland. *This volume.*

Huntley, B. (1981). The past and present vegetation of the Caenlochan National Nature Reserve, Scotland II. Palaeoecological investigations. *New Phytologist*, **87**, 189–222.

Imbrie, J. (1985). A theoretical framework for the Pleistocene ice ages. *Journal of the Geological Society of London*, **142**, 417–432.

Innes, J.L. (1983). Lichenometric dating of debris-flow deposits in the Scottish Highlands. *Earth Surface Processes and Landforms*, **8**, 579–588.

Iversen, J. (1958). The bearing of glacial and interglacial epochs on the formation and extinction of plant taxa. *Uppsala Universitets Arsskrift*, **6**, 210–215.

Jacobi, R.M., Tallis, J.H. & Mellars, P.A. (1976). The Southern Pennine Mesolithic and the ecological record. *Journal of Archaeological Science*, **3**, 307–320.

Johansen, J. (1985). Studies in the vegetational history of the Faroe and Shetland Islands. *Annales Societatis Scientiarum Færoensis Supplementum*, **XI**, 1–117.

Jones, V.J., Stevenson, A.C, & Battarbee, R.W. (1986). Lake acidification and the land-use hypothesis: a mid-post-glacial analogue. *Nature*, **322**, 157–158.

Keatinge, T.H. & Dickson, J.H. (1979). Mid-Flandrian changes in vegetation on Mainland Orkney. *New Phytologist*, **82**, 585–612.

Kutzbach, J.E. & Guetter, P.J. (1986). The influence of changing orbital parameters and surface boundary conditions on climatic simulations for the past 18,000 years. *Journal of Atmospheric Sciences*, **43**, 1726–1759.

Lamb, H.H. (1964). Trees and climatic history in Scotland. *Quarterly Journal of the Royal Meteorological Society*, **90**, 382–394.

Lee, J.A., Tallis, J.H. & Woodin, S.J. (1988). Acid deposition and British upland vegetation. *This volume.*

Mackereth, F.J.H. (1965). Chemical investigations of lake sediments and their interpretation. *Proceedings of the Royal Society of London B*, **161**, 295–309.

McVean, D.N. (1961). Post-glacial history of juniper in Scotland. *Proceedings of the Linnean Society of London*, **172**, 53–55.

McVean, D.N. & Ratcliffe, D.A. (1962). *Plant Communities of the Scottish Highlands.* HMSO, London.

Maguire, D.J. & Caseldine, C.J. (1985). The former distribution of forest and moorland on northern Dartmoor. *Area*, **17**, 193–203.

Maher, L.J. (1964). *Problems and possibilities of pollen analysis in mountainous regions.* Unpublished manuscript prepared for the American Association for the Advancement of Science Annual Meeting, Montreal, December 1964.

Maher, L.J. (1972). Absolute pollen diagram of Redrock Lake, Boulder County, Colorado. *Quaternary Research*, **2**, 531–553.

Merryfield, D.L. & Moore, P.D. (1974). Prehistoric human activity and blanket peat initiation on Exmoor. *Nature*, **250**, 439–441.

Miller, G.R. & Cummins, R.P. (1982). Regeneration of Scots pine *Pinus sylvestris* at a natural tree-line in the Caingorm Mountains, Scotland. *Holarctic Ecology*, **5**, 27–34.

Moore, P.D., Merryfield, D.L. & Price, M.D.R. (1984). The vegetation and development of British mires. *European Mires* (Ed. by P.D. Moore), pp. 203–235. Academic Press, London.

Mottershead, D.N. (1978). High altitude solifluction and post-glacial vegetation, Arkle, Sutherland. *Transactions of the Botanical Society of Edinburgh*, **43**, 17–24.

O'Sullivan, P.E. (1974). Two Flandrian pollen diagrams from the east-central Highlands of Scotland. *Pollen et Spores*, **16**, 33–57.

Parry, M.L. & Carter, T.R. (1985). The effect of climatic variations on agricultural risk. *Climatic Change*, **7**, 95–110.

Pears, N.V. (1968). Post-glacial tree-lines of the Cairngorm Mountains, Scotland. *Transactions of the · Botanical Society of Edinburgh*, **40**, 361–394.

Peglar, S.M. (1979). A radiocarbon-dated pollen diagram from Loch of Winless, Caithness, north-east Scotland. *New Phytologist*, **82**, 245–263.

Pennington, W. (1970). Vegetation history in the north-west of England: a regional synthesis. *Studies in the Vegetational History of the British Isles* (Ed. by D. Walker & R.G. West), pp. 41–79. Cambridge University Press, Cambridge.

Pennington, W., Haworth, E.Y., Bonny, A.P. & Lishman, J.P. (1972). Lake sediments in northern Scotland. *Philosophical Transactions of the Royal Society of London B*, **264**, 191–294.

Pye, K. & Paine, A.D.M. (1984). Nature and source of aeolian deposits near the summit of Ben Arkle, Northwest Scotland. *Geologie en Mijnbouw*, **63**, 13–18.

Ratcliffe, D.A. (Ed.) (1977). *A Nature Conservation Review, Vol. I.* Cambridge University Press, Cambridge.

Ritchie, J.C., Cwynar, L.C. & Spear, R.W. (1983). Evidence from north-west Canada for an early Holocene Milankovitch thermal maximum. *Nature*, **305**, 126–128.

Simmons, I.G. (1969). Pollen diagrams from the North York Moors. *New Phytologist*, **68**, 807–827.

Sugden, D.E. (1971). The significance of periglacial activity on some Scottish mountains. *Geographical Journal*, **137**, 388–392.

Sugden, D.E. (1977). Did glaciers form in the Cairngorms in the 17th–19th centuries? *Cairngorm Club Journal*, **18**, 189–201.

Tallis, J.H. (1964a). The pre-peat vegetation of the Southern Pennines. *New Phytologist*, **63**, 363–373.

Tallis, J.H. (1964b). Studies on Southern Pennine peats III. The behaviour of *Sphagnum*. *Journal of Ecology*, **52**, 345–353.

Tallis, J.H. (1985). Mass movement and erosion of a Southern Pennine blanket peat. *Journal of Ecology*, **73**, 283–315.

Tallis, J.H. & Switsur, V.R. (1983). Forest and moorland in the South Pennine uplands in the mid-Flandrian period I. Macrofossil evidence of the former forest cover. *Journal of Ecology*, **72**, 585–600.

Taylor, J.A. (1976). Upland climates. *The Climate of the British Isles* (Ed. by T.J. Chandler & S. Gregory), pp. 264–287. Longman, London.

Tinsley, H.M. & Derbyshire, E. (1976). Late-glacial and postglacial sedimentation in the Peris–Padarn Rock Basin, North Wales. *Nature*, **260**, 234–238.

Turner, J. (1965). A contribution to the history of forest-clearance. *Proceedings of the Royal Society of London B*, **161**, 343–354.

Turner, J. (1984). Pollen diagrams from Cross Fell and their implications for former tree-lines. *Lake Sediments and Environmental History* (Ed. by E.Y. Haworth & J.W.G. Lund), pp. 317–357. Leicester University Press, Leicester.

Turner, J. & Hodgson, J. (1979). Studies in the vegetational history of the Northern Pennines I. Variations in the composition of the early Flandrian forests. *Journal of Ecology*, **67**, 629–646.

Turner, J. & Hodgson, J. (1981). Studies in the vegetational history of the Northern Pennines II. An atypical pollen diagram from Pow Hill, Co. Durham. *Journal of Ecology*, **69**, 171–188.

Turner, J. & Hodgson, J. (1983). Studies in the vegetational history of the Northern Pennines. III. Variations in the composition of the mid-Flandrian Forests. *Journal of Ecology*, **71**, 95–118.

Turner, J., Hewetson, V.P., Hibbert, F.A., Lowry, K.H. & Chambers, C. (1973). The history of the vegetation and flora of Widdybank Fell and the Cow Green reservoir basin, Upper Teesdale. *Philosophical Transactions of the Royal Society of London B*, **265**, 327–408.

Walker, M.J.C. (1984a). Pollen analysis and Quaternary research in Scotland. *Quaternary Science Reviews*, **3**, 369–404.

Walker, M.J.C. (1984b). A pollen diagram from St Kilda, Outer Hebrides, Scotland. *New Phytologist*, **97**, 99–113.

Walker, R. (1978). Diatom and pollen studies of a sediment profile from Melynllyn, a mountain tarn in Snowdonia, North Wales. *New Phytologist*, **81**, 791–804.

Whittington, G. (1978). A sub-peat dyke on Shurton Hill, Mainland, Shetland. *Proceedings of the Society of Antiquaries of Scotland*, **109**, 30–35.

Wilkins, D.A. (1984). The Flandrian woods of Lewis. *Journal of Ecology*, **72**, 251–258.

Vegetation and soil change in the uplands

J. MILES

Institute of Terrestrial Ecology, Hill of Brathens,
Banchory, Kincardineshire AB3 4BY

SUMMARY

1 The paper reviews the gross successional changes that occur in semi-natural vegetation in the upland forest zone of Britain, certain factors that promote or inhibit change, the changes in labile soil properties resulting from succession, and the floristic reaction to such soil changes.

2 Most successional transitions are initiated by the death of dominant species or by changes in management, but vegetation structure, soil fertility and the availability of propagules of successional species also influence the rate of succession.

3 There is considerable variety in successional transitions, and there seem to be few if any fixed directions of change. The availability of propagules of successional species (species accessibility) seems to be the most important factor determining which transition or transitions occur (i.e. the direction of change) at any site.

4 Transitions between many vegetation types with contrasting pedogenic effects cause marked changes in labile soil properties in decades or less.

5 Such plant-induced changes in soil 'fertility' have been shown largely to drive the succession of the field layer when *Betula* spp. colonize *Calluna* moorland. It is suggested that similar changes are widespred among dynamic mosaics of upland vegetation.

INTRODUCTION

This paper briefly reviews (i) the characteristics that promote and inhibit successional change in semi-natural vegetation within the upland forest zone of Britain, (ii) the gross successional changes that occur, (iii) the associated plant-induced changes in soil properties, and (iv) the consequential floristic responses to soil change. Vegetation above the tree-line is not considered. Although not unaffected by man, the mosaics of montane and sub-montane vegetation are determined mainly by climate, topography and soil, and are relatively stable compared with forest-zone vegetation. In contrast, the present patterns of forest-zone vegetation owe more to land-use and management history than to natural environmental factors, and, given a varied seed rain, are relatively dynamic.

Plant nomenclature follows Clapham, Tutin & Moore (1987).

FACTORS DETERMINING CHANGE IN UPLAND VEGETATION

The crucial feature of upland vegetation within the natural forest zone in Britain is that it is largely man-made, directly or indirectly. The predominant woodland cover that developed after the last glaciation was progressively changed and reduced by man's activities, a process which in some areas perhaps began as the woods first developed. Now, all but fragments have been destroyed by burning, felling and grazing by domestic livestock, and by wild herbivores such as red deer (*Cervus elaphus* L.) whose other predators man has exterminated. In place of woodland, the familiar moorland landscapes developed, including the forest-zone blanket bogs (Moore 1973; Merrifield & Moore 1974). Indeed, even the Caithness peatlands may owe their origins to man's early activities (Moore 1987).

Most forest-zone moorland vegetation is, like most man-made vegetation, intrinsically unstable. This results in a general tendency for secondary succession to scrub and woodland. However, the tendency for successional change varies greatly in the uplands. In some places, the land manager is hard put to it to contain the unwanted spread of trees, for example on certain grouse moors, heathland nature reserves and limestone-dale sheepwalks. Yet in other areas there are few signs of a reversion to woodland, despite low grazing pressures and infrequent burning. The strength of this tendency for succession and the speed of succession depend in particular on four factors: species diversity and accessibility, soil fertility, the structure of the vegetation and the effects of management.

Species diversity and accessibility

Succession can only occur if there are available seeds, fruits or other propagules of species that are more competitive under the prevailing environment than those currently predominant. Sources may already be present in the soil seed bank or in the vegetation. For example, a grassland that is trampled by livestock may develop a dense stand of *Cytisus scoparius* from long-buried seeds. Another may harbour enough grazed-down, small and inconspicuous plants of *Calluna vulgaris, Betula pendula* or *B. pubescens* for these to become dominant in 8–10 years if grazing is largely prevented.

However, because deforestation, with subsequent repeated burning or heavy grazing, or both, tended to extinguish all woodland and fire-sensitive plants, successional species are commonly lacking in upland vegetation. Species accessibility (Heimans 1954), i.e. the possibility of propagules reaching that spot, is then critical. Although moorland edges in particular can have intricate mosaics of varied vegetation, large expanses of relatively uniform moorland commonly lack appreciable sources of successional species over much of their area, in the vegetation, in the soil seed bank and in the dispersal 'rain' of propagules. This species poverty gives such areas a marked successional inertia akin to that of deserts and tundra, such that the existing vegetation redevelops directly after disturbance simply because no other course is possible. Miles (1973) demonstrated the effect of species poverty at three *Calluna* moorland sites in northeast Scotland. When patches of vegetation were killed and the soil was bared, recolonization just reflected the composition of the surrounding vegetation and the soil seed bank. However, when seeds or fruits of 107 species, most of which did not currently grow at the sites, were sown on the bared soil, from 28 per cent to 67 per cent of the species established, depending on the soil fertility (Miles 1974a). These new colonists included many species that probably grew at the sites when they were wooded.

A wide range of species colonize after disturbance or during succession in the uplands. Table 1 lists just those vascular species that I have recorded as colonizers since 1969; a literature search would doubtless add to the roll. The main inference is that, at the colonizing, pre-competitive stage, almost anything can grow on at least the less acid upland soils. A second point is that 15 species which in eastern England are confined mainly to ancient woodland, apparently because they are poor colonists there, are much more mobile in the northern uplands. Indeed, two of them, *Lathyrus montanus* and *Viola riviniana*, are not even woodland indicators in the north. However, the remaining 13 species, and a handful of others, do indicate the likelihood of old or former woodland sites in the uplands, especially if several are present. Although they can colonize secondary woodland and quarries, I have never found most doing so further than about 100–200 m from a seed source.

Soil fertility

Most non-calcareous upland soils are relatively infertile, partly intrinsically, because of the prevalence of acidic rocks, but partly because of the loss of the natural woodland cover and the subsequent acidifying effects of many moorland species, especially *Calluna* (Miles 1985). However, soil fertility varies greatly, depending in particular on differences in the soil parent material, on the degree of podzolization and on the nature of the recent vegetation cover. For example, the calcium content of C horizon soil, at a sample of twelve *Calluna* moorland sites that I have studied, varied from less than 0·01 per cent to 1·6 per cent and was reflected in differences in podzolization, the species richness of the vegetation and the growth of test plants. There was a strong positive correlation at these sites between soil pH, which is a good general indicator of fertility in moorland soils, and the species richness in stands of *Calluna* and in successional stands of *Betula* spp. Even the growth of species characteristic of poor, acid soils, e.g. *B. pendula*, *Calluna* and *Pinus sylvestris*, responds to increasing soil nutrient content (Miles 1974a). The effect of soil fertility on species richness was shown at the three *Calluna* moorland sites noted earlier (Miles 1973, 1974a). When soil was bared experimentally, only one species of flowering plant that was not already growing in the surrounding vegetation successfully colonized at any of the sites. In contrast, when the soil was given a dressing of fertilizer, ten additional species colonized. Also, in the experiments in which seeds or fruits of 107 species were sown, from 68 to 86 per cent of the species established on fertilized soil at the different sites, compared with only 28 to 67 per cent on unfertilized soil. The effects of vegetation on soil fertility are discussed later.

TABLE 1. Vascular species known or inferred to colonize after disturbance or during succession in upland habitats (excluding those on limestone) in Scotland and northern England (* and † respectively denote species with an affinity with ancient woodland in central Lincolnshire (Peterken 1981) and in eastern England (Rackham 1980), and § denotes species characteristic of old or former woodland, especially broadleaved sites, in the Scottish Highlands and Islands)

Acer pseudoplatanus
Betula pendula
B. pubescens
§Corylus avellana
Crataegus monogyna
Fagus sylvatica
Fraxinus excelsior
Ilex aquifolium
§Juniperus communis
Larix decidua
Picea abies
Pinus sylvestris
Populus tremula
Prunus avium
P. padus
Quercus robur
Salix aurita
S. cinerea
S. purpurea
Sorbus aucuparia

Arctostaphylos uva-ursi
Calluna vulgaris
Cytisus scoparius
Erica cinerea
E. tetralix
Genista anglica
Myrica gale
Rosa canina
Rubus fruticosus agg.
R. idaeus
Salix repens
Ulex europaeus
Vaccinium myrtillus
V. vitis-idaea

Athyrium filix-femina
Blechnum spicant
†Dryopteris affinis
D. carthusiana
D. dilatata
D. filix-mas
Equisetum arvense
§Phegopteris connectilis
Pteridium aquilinum

Agrostis capillaris
A. stolonifera
A. vinealis
Aira praecox
Briza media
Carex binervis
C. demissa
C. echinata
C. flacca
C. nigra
*†§C. pallescens
C. panicea
C. pilulifera
C. rostrata
Dactylis glomerata
Dactylorhiza maculata
D. majalis subsp. purpurella
Deschampsia cespitosa
D. flexuosa
Eriophorum angustifolium
Festuca ovina
F. rubra
Holcus lanatus
H. mollis
*†§Hyacinthoides nonscripta
Juncus acutiflorus

J. articulatus
J. bulbosus
J. conglomeratus
J. effusus
J. squarrosus
Listera cordata
Luzula campestris
L. multiflora
*†§L. pilosa
*†§L. sylvatica
*†§Milium effusum
Molinia caerulea
Nardus stricta
Narthecium ossifragum
Platanthera bifolia
Poa annua
P. nemoralis
P. pratensis
Trichophorum cespitosum

Achillea millefolium
A. ptarmica
*§Adoxa moschatellina
Ajuga reptans
*†§Anemone nemorosa
Angelica sylvestris
Antennaria dioica
Anthoxanthum odoratum
Bellis perennis
Campanula rotundifolia
Cardamine pratensis
Centaurea nigra
Cerastium fontanum
Chamaenerion angustifolium
§Circaea lutetiana
Cirsium palustre

C. vulgare
*†§Conopodium majus
Crepis capillaris
Digitalis purpurea
Epilobium brunnescens
E. montanum
E. palustre
Euphrasia arctica ssp. borealis
Filago minima
Filipendula ulmaria
*Fragaria vesca
Galium palustre
G. saxatile
G. verum
Geranium robertianum
Gnaphalium sylvaticum
Heracleum sphondylium
Hieracium pilosella
Hypericum pulchrum
Hypochoeris radicata
*†Lathyrus montanus
L. pratensis
Linum catharticum
Lonicera periclymenum
Lotus corniculatus
*†§Lysimachia nemorum
*†§Melampyrum pratense
Montia fontana
Myosotis arvensis
*†§Oxalis acetosella
Plantago lanceolata
P. major
Polygala serpyllifolia
Polygonum viviperum
Potentilla anserina
P. erecta

*§Primula vulgaris
Prunella vulgaris
§Pyrola rotundifolia
Ranunculus acris
R. flammula
R. repens
Rhinanthus minor
Rumex acetosa
R. acetosella
R. obtusifolius
Sagina procumbens
Serratula tinctoria
Solidago virgaurea
Sonchus oleraceus
Stellaria alsine
§S. graminea
S. media
Succisa pratensis
Taraxacum officinale agg.
Teucrium scorodonia
Thymus praecox
§Trientalis europaea
Trifolium dubium
T. pratense
T. repens
Tussilago farfara
Urtica dioica
Veronica arvensis
V. chamaedrys
V. officinalis
V. serpyllifolia
Viola lutea
†V. riviniana

Vegetation structure

The structure of vegetation often determines whether or not other species can colonize. Many species establish poorly from seed in undisturbed grassland and dwarf shrub stands (Cavers & Harper 1967; Miles 1972, 1974b). Some species can form very dense stands that resist invasion by trees, e.g. *Pteridium aquilinum* and *Rhododendron ponticum*, which makes them very stable.

I know of no moorland phanerogams whose seedlings can tolerate heavy shade. Most establish from seeds or fruits only in gaps, though the parameters of a 'gap' vary from species to species. Removal of different layers of the canopy at four *Calluna*-dominant sites, and the creation of varying-sized patches of bared soil at one of them, had differential effects on species establishment (Miles 1974b).

Effects of management

The most important management practices in the uplands influencing vegetation are grazing and burning. The former is ubiquitous and the latter almost so. They variously constrain succession in the uplands and maintain the bulk of upland, non-woodland vegetation. Any variations in the intensity of grazing and the frequency of burning can drive or allow change. For example, although the response of any vegetation to fire is markedly influenced by the vegetation's composition beforehand, burning *Calluna*-dominant stands on mineral soils at about 3–6-year intervals shifts dominance to grasses, especially *Deschampsia flexuosa* on well drained soils and *Molinia caerulea* on poorly drained soils, whereas a frequency of about 6–10 years favours *Erica cinerea* and *E. tetralix*, and one of about 10–20 years favours *Calluna*. On peat, however, a 20-year frequency favours *Eriophorum vaginatum*, and *Rubus chamaemorus* where present, at the expense of *Calluna* (Taylor & Marks 1971; Hobbs 1984). Further details about vegetation changes caused by management are given in the next section and by Murray (1985), Hobbs & Gimingham (1987) and Miles (1987b).

SUCCESSIONAL CHANGES IN THE UPLANDS

There is considerable variety in successional transitions in the uplands and, although some transitions occur more frequently than others, there seem to be few, if any, fixed directions of change. However, in having multiple pathways of change, upland vegetation is no different from other classes of vegetation that have been examined critically (Miles 1987c). Fig. 1 summarizes the known successional transitions between eight common types of semi-natural vegetation on well-drained, acid mineral soils, divided according to three broad levels of grazing intensity and burning frequency under which they occur. Although it is not currently possible to indicate the relative frequency of occurrence of the different transitions for any specific upland area, the diagrams suggest two particular points of interest, even at this general level.

Vegetation and soil change

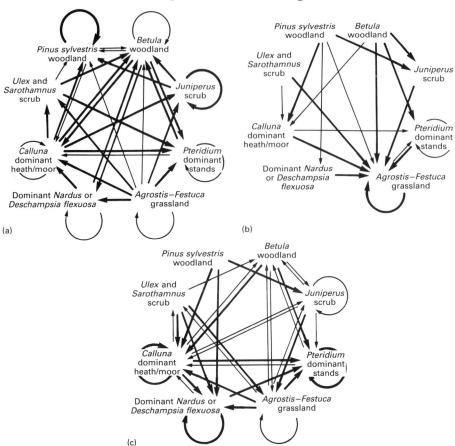

FIG. 1. Successional transitions in the uplands (particularly north-east Scotland) between eight vegetation types given (a) low grazing pressures (<1 sheep equivalent ha^{-1} year^{-1}) and no burning, (b) high grazing pressures (>2–3 sheep equivalents ha^{-1} year^{-1}) and frequent burning, and (c) intermediate levels of grazing (1–2 sheep equivalents ha^{-1} year^{-1}) and occasional burning. Broad arrows represent common transitions, thin arrows apparently less frequent transitions, and curved arrows self-replacement. The vegetation types are arranged so that types tending to podzolize and/or acidify soils are on the left, and types with contrasting effects on the soil are on the right (from Miles 1985, courtesy of the British Society of Soil Science).

First, Figs 1a and 1c suggest that, at low grazing pressures, several vegetation types may be able to coexist in a given tract, forming a mosaic that will change with time as new stands establish and old ones die. This is effectively the concept of the shifting-mosaic steady state (Bormann & Likens 1979) or of multiple stable points (Sutherland 1974). Second, Fig. 1 forces us to ask why particular transitions occur at any site. From the inference derived from Table 1, and from experience in the field, I suggest that, for any given site and management regime, the accessibility of alternative dominant species is usually the main reason why any particular transition (other than self-regeneration) occurs.

Accessibility variously means the presence or nearness and the size of sources of seeds and fruits, including buried seeds, and of colonizing rhizomes or other organs of vegetative reproduction. Table 2 lists modes of accessibility of various species that can be dominant or co-dominant in the uplands. It gives much of the information needed to predict the changes that will occur in any mosaic of vegetation given changes in its management (Miles 1987c).

Although the minimum area for a system approaching a shifting-mosaic steady state is unknown, many different transitions can occur concurrently in quite small areas. Fig. 2 shows the transitions currently occurring in a 3 ha patch at a moorland edge in Deeside. Including self-regeneration, there are 17 transitions, but potentially there are even more. For example, all the seed-bearing trees of *P. sylvestris* are as yet too far from any patches of *Cytisus* scrub, *Agrostis–Festuca* grassland or *Pteridium* to have much chance of colonizing them, while none are approaching senescence.

Successional transitions are influenced by soil type (Ratcliffe 1959; Miles, Welch & Chapman 1978), and there are fewer alternatives on wet or waterlogged soils. Fig. 3 shows the variability on different soil types in the succession of the ground layer under *Betula* spp. colonizing *Calluna* moorland, and in the range of possible successional transitions when the *Betula* wood dies of old age.

Rates of change in upland vegetation vary greatly. The subject is too large to be dealt with here, but is discussed by Miles *et al.* (1978). Perhaps the main point is that change càn be rapid. For example, a *Calluna*-dominant dwarf-shrub heath can change to *Agrostis–Festuca* grassland in only 2–3 years with heavy grazing and trampling by livestock, or to *Betula* or *P. sylvestris* woodland in 15–20 years where there is a heavy seed rain.

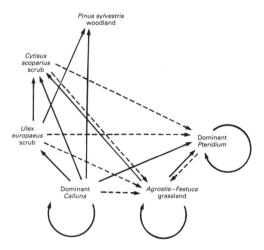

FIG. 2. Successional transitions occurring in 1987 in a 3 ha patch at the edge of moorland in Glen Cat, Deeside, bearing a mosaic of vegetation types. Continuous straight arrows represent transitions involving colonization of healthy vegetation by taller growing species, dashed arrows are transitions resulting from the death of stands of one dominant, and curved arrows are self-replacement.

TABLE 2. Ways by which groups of certain common and abundant upland plant species can become sufficiently abundant at a site for a successional transition to occur under favourable conditions

Species	Description
Betula pendula, *B. pubescens*	Mass colonization by wind-dispersed fruits within about 50 m of parent trees, but establishment at low density for 500+ m. Trees fruit from 5–10+ years. Seed banks short-lived (1–2 years), with up to 1000+ fruits m^{-2}
Pinus sylvestris	Mass colonization by wind-dispersed fruits within about 100 m of fruiting trees, but establishment at low density for 500+ m. Trees fruit from 15–20+ years. No apparent seed bank
Cytisus scoparius, *Ulex europaeus*	Rapid mass establishment from exposure of buried seeds, usually by livestock trampling. Stands can spread by seedling establishment at the margins. Plants fruit from 2–3+ years. Seed banks persistent for decades, with up to 10,000+ seeds m^{-2} of *C. scoparius* and 1000+ seeds m^{-2} of *U. europaeus*
Calluna vulgaris, *Erica cinerea,* *E. tetralix*	Rapid mass establishment from buried seeds after fire or soil disturbance. Can also colonize from wind-blown seeds. *Calluna* for up to about 100 m, the *Erica* spp. probably for similar distances. *Calluna* seeds are also transported in the dung of mammalian herbivores and red grouse (*Lagopus lagopus scoticus*). Also, *Calluna* can become dominant in 8–10 years by vegetative growth after grazing is prevented in grasslands with even a few per cent cover of inconspicuous, grazed-down *Calluna* plants. Plants fruit from 1–2+ years. Seed banks persistent, perhaps for 80+ years, with up to 90,000+ seeds m^{-2} of *Calluna* and *E. tetralix*, and up to about 20,000 m^{-2} of *E. cinerea*
Deschampsia flexuosa, *Festuca ovina,* *F. rubra,* *Molinia caerulea,* *Nardus stricta*	Fruits dispersed for a few metres by wind but over long distances in the dung of mammalian herbivores, and perhaps on occasion by red grouse. Slow vegetative spread of individual plants, but *D. flexuosa* can increase its cover a hundredfold in 2 years from seed after fire, felling of conifers or ploughing for afforestation. Plants can fruit in their first year. Seed banks short-lived; usual maximum sizes (seeds m^{-2}) are *Nardus*, 3000+; *F. rubra*, 500+; *D. flexuosa* & *F. ovina*, 200; and *Molinia*, less than 50
Agrostis capillaris, *A. vinealis*	Fruit dispersal as in the previous group. Can spread quite rapidly by vegetative growth, especially the rhizomatous *A. vinealis*. Plants can fruit in their first year. Seed banks persistent, commonly 2000–4000+ seeds m^{-2}
Vaccinium myrtillus, *V. vitis-idaea*	Seed dispersal via dung and bird droppings. Plants fruit from 2–3+ years. Seed banks perhaps persistent but very small (<200 seeds m^{-2}). Main increase is by rhizome growth. *V. myrtillus*, often with *V. vitis-idaea*, can replace a *Calluna*-dominant sward within 10 years under a developing tree canopy
Crataegus monogyna, *Juniperus communis,* *Sorbus aucuparia*	Dispersal mainly via bird droppings, which are often deposited preferentially under perches. No seed banks recorded. *C. monogyna* and *S. aucuparia* fruit from 5–10+ years, *J. communis* usually from 20+ years. Seeds of *S. aucuparia* quickly rot in the soil; those of *C. monogyna* and *Juniperus* are soon eaten by small mammals. Plants of *J. communis* commonly regenerate by layering of branches
Corylus avellana, *Quercus petraea,* *Q. robur*	Mainly dispersed by various mammals and birds, with uneaten fruits either cached or accidentally dropped. Most seedlings are found within 200–300 m of parent trees. *C. avellana* nuts may survive for a few years in soil or in caches, but *Quercus* acorns do not. *C. avellana* fruits from 5–10+ years, *Quercus* spp. after 35–45+ years
Pteridium aquilinum	Establishment of prothalli is rare. Main spread is by rhizomes. An invading front may advance 0·25–0·5 m year^{-1}, and in exceptionally good soils by 1·5 m year^{-1}. Many transitions to *Pteridium*-dominant stands are by increase from existing, low-density frond populations rather than by invasion
Rhododendron ponticum	Seeds dispersed several hundred metres by wind. Plants fruit from about 12 years. No information on seed banks. Plants can regenerate by layering of branches

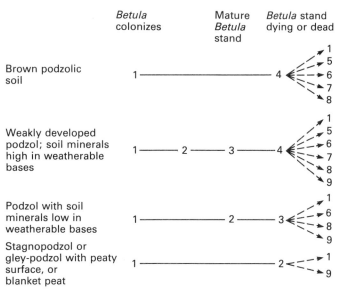

FIG. 3. Generalized sequences of vegetation change during the life cycle of an upland birchwood on different soil types. Continuous arrows represent observed transitions in the field layer, dashed arrows observed or likely potential transitions with death of the birch stand. Numbers represent: 1, *Calluna* heath; 2, *Vaccinium* heath; 3, co-dominant *Deschampsia flexuosa* and *Vaccinium* spp.; 4, species-rich grassy field layer; 5, *Agrostis–Festuca* grassland; 6, *Pteridium* stands; 7, *Juniperus communis* scrub; 8, woodland dominated by *Quercus* and other broadleaved spp.; 9, *Pinus sylvestris* woodland.

CHANGES IN SOIL UNDER VEGETATION

Many different upland plant species and vegetation types can markedly influence labile soil properties, which tend to change in a cyclic way during the life of a stand of vegetation (Miles 1986). Change can be rapid in the poorly buffered siliceous soils prevalent in much of the uplands. For example, increases and decreases in pH of 1–2 units within the range 3·5–6 can occur in decades or less (Miles 1985). The broad pedogenic effects of common dominant species and vegetation types are shown in Fig. 1. This is arranged so that, with a simple vertical division into two equal halves, those types tending to acidify soil, and in some cases to promote podzolization, are on the left, and those with contrasting tendencies on the right. It suggests that soil regularly changes during succession.

By examining chronosequences, we are currently assessing the rates of soil change under examples of these vegetation types in north-east Scotland in order to test and refine Fig. 1. Table 3 gives means and ranges of topsoil pH under different types from successional sequences. It suggests that transitions to *Agrostis–Festuca* grassland from any of other eight types will tend to increase soil pH, but that stands of *Betula* spp., *Pteridium* and *Juniperus* will tend to acidify the soil on colonizing *Agrostis–Festuca* grassland, but increase pH if replacing

TABLE 3. Mean soil pH values at 0–5 cm in stands of upland vegetation in north-east Scotland that were all either the precursors or the result of recent successional transitions. Successional transitions between vegetation types having mean pH values with the same subscripts tended not to produce significant changes in soil pH. Transitions between types with different subscripts consistently produced pH changes; changes to subscript letters later in the alphabet resulted in increased pH values, and *vice versa*

Vegetation type	Number of stands sampled	Mean pH of stands	Range of mean stand pH
Dominant *Calluna vulgaris*	16	$4{\cdot}2_a$	(4·2–4·6)
Cytisus scoparius scrub	2	$4{\cdot}4_a$	(4·2–4·6)
Nardus-dominant grassland	1	$4{\cdot}2_a$	
Pinus sylvestris plantation	5	$4{\cdot}0_a$	(3·7–4·8)
Ulex europaeus scrub	5	$4{\cdot}4_a$	(4·2–4·6)
Betula pendula and B. pubescens woodland	24	$4{\cdot}5_b$	(3·7–5·3)
Juniperus communis scrub	5	$4{\cdot}7_b$	(4·5–5·1)
Dominant *Pteridium aquilinum*	5	$4{\cdot}8_b$	(4·4–5·3)
Agrostis–Festuca grassland	8	$5{\cdot}1_c$	(4·7–5·5)

any of the other five types. Transitions between types with similar effects may result in minimal soil change, but Fig. 1 shows that there are many successional transitions which will bring about soil change. Indeed, the cyclic relationship between *Calluna* and *Pteridium* shown on one Breckland heath by Watt (1955) will have tended to produce a regular alternation of soil properties. Although phasic change at this site has now broken down (Marrs 1986), *Pteridium* stands do show similar fluctuations in extent in the uplands (Birnie & Miller 1986).

Perhaps the most dramatic and commonly occurring cycles of soil change are those occurring in boreal forests as a result of periodic fires caused by lightning. The natural *P. sylvestris* forests of the Scottish Highlands are part of this system (Miles 1985). The evidence suggests that *Betula* spp. and *Populus tremula* (which has similar effects on the soil) tended to establish more rapidly after fire than *P. sylvestris* (which has contrasting effects), but that the longer-lived pines gradually became predominant until the next fire (Fig. 4). These soils will thus, for several thousand years, have been subjected after fire to initial influences tending to create brown profiles with mull humus, giving way to acidifying and podzolizing effects as the pines gradually became predominant. Mor humus will have accumulated under the pines until it was destroyed by the next fire, releasing nutrients. The soils over large parts of the Highlands will have cycled in 'fertility' along with the forest cycles shown in Fig. 4. Similar alternations of species and soil properties are caused by windthrow, which is also a common natural occurrence in the uplands.

FLORISTIC RESPONSES TO SOIL CHANGE

Soil changes of the magnitude of those reviewed by Miles (1985, 1986, 1987a) would be expected to influence the growth of many plant species. It is thus

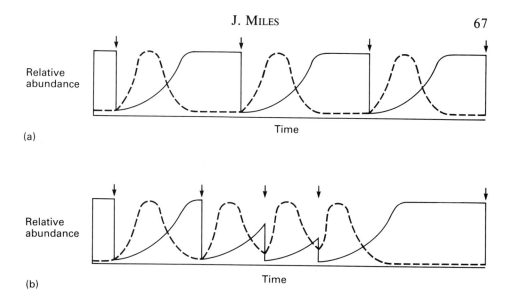

FIG. 4. Diagrammatic representation of the trend of natural changes caused by lightning fires in the abundance of *Pinus sylvestris* (continuous lines) and of *Betula* spp. and *Populus tremula* (dashed lines) in boreal forest in the Scottish Highlands with (a) fires at regular intervals and (b) fires at irregular intervals. Arrows represent fires.

reasonable to assume that marked floristic changes will often occur during upland successions, although as yet there is little information about this. However, the soil changes that occurred when *B. pendula* colonized *Calluna* moorland and felled *P. sylvestris* plantation at 11 upland sites were always associated with an increase in the numbers of field layer species and in the growth of test plants (Miles & Young 1980; Miles 1986).

In experiments and in chronosequence studies of moorland edge vegetation, I have found all the species in Table 1 acting as colonists, although with varying frequency and to a varying extent. Of the eighteen species noted in Table 1 as characteristic of old or former woodland sites in the Highlands, three, *Anemone nemorosa*, *Oxalis acetosella* and *Trientalis europaea*, have commonly appeared when dwarf-shrub heathland or grassland has given way to *Pteridium* or to *Betula*, *Cytisus* or *Juniperus* scrub. It is difficult to separate the effects of differences in shading and competition from those of soil change. However, experimentally sowing a variety of field layer species at one site showed a sequence of species progressively able to establish as the soil changed towards mull conditions (Table 4), indicating that soil change can be essential.

Thus, although plant-induced soil changes can produce a marked floristic reaction, it is not known how widespread such responses are. I suspect that the tendency for alternating dominance of *Betula* and *P. sylvestris* in the boreal forest of the Highlands with frequent lightning fires will have been associated with a corresponding change in the composition of the field layer as a result of the alter-

Vegetation and soil change

TABLE 4. Persistence of species 2 years after germinating from experimental sowings of seed on bared ground in *Calluna* moorland and adjacent, first generation, successional stands of *Betula pendula* of different ages, near Advie, Morayshire; brackets indicate that all plants were weak and unhealthy-looking (from Miles 1986)

	Calluna moorland	Age of *Betula pendula* 18 years	Age of *Betula pendula* 26 years	Age of *Betula pendula* 38 years
Calluna vulgaris	+	+	+	+
Deschampsia flexuosa	+	+	+	+
Luzula sylvatica	+	+	+	+
Festuca rubra	(+)	+	+	+
Holcus lanatus	(+)	(+)	+	+
Galium saxatile		(+)	+	+
Rumex acetosa		(+)	+	+
Ranunculus acris			(+)	+
Rubus idaeus			(+)	+
Geranium sylvaticum				+
Primula vulgaris				+
Prunella vulgaris				+
Viola riviniana				+

nating trends of soil development and fertility. The changes expected would be similar to those occurring when *Betula* spp. colonize *Calluna* moorland. Unfortunately, this hypothesis cannot now be tested in Scotland because the natural boreal woodlands are largely destroyed.

I have suggested elsewhere (Miles 1987a), on admittedly slender circumstantial evidence, that some of the changes in species composition that occur, when grazing of *Agrostis–Festuca* grassland is increased or decreased, reflect changes in pH and nutrient cycling rates with different degrees of sward tussockiness and litter accumulation. It is also possible that some of the characteristic associations of grassland species (de Vries 1954; Aarsen, Turkington & Cavers 1979) and some correlations between species abundance and soil conditions (Rorison 1969) are the result of particular soil conditions brought about by the presence of the plants rather than the converse. Thus Ernst (1978) suggested that the discrepancy between the physiological and ecological optima of *Deschampsia flexuosa* to pH was because the plants acidify the soil (Pigott 1970) and are thus inevitably associated with acid soils where they are growing well.

IMPLICATIONS FOR CONSERVATION MANAGEMENT OF VEGETATION

There are a number of general implications for managing vegetation for its conservation and wildlife interest. It is clear that all semi-natural vegetation now existing in place of natural woodland within the forest zone is intrinsically unstable because there is a general tendency for succession back to woodland.

Thus, wherever sources of successional species are present, whether in the vegetation, the soil seed bank or the seed rain, active management will usually be needed to prevent succession.

Most sites can potentially support a range of different kinds of vegetation, including natural, semi-natural and directly man-made types. This gives the creative manager great scope for positive, direct intervention either to recreate putative former kinds of habitat or to create new habitat types to increase wildlife diversity (e.g. by modifying the design and composition of planted forests).

It seems that upland vegetation mosaics can be dynamic, and that many plant species, perhaps even most, are capable of colonizing new sites, at least in the Scottish Highlands. The mobile species include many indicators of old or former woodland sites. Thus the effects on vegetation of small-scale habitat loss within a mosaic of varied vegetation types (e.g. afforestation, agricultural reclamation) may commonly be reversible.

Woodlands commonly show regeneration cycles (Peterken 1981; Miles 1985). These spatial fluctuations in composition are particularly marked in the boreal zone, which includes the Scottish Highlands. The concept of the static ancient woodland in lowland England is quite inappropriate for the Scottish Highlands where birch, pine and perhaps even oak woods effectively move about with time during natural regeneration cycles and succession. Given this, and also the dynamic nature of many moorland vegetation mosaics, it is clear that upland nature reserves need to be large if there is to be any chance of maintaining vegetation diversity through natural or semi-natural processes. Individual reserves should preferably be in thousands of hectares, and certainly in hundreds, rather than in just tens of hectares.

REFERENCES

Aarsen, L.W., Turkington, R. & Cavers, P.B. (1979). Neighbour relationships in grass–legume communities. III. Temporal stability and community evolution. *Canadian Journal of Botany*, 57, 2695–2703.

Birnie, R.V. & Miller, D.R. (1986). The bracken problem in Scotland: a new assessment using remotely sensed data. *Bracken, Ecology, Land Use and Control Technology* (Ed. by R.T. Smith & J.A. Taylor), pp. 43–55. Parthenon Publishing Group, Carnforth.

Bormann, F.H. & Likens, G.E. (1979). *Pattern and Process in a Forested Ecosystem.* Springer-Verlag, New York.

Cavers, P.B. & Harper, J.L. (1967). Studies in the dynamics of plant populations. I. The fate of seed and transplants introduced into various habitats. *Journal of Ecology*, 55, 59–71.

Clapham, A.R., Tutin, T.G. & Moore, D.M. (1987). *Flora of the British Isles,* 3rd edn. Cambridge University Press, Cambridge.

de Vries, D.M. (1954). Constallation of frequent herbage plants, based on their correlation in occurrence. *Vegetatio*, 5–6, 105–111.

Ernst, W. (1978). Discrepancy between ecological and physiological optima of plant species. A reinterpretation. *Oecologia Plantarum*, 13, 175–188.

Heimans, J. (1954). L'accessibilité, terme nouveau en phytogéographie. *Vegetatio*, 5–6, 142–146.

Hobbs, R.J. (1984). Length of burning rotation and community composition in high-level *Calluna–Eriophorum* bog in N. England. *Vegetatio*, 57, 129–136.

Hobbs, R.J. & Gimingham, C.H. (1987). Vegetation, fire and herbivore interactions in heathland. *Advance in Ecological Research*, 16, 87–173.

Marrs, R.H. (1986). Study of vegetation change at Lakenheath Warren: a re-examination of A.S. Watt's theories of bracken dynamics in relation to succession and vegetation management. *Journal of Applied Ecology*, **23**, 1029–1046.

Merrifield, D.L. & Moore, P.D. (1974). Prehistoric human activity and blanket peat initiation on Exmoor. *Nature*, **250**, 439–441.

Miles, J. (1972). Experimental establishment of seedlings on a southern English heath. *Journal of Ecology*, **60**, 225–234.

Miles, J, (1973). Natural recolonization of experimentally bared soil in Callunetum in north-east Scotland. *Journal of Ecology*, **61**, 399–412.

Miles, J. (1974a). Experimental establishment of new species from seed in Callunetum in north-east Scotland. *Journal of Ecology*, **62**, 527–551.

Miles, J. (1974b). Effects of experimental interference with stand structure on establishment of seedlings in Callunetum. *Journal of Ecology*, **62**, 675–687.

Miles, J. (1985). The pedogenic effects of different species and vegetation types and the implications of succession. *Journal of Soil Science*, **36**, 571–584.

Miles, J. (1986). What are the effects of trees on soils? *Trees and Wildlife in the Scottish Uplands* (Ed. by D. Jenkins), pp. 55–62. Institute of Terrestrial Ecology, Huntingdon.

Miles, J. (1987a). Soil variation caused by plants: a mechanism of floristic change in grassland? *Disturbance in Grasslands: Causes, Effects and Processes* (Ed. by J. van Andel, R.W. Snaydon & J.P. Bakker), pp. 37–49. Junk Dordrecht.

Miles, J. (1987b). Effects of man on upland vegetation. *Agriculture and Conservation in the Hills and Uplands* (Ed. by M. Bell & R.G.H. Bunce), pp. 7–18. Institute of Terrestrial Ecology, Huntingdon.

Miles, J. (1987c). Vegetation succession: past and present perceptions. *Colonization, Succession and Stability* (Ed. by A.J. Gray, M.J. Crawley & P.J. Edwards), pp. 1–29. Blackwell Scientific Publications, Oxford.

Miles, J. & Young, W.F. (1980). The effects on heathland and moorland soils in Scotland and northern England following colonization by birch (*Betula* spp.). *Bulletin d'Écologie*, **11**, 233–242.

Miles, J., Welch, D. & Chapman, S.B. (1978). Vegetation and management in the uplands. *Upland Land Use in England and Wales* (Ed. by O.W. Heal), pp. 77–95. Countryside Commission, Cheltenham.

Moore, P.D. (1973). The influence of prehistoric cultures upon the initiation and spread of blanket bog in upland Wales. *Nature*, **241**, 350–353.

Moore, P.D. (1987). A thousand years of death. *New Scientist*, **113**, 46–48.

Murray, R.B. (Ed.) (1985). *Vegetation Management in Northern Britain*. British Crop Protection Council, Croydon.

Peterken, G. (1981). *Woodland Conservation and Management*. Chapman & Hall, London.

Pigott, C.D. (1970). Soil formation and development on the carboniferous limestone of Derbyshire. II. The relation of soil development to vegetation on the plateau near Coombs Dale. *Journal of Ecology*, **58**, 529–541.

Rackham, O. (1980). *Ancient Woodland: Its History, Vegetation and Uses in England*. Edward Arnold, London.

Ratcliffe, D.A. (1959). The vegetation of the Carneddau, North Wales. I. Grasslands, heaths and bogs. *Journal of Ecology*, **47**, 371–413.

Rorison, I.H, (1969). Ecological inferences from laboratory experiments on mineral nutrition. *Aspects of the Mineral Nutrition of Plants* (Ed. by I.H. Rorison), pp. 155–175. Blackwell Scientific Publications, Oxford.

Sutherland, J.P. (1974). Multiple stable points in natural communities. *American Naturalist*, **108**, 859–873.

Taylor, K. & Marks, T.C. (1971). The influence of burning and grazing on the growth and development of *Rubus chamaemorus* L. in *Calluna–Eriophorum* bog. *Scientific Management of Animal and Plant Communities for Conservation* (Ed. by E. Duffey & A.S. Watt), pp. 153–166. Blackwell Scientific Publications, Oxford.

Watt, A.S. (1955). Bracken versus heather, a study in plant sociology. *Journal of Ecology*, **43**, 490–506.

Vegetation succession under birch: the effect of shading and fertilizer treatments on the growth and competitive ability of *Deschampsia flexuosa*

A. J. HESTER

Department of Plant Science, University of Aberdeen, Aberdeen AB9 2UD

INTRODUCTION

Changing environmental conditions have complex and often interactive effects on the species involved in any successional process (Miles 1979). Experimental examination of such effects is vital to our understanding of the mechanisms behind directional vegetation changes, such as those shown in Fig. 1 beneath increasing ages of birch (*Betula pubescens*). These species changes are common to several developing birch woodlands on moorland areas in the north and east of Britain (Miles & Young 1980). Precise causal factors relating to the changes in dominance of these species are not known.

Field and pot experiments were designed to examine in some detail how changes in two important environmental factors, light intensity and soil nutrient status, affect the growth and relative competitive abilities of the main species in-volved in this succession. The disappearance of *Calluna vulgaris*, frequently the dominant species before colonization by birch, can almost wholly be attributed to the increase in shade as the birch closes canopy (Hester 1987). This paper gives the results of a pot experiment examining the growth of *Deschampsia flexuosa*, both alone and in competition with the previous and subsequent dominants,

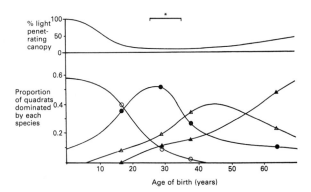

FIG. 1. Generalized changes in dominance of main species beneath increasing ages of birch at Glenlivet, Grampian Region (National Grid reference NJ190322). Vegetation data were collected from 432, 5 × 5 cm squares within each aged stand. ○ = *Calluna vulgaris*; ● = *Vaccinium myrtillus*; △ = *Deschampsia flexuosa*; ▲ = *Agrostis tenuis*. * Denotes stage of greatest increase in soil nutrients (Miles & Young, 1980). Light levels were recorded by J. Miles.

71

Vaccinium myrtillus and *Agrostis tenuis*. Nomenclature follows that of Clapham, Tutin & Warburg (1981).

METHODS

17·8 × 17·8 × 14 cm deep square plant pots were filled with a 'poor' soil (pH 4·0) made up from 2 parts birch wood soil, 1 part *Sphagnum* moss peat and 1 part sharp sand. Plants were collected from the field and split into individual uniformly sized rooted tillers or rooted shoots (*Vaccinium myrtillus*). Four plants were placed in each pot in the following five mixtures: *V. myrtillus* alone and together with *Deschampsia flexuosa*; *D. flexuosa* alone and together with *Agrostis tenuis*; *A. tenuis* alone. Five replicate pots of each species mixture were given the following six treatment combinations: no shade, 65 per cent shade and 87·75 per cent shade, each with or without fertilizer.

Four 50 cm high wooden shade frames were constructed using 1 or 2 layers of Lobrene KX02 nylon shade netting (neutral shade), giving 35 per cent and 12·25 per cent light penetration respectively. The latter light intensity was similar to that under a young dense birch canopy and the former to that under a mature stand where the canopy has opened out (see Fig. 1). Relative humidity levels measured beneath the shades were comparable to those beneath birch.

The nutrient concentrations applied were selected so as to relate broadly to the nutrients released into the soil annually from birch leaf litter, throughfall, etc. (Ovington & Madgwick 1959; Forbes 1973; Miller 1984; J. Miller pers. comm.) and to the increases in soil nutrients recorded beneath developing birch (Miles & Young 1980). The following fertilizers and amounts of each element were applied to each pot: sulphate of ammonia (32 mg N), Nitram (105 mg N), muriate of potash (150 mg K), nitro-chalk (15 mg Ca, 39 mg N), phosphate rock (105 mg Ca, 42 mg P), magnesium sulphate (30 mg Mg). Two applications were given each year (rather than one large dose) to avoid damaging the plants.

All pots were randomized both beneath the frames and in the open at the Banchory Research Station (Institute of Terrestrial Ecology). The experiment was run from April 1985 to April 1987. The shades were put in place from May to November each year, corresponding to the unfolding and fall of birch leaves in the nearby woodlands, and the fertilizer was applied in June and in August, during the main period of growth. All plants were harvested, dried and weighed in April 1987. Only the results for *D. flexuosa* are considered in this paper. Data were analysed using two-way analyses of variance (plus Bartlett-Box F tests for homogeneity of variances) and least significant difference comparisons among means (Sokal & Rohlf 1981). Fertilized and unfertilized pots were analysed separately due to unequal variances (independent of the means).

RESULTS

Fertilizer addition improved the yield of *D. flexuosa* under all conditions of shade and species combination (Table 1). *D. flexuosa* consistently grew better

TABLE 1. Above-ground dry weight of *Deschampsia flexuosa* under different conditions of shade, soil fertility and species combination. The table shows dry weight (g) per plant (mean ±S.E., n = 5). Abbreviations are V.m., *Vaccinium myrtillus*; D.f., *Deschampsia flexuosa*; and A.t., *Agrostis tenuis*. Analysis of variance was used, plus LSD comparisons among means (unfertilized and fertilized pots were analysed separately due to unequal variances). Numerical values with the same superscript are not significantly different at $P < 0.05$

Treatment		Species combination		
Fertilizer	Shade	D.f. + V.m.	D.f. + D.f.	D.f. + A.t.
None	0	2·3 (±0·10)[d]	1·5 (±0·15)[c]	0·5 (±0·04)[a]
None	65%	1·7 (±0·04)[c]	1·1 (±0·05)[b]	0·3 (±0·07)[a]
None	87·75%	1·7 (±0·09)[c]	1·0 (±0·04)[b]	0·6 (±0·10)[a]
Added	0	8·7 (±1·09)[z]	5·6 (±0·39)[y]	2·0 (±0·37)[ywx]
Added	65%	7·1 (±0·57)[y]	3·4 (±0·36)[x]	2·0 (±0·63)[ywx]
Added	87·75%	2·8 (±0·55)[wx]	1·5 (±0·23)[vw]	1·1 (±0·22)[v]

when in mixture with *V. myrtillus* than when alone and consistently grew less well in competition with *A. tenuis*, although these differences were smaller at the denser shade level.

Interactions between shade and species combination were significant, both with and without fertilizer ($P < 0.01$). In the absence of interspecific competition, shading at either level caused about the same decrease in yield in the unfertilized soil; when fertilizer was added, yield decreased significantly with each increasing level of shade. The presence of *V. myrtillus* had little effect on the nature of these responses. However, competition with *A. tenuis* greatly decreased the positive effect of fertilizer, and obscured any responses to the different shade levels.

DISCUSSION

As indicated earlier, *D. flexuosa* begins to increase under young birch as soon as *C. vulgaris* starts to decline, eventually attaining dominance under 35-year-old birch (Fig. 1). *D. flexuosa* subsequently declines and *A. tenuis* takes over as the birch matures and the canopy becomes more open.

The results of this experiment suggest that *D. flexuosa* will respond positively to the marked increases in availability of soil nutrients measured under 25 to 35-year-old birch (Miles & Young 1980), becoming dominant to *V. myrtillus*, which seems less able to exploit these nutrient increases. The results of other experiments suggest that, at least at lower light intensities, increases in nutrient supply are generally only beneficial to *D. flexuosa* in limited quantity and on soils of low pH (Watt & Fraser 1933; Hackett 1965; Rorison 1985).

Wild herbivores often select vegetation with a higher nutrient content (Miller 1968; Iason, Duck & Clutton-Brock 1986). Grazing has a much more detrimental effect on *V. myrtillus* than on *D. flexuosa* (Scurfield 1954; Pigott 1983). Therefore, if grazing pressure increases as nutrients increase in the herbage beneath birch, then this is also likely to contribute to the change in dominance from *V. myrtillus* to *D. flexuosa*.

The detrimental effect of *A. tenuis* on the growth of *D. flexuosa* under all conditions in this experiment suggests that the decline of *D. flexuosa* beneath *c.* 45-year-old birch can be attributed primarily, if not wholly, to increased competition by *A. tenuis* (cf. work by Mahmoud & Grime 1976).

The results discussed in this paper suggest some ways in which soil fertility and shading by birch may affect the abundance of *D. flexuosa* during this succession. Other experiments in progress should give further insights into the factors controlling the changes in species dominance observed and should thus increase our understanding of the processes behind a successional system of this sort.

ACKNOWLEDGMENTS

I am grateful to my two supervisors, Professor C. H. Gimingham and Dr J. Miles, for their advice during this work, which was carried out whilst in receipt of a Natural Environment Research Council research studentship.

REFERENCES

Clapham, A.R., Tutin, T.G. & Warburg, E.F. (1981). *Excursion Flora of the British Isles*, 3rd edn. Cambridge University Press, Cambridge.

Forbes, J.C. (1973). *Production and nutrient cycling in two birchwood ecosystems*. Ph.D. Thesis, University of Aberdeen.

Hackett, C. (1965). Ecological aspects of the nutrition of *Deschampsia flexuosa* (L.) Trin. II. The effects of Al, Ca, Fe, K, Mn, N, P and pH on the growth of seedlings and established plants. *Journal of Ecology*, 53, 315–333.

Hester, A.J. (1987). Successional vegetation change: the effect of shading on *Calluna vulgaris* (L.) Hull. *Transactions of the Botanical Society of Edinburgh*, 45, 121–126.

Iason, G.R., Duck, C. & Clutton-Brock, T.H. (1986). Grazing and reproductive success of red deer. The effect of local enrichment by gull colonies. *Journal of Animal Ecology*, 55, 507–515.

Mahmoud, A. & Grime, J.P. (1976). An analysis of competitive ability in three perennial grasses. *New Phytologist*, 77, 431–435.

Miles, J. (1979). *Vegetation Dynamics*. Chapman & Hall, London.

Miles, J. & Young, W. (1980). The effects on heathland and moorland soils in Scotland and northern England following colonisation by birch (*Betula* spp.). *Bulletin of Ecology*, 11, 233–242.

Miller, G.R. (1968). Evidence for selective feeding on fertilised plots by red grouse, hares and rabbits. *Journal of Wildlife Management*, 32, 849–853.

Miller, H. (1984). Nutrient cycles in birchwoods. *Proceedings of the Royal Society of Edinburgh, B*, 85, 83–96.

Ovington, J.D. & Madgwick, H.A.I. (1959). The growth and composition of natural stands of birch. I. Dry matter production. *Plant and Soil*, 10, 271–283.

Pigott, C.D. (1983). Regeneration of oak–birch woodland following exclusion of sheep. *Journal of Ecology*, 71, 629–646.

Rorison, I.H. (1985). Nitrogen source and the tolerance of *Deschampsia flexuosa, Holcus lanatus* and *Bromus erectus* to aluminium during seedling growth. *Journal of Ecology*, 73, 83–90.

Scurfield, G. (1954). Biological flora of the British Isles: *Deschampsia flexuosa* (L.) Trin. *Journal of Ecology*, 42, 225–233.

Sokal, R.R. & Rohlf, F.J. (1981). *Biometry*, 2nd edn. Freeman, San Francisco.

Watt, A.S. & Fraser, G.K. (1933). Tree roots and the field layer. *Journal of Ecology*, 35, 404–414.

Animal communities in the uplands: how is naturalness influenced by management?

M. B. USHER AND S. M. GARDNER

Department of Biology, University of York, York YO1 5DD

SUMMARY

1 The faunas of three upland habitats are selected for discussion. The paper focuses attention on the trophic structure of the animal communities of upland grasslands, on faunal taxonomic diversity in peatlands, and on the ecological processes influencing animal species occurrence in dwarf-shrub heaths. Although all of these habitats have been affected by management activities, the examples quoted demonstrate semi-naturalness.

2 The aim of this paper is to raise questions about the animal communities of the uplands. The questions asked derive both from a literature review and from planning research on the North York Moors.

3 Management is seen as having two classes of effects. Operations such as drainage, burning, grazing and cutting largely influence the relative abundances of the native species. Changing land use, such as afforestation or agricultural improvement, drastically alters the species complement and hence many of the ecosystem functions.

4 Conservation objectives for upland ecosystems are often ill-defined, and hence questions such as 'what is to be conserved?' are difficult to answer. Two criteria, rarity and naturalness, are considered to be important in selecting areas of the British uplands for conservation.

INTRODUCTION

There is no precise dividing line between the 'uplands' and the 'lowlands' of Britain, although about 30 per cent of the land above the limits of arable farming is generally considered to be upland (Ratcliffe & Thompson 1988). In considering a series of sites in the uplands for conservation purposes, Ratcliffe (1977) recognized a total of 105 vegetation types (summarized in Fig. 1). A classification of the upland vegetation, although unlikely to be completely reflected in the animal species, provides a framework within which to review the animal communities.

Two of the primary divisions of the diagram in Fig. 1 contain dwarf-shrub heaths, grasslands and at least one other community type. The following discussions will focus on the fauna of these two major community types, concentrating primarily on the ecological processes in dwarf-shrub heaths and the trophic relationships in grasslands. There is considerably less literature that relates to the fauna of the forb, fern and bryophyte communities of the uplands.

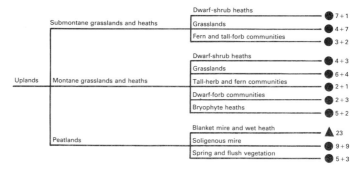

FIG. 1. The classification of 105 upland communities adopted by Ratcliffe (1977). The third level of division of the diagram is indicated symbolically; for those indicated by ● it is according to whether the substrate is acidic or basic (the number of communities is given as *A* + *B*, where *A* is the number of acidic communities and *B* the number of basic communities). For one of the peatland groups, indicated by ▲, there are 23 communities, not based on the distinction between acidic and basic substrates.

The other primary division in Fig. 1 relates to peatlands; this review concentrates on the taxonomic diversity of some of the 49 peatland types (not all of Ratcliffe's peatland types are necessarily upland, or exclusively upland, in distribution).

There are problems in defining the term 'natural' (Usher 1986), since this implies that the community type in the absence of man is known, although in practice some small amount of anthropogenic alteration is tolerated (see Gjaerevoll 1988). Because the British Isles have been populated by a variety of cultures during the period since the last glaciation, it is doubtful whether any upland area is entirely free from the influence of man, either through the use of biological resources (forest felling, peat digging) or through agriculture (crop growing, grazing by domestic stock). Hence the ecosystems discussed in this review are more correctly described as 'semi-natural' because there will have been some modification from the unknown 'natural' state. Management operations affect the degree of this semi-naturalness. Some forms of management manipulate the natural vegetation by burning, cutting, grazing or drainage; others change the natural vegetation by fertilizing and reseeding or by afforestation. One aim of this review is to explore the different effects of managing natural vegetation on the upland animal communities. Another aim is to ask questions about upland animal communities which still need answers.

There are comparatively few studies of upland animal communities in Britain. The majority of studies have concentrated either on a single species of wild upland animal, for example *Charadrius morinellus* (dotterel), *Tringa nebularia* (greenshank) or *Lepus timidus* (mountain hare), or on species of economic importance, such as *Lagopus lagopus scoticus* (red grouse), *Cervus elaphus* (red deer) or *Ovis aries* (sheep). In Britain, there are two approaches to community-based studies. First, there is the work that focuses attention on a single upland habitat. An example is the Moor House National Nature Reserve, reviewed by Coulson & Whittaker (1978). Second, following the classical zonation approach to field ecology, there are studies of the altitudinal distribution of species. Greenslade's (1968) study of Carabidae (beetles) on Ben Cruachen (Argyll) showed that there were 39 species in the lowlands (below 150 m), 25 species in the moorland zone between 150 and 760 m (all except 2 of them also occurred in the lowlands) and only 3 species in the montane zone above 760 m (all of which occurred in the moorland zone, and 2 in the lowland zone as well). The Araneae (spiders) along an altitudinal gradient in western Norway were studied by Otto & Svensson (1982); the community at the maximum altitude of 800 m was similar to the beetle community in Scotland, being species-poor and with most species also occurring at lower altitudes. Although altitudinal studies tend to be descriptive, Coulson et al. (1976) related the development of two species of Tipulidae to the temperature differences along an altitudinal gradient. It was concluded that emergence was later at higher altitudes, that larval growth was independent of temperature and that the species were adapted to life under montane conditions.

Such community studies raise three questions. First, are upland animal communities always species-poor compared with their lowland counterparts? Second, is it true that the proportion of specialist upland species in the upland community is smaller than the proportion of specialist lowland species in the lowland community? Greenslade's (1968) study showed that 41 per cent of species of Carabidae found in the lowland sites were confined to those sites, i.e. were lowland specialists, whereas there were no upland specialists (there are ten species of Carabidae in Britain that are predominantly upland in distribution; two of them, *Nebria nivalis* and *Amara alpina,* are strictly montane, but neither species was found by Greenslade). Third, do species which occur in the uplands show eco-physiological adaptations to the upland environment? Studies such as those of Leinaas & Sømme (1984) or Sømme (1986) indicate that there are a number of physiological adaptations to upland environments that are, as yet, poorly researched and poorly understood.

THE FAUNA OF UPLAND GRASSLANDS

The upland grasslands, whether montane or sub-montane, are divided by Ratcliffe (1977) on the basis of whether the substrate is basic or acidic. The semi-natural plant communities of acidic strata in Britain are dominated by *Nardus stricta, Molinia caerulea, Agrostis* spp., *Festuca* spp., *Juncus squarrosus,* etc., often reflecting the moisture condition of the site. On basic strata the main plant dominants can be the same species, often with *Deschampsia cespitosa,* but always mixed with appreciable numbers of forb species. The taxonomic diversity of the fauna of grassland habitats is reviewed by Curry (1987).

Perhaps the most intensively studied upland grassland area in Britain is adjoining Malham Tarn, North Yorkshire (the tarn surface is at 375 m above sea-level). The surrounding land, mostly 400–500 m above sea-level, contains a mosaic of acidic, neutral and basic grasslands (Usher 1980), and hence the majority of studies of animals has compared the communities of different grassland types. In this environment the birds apparently showed no preference for the grassland type (Williamson 1968), and indeed the grassland and other open communities (mires and a raised bog) were basically similar in species complement, except for *L. lagopus scoticus*, which was confined to Tarn Moss. The commonest species of the grasslands are *Alauda arvensis* (skylark) and *Anthus pratensis* (meadow pipit) with densities of 1·38 and 0·54 breeding pairs ha^{-1}, respectively. The next most abundant species of the grasslands are *Motacilla flava* (yellow wagtail), *Numenius arquata* (curlew) and *Gallinago gallinago* (snipe), all with densities of between 0·13 and 0·16 breeding pairs ha^{-1} (Williamson 1968). The trophic position of these birds is interesting; none is a complete herbivore, though the skylark is predominantly herbivorous, supplementing its food by a variety of invertebrate foods especially in the summer (Green 1978). Both the curlew and snipe are predominantly invertebrate predators, although both will eat some seeds. The meadow pipit and yellow wagtail are essentially invertebrate predators, the former tending to feed catholically whereas the latter tends to catch flying insects. The abundance of invertebrate food alone is therefore sufficient to support these breeding densities of avian predators.

The invertebrate communities of Malham have been documented by Flint (1963) for insects, Duffey (1963) for spiders and Cameron & Redfern (1972) for molluscs. Flint's list shows that there is a large variety of insects, with the usual range of herbivores, predators, parasites, parasitoids and detritivores. More recent work has shown that surveys are never complete; for example, Flint's (1963) list of 436 species of Diptera had been extended to more that 1000 species by 1983 (Disney 1986), and will probably reach about 2000 species when all species in the Malham area have been discovered (Disney 1987). Vertebrate carnivores are also present; Holmes (1960) documented the occurrence of many birds of prey, including three *Falco* species (merlin, peregrine falcon and kestrel; *F. columbarius, F. peregrinus,* and *F. tinnunculus* respectively) and four owls (Strigidae); the abundances of *F. tinnunculus* (kestrel) and *Asio flammeus* (short-eared owl) are particularly dependent upon the abundance of small mammals in the grasslands.

From the example of the above-ground fauna of the Malham area there seem to be similarities between the trophic structure of upland and lowland communities. However, how similar are they? Is the mean length of an upland food chain the same as that in a lowland community? Is the connectance in an upland food web the same as in its lowland counterpart? There are too few studies of the uplands, as shown by Pimm (1982), to be able to answer such questions. However, as the studies of terrestrial communities in the Arctic and Antarctic tend to illustrate, it could be hypothesized that food chains might become shorter with increasing altitude whereas the connectance of food webs might not be radically changed.

TABLE 1. Comparative data for the soil arthropods in six limestone, neutral
and acidic grasslands at Malham, North Yorkshire (from Wood 1967b). Mean
species richnesses (S) are rounded to the nearest integer number of species,
and mean densities (D) are given to the nearest thousand. The soils of the
three limestone grasslands include a mull-like rendzina, a protorendzina
and a skeletal soil on limestone boulders. The neutral grassland was on a
strongly leached brown earth, and the two acidic grasslands were on gleyed,
podzolic brown earths (degree of gleying and podzolization differing)

Taxa	Limestone grasslands		Acidic and neutral grasslands	
	S	D	S	D
Collembola	16	33,000	24	59,000
Cryptostigmata	26	110,000	28	56,000
Astigmata	2	1,000	3	8,000
Prostigmata	27	95,000	32	68,000
Mesostigmata	9	7,000	20	18,000

Below ground, the length of a food chain is much less clearly defined
(Coleman 1985). Wood (1967a, b) analysed the collembolan and acarine
populations of six grassland sites at Malham (Table 1). Three of these sites could
be classified as limestone grasslands with soils that were either rendzinas or
skeletal, and three sites , on deeper glacial deposits, were neutral or acidic, with
soils that were leached, gleyed or podzolized brown earths. The comparison of
species richnesses in Table 1 shows that the fauna of limestone grassland soils was
less rich, particularly the Collembola (mostly fungivores and detritivores) and
Mesostigmata (mostly predators). Population densities of these grasslands were
fairly similar; the maximum was on the protorendzina (281,000 m^{-2}) and the
minimum on a podzolized brown earth (167,000 m^{-2}).

Without comparative studies on the bacteria, fungi, protozoa and other
invertebrate animals of the same grassland soils, it is impossible to assess these
upland, below-ground food chains or food webs. However, data such as these
indicate that upland sites can be very species-rich and can have large invertebrate
population densities. The data in Curry's (1987) review of grassland fauna
suggest that upland grasslands and tundra may have appreciably larger soil
arthropod biomasses than lowland grasslands. Such observations lead to three
further questions about the grassland fauna of the uplands. First, are the soil communities of upland grasslands more diverse than the soil communities of lowland
grasslands? (And, if they are, is it a function of the lesser dominance of some
animal groups, particularly the earthworms, in the uplands?) Second, is the soil
animal biomass in the upland grasslands greater than that in the lowland
grasslands? Third, if the answers to the first two questions are affirmative, the
next question to ask is: what are these species doing in the soil? Primary
autotrophic productivity is lower in the uplands, and this would usually suggest
that the heterotrophic component of the food webs should also decrease. How are

upland soils able to support large biomasses of animals, many of which are likely to be detritivores, fungivores, bacterivores and predators? A partial answer to this question may relate to differences in bite size among the herbivores (Illius & Gordon 1987).

THE FAUNA OF UPLAND PEATLANDS

In terms of the larger vertebrate species, the faunas of upland non-forested habitats tend to be similar, as demonstrated for the birds of the Malham Tarn area (Williamson 1968). However, surveys of the spiders (Duffey 1963) and insects (Flint 1963) indicated that there were considerable differences between the grasslands and the peatlands.

More recent studies have surveyed the fauna of a series of upland peatland and grassland habitats in the northern Pennines and North York Moors. Houston's (1971) study of the Carabidae of Moor House National Nature Reserve and Cow Green, separated by a distance of 6 km, showed that the grassland species were similar between the two locations, as were the peatland species, but that the main differences were between the grassland and peatland faunas. The species richnesses were similar, with a mean of 21·5 species per 'dry' habitat and 18·7 species per 'wet' habitat. The more extensive data available to Butterfield & Coulson (1983) enabled them to recognize 5 different communities of Carabidae on peat with species richnesses between 12 and 24.

A similar study (Coulson & Butterfield 1985, 1986; Coulson 1988) showed that the major division of the spider fauna was related to the distinction between mineral soils and peat soils. Species richness does not appear to differ greatly between the peatland and non-peatland habitats (see lists of species in Coulson *et al.* 1984), but the composition of the fauna varies with altitude. Coulson & Butterfield (1986) showed that although the number of linyphiid species is virtually independent of altitude, the number of species of all other families decreases. Thus, the proportion of Linyphiidae in the spider fauna increases, a situation paralleling the distribution of the family with increasing latitude in northern Europe (Bristowe 1958). The family Lycosidae also occurs at high altitudes; Thompson, Galbraith & Horsfield (1987) record that *Tricca alpigena* comprised 24 per cent of the spider catch at 975 m and 1000 m on Creag Meagaidh (Inverness-shire). Both Coulson & Butterfield (1986) and Ratcliffe (1977) provide lists of spider species confined to high altitudes in Britain.

Peatlands can be viewed in two ways. One is the extensive area of peat; many of the comparisons quoted above relate to those peat habitats of large extent. The other is the small, isolated area of peat, often a small flush, in an otherwise different habitat. It is these small areas that Pearsall (1950) drew attention to, and it is these areas that are selected by adult grouse with young broods (Hudson 1986). Do the broadly based studies of peatland animal communities reflect the communities of the small flushes? Are the invertebrate communities of the flushes important for the animal communities of the wider upland environment?

And, how, when these small areas of wetland are subject to successional or cyclical changes (Murphy 1955), do their dynamics relate to the longer-term faunal stability of upland environments? These are all important questions touching on the equilibrium of upland communities and having important implications for moorland management.

THE FAUNA OF UPLAND DWARF-SHRUB HEATHS

The dominance of ericaceous dwarf shrubs is a key factor in shaping the faunal community. Watt (1955) applied the terms 'pioneer', 'building', 'mature' and 'degenerate' to the four growth phases of *Calluna vulgaris*, the characteristics of which have been described by Gimingham (1972). The extent to which this cyclical growth pattern of *Calluna* can be maintained in the absence of moorland management is open to conjecture (Hobbs & Gimingham 1987), but the growth phases provide a useful framework in which to review the development of the heathland fauna. Changes in the microclimate, for example, will be important in the establishment of populations of the ground-dwelling, litter and soil animals, which have mostly been studied on lowland as opposed to upland heaths in Britain.

As the *Calluna* life cycle proceeds, changes also occur in the nutrient content of the foliage, which may in turn determine the pattern of utilization of *Calluna* by herbivores. Similar changes can be seen in the production of litter; the maximum occurs towards the end of the mature phase (Chapman & Webb 1978) and the quantity can influence the development of the soil and ground-dwelling fauna. Although there is an extensive literature on the development of *Calluna*, little information is available for other dwarf-shrub species, e.g. *Erica cinerea*, *E. tetralix* and *Vaccinium myrtillus*. Similarly, there is little information on how these species influence the nature of the faunal community, particularly during the pioneer and degenerate phases of the *Calluna* cycle when *Calluna* occupies less than 50 per cent ground cover (Barclay-Estrup & Gimingham 1969). The heathland fauna itself may also have a profound effect on the vegetational development of an existing heath. This has been dramatically demonstrated on Dutch heathlands where infestations of *Lochmaea suturalis* (heather beetle) have severely restricted the regeneration of *Calluna* (Diemont & Heil 1984).

The development of a soil fauna is primarily influenced by the growth phase of *Calluna*. During the pioneer phase the microclimate is harsh because much of the litter is blown away. As the canopy begins to close towards the end of the building phase, the microclimate beneath the plants stabilizes and the developing litter layer retains more moisture. Under these conditions the soil and litter faunas develop rapidly (Chapman & Webb 1978). Litter production begins to decline as the plants reach the degenerate phase; the canopy opens out and the microclimate beneath the plants becomes less humid. Sufficient moisture is, however, retained by the existing litter and mosses to enable the soil fauna to retain its diversity (Chapman & Webb 1978).

While the majority of the soil fauna occurs in the surface zone, seasonal migrations to deeper zones have been observed for some groups. Tipulids (Hadley 1971), enchytraeids (Springett 1970) and Collembola (Hale 1966) in the uplands all show a tendency to move into deeper layers in winter when the surface zone is subject to freezing. Are such migrations more characteristic of those growth phases of *Calluna* when the canopy is open, and are they more pronounced in the uplands than in the lowlands?

A feature of the managed heathland community is the pattern of succession that is associated with the growth phases of *Calluna*. Studies of ants (Brian *et al.* 1976) and spiders (Merrett 1976) showed marked changes in species composition and abundance as *Calluna* progressed from the pioneer to the mature phase. The ground-dwelling and ground-feeding taxa, such as Carabidae, Collembola and some Acarina and Araneae, were particularly abundant in upland pioneer stands in north-east Scotland, while the building and mature phases were dominated by shoot and sap feeders such as the Cercopidae, Psyllidae and Curculionidae (Miller 1974; Gimingham 1985).

Habitat heterogeneity appears to be an important factor in determining the diversity of the invertebrate fauna. Pioneer and degenerate stands of *Calluna*, which exhibit greater floristic diversity in comparison with building and mature phases, showed greatest invertebrate diversity (Miller 1974; Gimingham 1985). Further evidence of the importance of habitat heterogeneity was shown in the comparison of the invertebrate fauna of unmanaged and managed *Calluna* stands (Miller 1974). Unmanaged stands included bushes of all ages whereas those in managed stands were of similar age; many taxa were more numerous in the unmanaged stands (Gimingham 1985). Studies such as these have implications for the conservation of invertebrate diversity on heathlands.

There is a paucity of species of mammals, reptiles and amphibians occurring on the British upland heaths (Ratcliffe 1977). Only *L. timidus* can be regarded as a typical montane heath species. The use of dwarf-shrub heaths by grazing mammals is largely influenced by the developmental stage and nutrition of *Calluna*. Comparison of the utilization of *Calluna* (both for feeding and resting) on three Scottish moors by mountain hares, sheep and red deer, as assessed from the distribution of faecal droppings, indicated that mountain hares spent most time on heather less than 15 cm high, sheep on heather less than 20 cm high and red deer on heather over 25 cm high (Savory 1986). Both red deer and mountain hares appear to be able to select nitrogen-rich patches within apparently uniform vegetation (Miller 1968; Moss, Welch & Rothery 1981). There may also be seasonal variation in feeding behaviour; Hewson (1976) found that mountain hares switched from feeding on pioneer heather in summer and autumn to feeding on building and mature heather in winter and spring. The utilization of heather by red deer and sheep is also much heavier in winter when few alternative sources of food are available (Welch 1984a). Does this seasonal pressure result in competition for food resources between domestic livestock and the wild heathland fauna?

Grazing by herbivores can have a significant effect on the regeneration and subsequent pattern of development of *Calluna*. Older stands grazed by deer are particularly susceptible to damage (Grant, Hamilton & Souter 1981), possibly because of the reduced ratio of overwintering shoots to wood and roots (Grant *et al.* 1978). Heavy grazing of young pioneer *Calluna* by mountain hares has resulted in the regeneration of short, lawn-like *Calluna* bushes, which retained a juvenile appearance, and with shoots that contained more nitrogen and phosphorus than the surrounding stands and were apparently more digestible (Moss & Hewson 1985). The difference produced by deer and hares when feeding on *Calluna* is largely a reflection of the different feeding preferences and the different regeneration capabilities of the preferred *Calluna* stands. It raises the question of whether grazing animals can actively or passively improve the nutritional quality of *Calluna*.

The principal mammalian predators on upland dwarf-shrub heaths are *Vulpes vulpes* (fox), *Mustela erminea* (stoat) and *Felis sylvestris* (wildcat). Although there is little information on the factors affecting the distributions of these species in upland areas, both the availability of prey species such as voles (*Microtus* spp.) and the degree of management of the heathland area are likely to be important.

Upland dwarf-shrub heaths are important breeding areas for birds including, in the North York Moors National Park, several species of conservation interest, e.g. *Tringa totanus* (redshank), *G. gallinago, Pluvialis apricaria* (golden plover) and *N. arquata*. Many of these species also nest in other habitats and cannot be regarded strictly as typical heathland birds. Other heathland birds are *L. lagopus scoticus, Falco columbarius* (merlin) and *Saxicola rubetra (*whinchat) (Hudson 1988; Bibby 1988). The economic importance of the red grouse is such that current regimes of heathland management are aimed at optimizing conditions to ensure its breeding success. Miller & Watson (1978) have shown that territory size in red grouse is related to both vegetation structure and the amount of nitrogen in heather shoot tips (Lance 1978). Although the nutritional quality of *Calluna* is one factor determining the population density of red grouse, it does not explain the cyclical fluctuations in population density frequently observed in this species (Hudson 1986). It is important to note that the young of several birds breeding on upland heaths, including red grouse, are dependent on invertebrates for food during part of their life. Any change in the abundance or diversity of the invertebrate fauna may, therefore, significantly affect the breeding success of such species.

THE EFFECTS OF UPLAND LAND MANAGEMENT

Upland habitats have traditionally been managed by burning, cutting or grazing the vegetation and, more recently, by drainage with the intention of improving or stimulating the vegetation production for livestock (Ratcliffe & Thompson 1988). The current emphasis in management has now focused on changing the traditional land-use patterns and introducing schemes for agricultural improvement and afforestation, as well as on encouraging tourism and conservation, in a bid to improve the economic viability of upland areas. This section of the review examines the impacts and possible implications of land management schemes on upland animal communities (see Bibby's (1988) specialized treatment of the birds).

Burning

Regular burning, on an approximate 15-year cycle in Britain, is the principal land management practice on upland dwarf-shrub heaths in western European countries (Gimingham 1981). The purpose is to provide a mosaic of regenerating heather to support the grazing requirements of domestic livestock (mainly sheep and cattle), to improve the breeding success of the red grouse and to maintain habitat heterogeneity. Burning reduces the diversity of the invertebrate fauna (Gimingham 1985), but initially the soil fauna is not adversely affected as apparently little heat is transferred beneath the soil surface (Hobbs & Gimingham 1984). However, the subsequent removal of litter by wind has been shown to cause a decrease in soil invertebrate population sizes (Chapman & Webb 1978).

During the post-burn phase faunal diversity may be quite high as the developing stand of pioneer heather provides a wide range of niches for ground- and plant-dwelling invertebrates (Gimingham 1985). Little is known about the rate of recolonization of burned areas by invertebrates (however, see Gardner & Usher (in press) and Usher & Smart (in press)) or about the effects of surrounding habitats, notably coniferous forest or improved agricultural grassland, on the species composition of the recolonizing fauna. Rotational burning of upland heaths may reduce the overall invertebrate diversity, since much of the area will be dominated by stands of building or mature *Calluna,* both of which support abundant populations of some herbivorous species but which lack the diversity of the other growth phases (Gimingham 1985).

Burning will favour populations of vertebrate herbivores, especially those species which prefer to feed on young heather (e.g. red grouse and mountain hare). It appears that heathland which is well managed for grouse also supports a rich bird community (Hudson 1986; Ratcliffe & Thompson 1988; Bibby 1988; Hudson 1988; R. W. Brown, pers. comm.).

Cutting

Cutting is one of the principal management practices for lowland grassland habitats; it has recently received more attention for managing upland dwarf-shrub heaths, having been introduced to the North York Moors National Park at the beginning of this decade. Although little information is available on the effects of cutting in the uplands, studies on lowland grasslands suggest that the timing of cutting is important in maintaining species richness and diversity in invertebrate communities (Morris & Lakhani 1979).

Heather-cutting is used particularly where there are extensive stands of degenerate *Calluna* which can prove hazardous for burning. Whilst little work has been done on the regeneration of heather in cut stands or on the effects of cutting on the animal communities (Gardner & Usher (in press) and Usher & Smart (in press)), the question of whether the impact of cutting differs significantly from that of burning remains unanswered. Since cut areas are usually smaller that burnt ones, one hypothesis could be that cut areas are recolonized more quickly by invertebrates than burnt areas; another hypothesis would be that they are less susceptible to invasion by 'non-heathland' species. There is clearly a need for further research to test such hypotheses.

Grazing by domestic livestock

The three principal effects of grazing in grass and heathland areas are vegetation defoliation, trampling and fertilization through dung and urine deposition. The principal grazing animal, the sheep, shows a clear preference for relatively rich grassland communities such as *Agrostis–Festuca* swards (Rawes & Welch 1969; Hewson & Wilson 1979), although species such as *Calluna* are important food resources in winter (Grant *et al.* 1976).

The selectivity of grazing animals is likely to have a positive effect on invertebrate populations of unpalatable plants such as aromatic herbs (Morris 1978). Light grazing of *Calluna* heathland has been shown to favour ericoids and lichens (Welch 1984b) and to prolong the pioneer phase of *Calluna* (Grant & Hunter 1966). This would have a positive effect in maintaining both invertebrate diversity and food resources for other herbivorous vertebrates, e.g. red grouse and mountain hare. Heavy grazing, however, tends to favour graminoids and forbs (Welch 1984b) and could in this case stimulate the replacement of the heathland animal species by grassland species. The question remains as to how to balance grazing requirements of domestic stock with the feeding and habitat preferences of the heathland fauna.

The interaction of grazing and burning is particularly important. Sheep tend to congregate in recently burned areas (Grant 1968); grazing of newly emerging shoots can significantly delay the regeneration of *Calluna* and may thus restrict the re-establishment of heathland invertebrate and vertebrate species. Alternatively, lengthening of the pioneer phase, once the *Calluna* has started to re-grow, might have a positive effect on invertebrate diversity.

Damage to vegetation by trampling is likely to accentuate the negative effects of grazing on heathland animal communities. The effects are likely to be particularly severe in fothering (gathering and winter feeding) areas for sheep. Further information is needed on the significance of trampling for the upland animal community.

Dunging can influence the composition of the vegetation, in particular assisting the introduction of grasses such as *Anthoxanthum odoratum*, *Holcus lanatus*, *Poa annua* and *P. pratensis* (Welch 1984c). The problem is associated particularly with cattle grazing, although the increase in graminoid cover was less than the decline in *Calluna* resulting from plant mortality below dung deposits (Welch 1984c). While the change in vegetation composition is likely to have a negative effect on the existing heathland fauna, dung from domestic livestock may provide additional niches.

Drainage

Moor-draining or 'gripping' has been undertaken on a variety of heathland soils. The immediate purpose is to increase the runoff and to lower the water-table so as to improve the vegetation and subsequent production of livestock and game animals. The impacts of moor-draining on the plant and animal communities of peatlands have been reviewed by Stewart & Lance (1983). They concluded that the effects of drainage on vegetation growth were very localized, confined to an area within 2–3 m of the drain's edge. Whilst the impacts of drainage on peatland animal communities would appear to be very localized (see Coulson 1988), such changes may have a significant effect on small, isolated flushes in areas dominated by dwarf-shrub heath. Given the apparent importance of these areas for feeding by chicks of several moorlands birds, further work on the effects of drainage, especially with respect to grazing pressure, is needed.

Change in land use

There appears to be little information on the effects of agricultural improvement practices on upland animal communities. Given the dramatic change in soil conditions and vegetation composition, it is likely that few heathland or peatland species would survive, although the impact on the upland grassland fauna may be less severe. Work by Moore (1962) on lowland heaths indicated that ploughing and subsequent fertilizer application resulted in the loss of virtually the entire heathland fauna, *A. pratensis* (meadow pipit) being one of the very few species that survived this radical treatment. Although some of the heathland fauna survived in the strips of heather that were left to form 'hedges', these were eventually replaced by woodland species.

The impacts of afforestation on upland animal communities have been reviewed by Ratcliffe (1986) and the Nature Conservancy Council (1986). In the first ten years after planting there is luxuriant growth of heathland and grassland plants although peatland bog species decline. Populations of *Microtus agrestis* (short-tailed field vole) increase in response to the luxuriant grass cover, and attract avian predators such as *A. flammeus, F. tinnunculus,* and, locally, *Circus cyaneus* (hen harrier). After about ten years the typical upland species are replaced by woodland species, principally of lowland origin, although remnants of the original population may survive in areas of open ground or along rides and roadsides.

Upland habitats adjacent to forests are likely to be prone to invasion by woodland animal and plant species, particularly in areas of recent burning, cutting or heavy grazing. To what extent does this invasion occur, and how will it affect the survival of the original upland animal communities? The increase in predatory species such as *Vulpes vulpes* (fox) and *Corvus corone* (carrion crow) is regarded as detrimental to the breeding success of birds nesting on surrounding heathland (Thompson, Stroud & Pienkowski 1988).

DISCUSSION: NATURALNESS OF THE UPLAND FAUNA

The preceding review has indicated that the majority of the uplands contain native species of fauna and flora. One of the two major exceptions is the replacement of upland vegetation with forest, largely by coniferous forest of non-native species. The development of a fauna and flora in such plantation forests has been reviewed in Ford, Malcolm & Atterson (1979); the studies have concentrated on the development of populations of pest species rather than on the establishment of a woodland animal community. The other major exception is agricultural intensification, mainly by ploughing, reseeding and the use of fertilizers. Woods (1984) documented the reduction in area of 'rough pasture' over the 30-year period from 1950 to 1980 in several of the British National Parks. The reduction in the North York Moors National Park was 25 per cent; the majority of this was due to either forestry or agricultural intensification (Thompson *et al.* 1988). These forests and fields, since they are dominated by plants that are not native to the upland sites, can no longer be considered as semi-natural.

The heathland, grassland and peatland communities that have been the focus of this review superficially appear to be far more natural. The species which occur *at the present time* are the species that one would expect to occur naturally in the uplands. But what one sees at the present time may not be natural; Hobbs & Gimingham (1987) argued that in the absence of management a substantial portion of heathland would revert to birch or pine woodland. Does the *Calluna* cycle of growth phases occur in unmanaged conditions; what effects could its loss have on the upland fauna; and, hence, can the animal communities of the open uplands be regarded as natural?

The relative proportions of the species in these unwooded upland communities are altered by management activities. The review has included a description of the relationship of vertebrate and invertebrate animals with the various developmental stages of *Calluna*. However, there have been other alterations in the animal communities of the uplands. Pearsall (1950) quoted the trapping returns for various birds and mammals in Glen Garry over the period 1837–1840 (Table 2). In the interests of game preservation, large numbers of upland predators were reported to have been killed. Although all of the mammals listed in Table 2 still occur in the uplands of Scotland, many of the birds have subsequently become extinct as breeding species or even as regular visitors; these include *Haliaëtus albicilla* (white-tailed sea eagle, 27 reported trapped in three years), *Accipiter gentilis* (goshawk, 63 reported trapped) and *Milvus milvus* (kite, 275 reported trapped). Such management activities in the uplands have certainly

TABLE 2. Animals trapped in Glen Garry between 1837 and 1840. The data are given on an annual basis, and are abstracted from more detailed information given by Pearsall (1950). Note that it is almost certain that the size of these trapping rates has been exaggerated (D. Nethersole-Thompson, pers. comm.)

Taxa	Annual trapping rate
Foxes and wildcats	70
Mustelids (pine marten, polecat, stoat, weasel, badger, otter)	256
Feral cats	26
Eagles (golden, white-tailed)	14
Hawks, etc. (osprey, goshawk, kite, hen-harrier, falcon, jerfalcon, hobby, merlin, kestrel, 3 species of buzzard)	585
Ravens and crows	635
Owls (short-eared, long-eared, brown)	36

altered the species complement; the effects of removing these predators on the remainder of the upland fauna are largely unknown.

The list in Table 2 contains only one invasive, introduced species, the feral cat (*Felis catus*). Brown's (1985) review of species introduced into the British Isles hardly contains any animal species that might occur in the uplands. This raises an interesting comparison between the lowlands and uplands; why are the uplands apparently so free from invasive, alien animal species? On the North York Moors alien plants have become widespread on managed heather moorland during the last 20 years. The most abundant is the moss *Campylopus introflexus*, a southern hemisphere species that was first found in Britain in Sussex in 1941. Although these invasive plants may be plentiful within the North York Moors, invasive animals are scarce, except for *Oryctolagus cuniculus* (rabbit) and, recently, *Mustela vison* (mink) on some of the river systems.

Conservation objectives for the uplands have often been ill-defined (but see Ratcliffe 1977), either because of a lack of understanding of species requirements or due to the extensive nature of the apparently semi-natural communities. The Countryside Commission's (1984) debate focused attention on the upland environment, challenging land managers to define the aims of their management and to devise multiple land-use systems for the benefit of the people who live and work in the uplands. From a conservation point of view, two features of the British uplands have become apparent.

First, all extensive upland tracts are to some extent modified by land management; only cliff ledges, steep gorges and some islands, etc., have not been modified, as demonstrated in Shetland by Spence (1979). Land management may be active at the present time, such as sheep grazing or moorland drainage, or it may be a historical legacy, as evidenced in Table 2 by the depletion of the predator populations. The animal communities are, to a large extent, man-induced; the natural species may be present, but the abundances of the various species reflect the current management practices as well as the history of site management for at least the last 100 to 200 years (for the birds, see also Ratcliffe & Thompson 1988).

Second, conservationists should value naturalness when comparing upland ecosystems that have been altered. Of the 10 criteria listed by Ratcliffe (1977) for Great Britain, or the 24 criteria used worldwide and listed by Usher (1986), only two seem to be of particular relevance to upland systems, although a third, typicalness, was advocated by Hopkins & Webb (1984) for lowland heaths. One of the two criteria is naturalness, which is a difficult criterion to use since the truly natural state of the British uplands is unknown, but pointers to naturalness would be the presence of large predators and the absence of alien species. The second criterion is rarity; this review has not concentrated on this aspect of the upland fauna, but several of the upland species are either rare or regarded as being of special conservation interest, as indicated by the schedules of the 1981 Wildlife and Countryside Act. Diversity, as a criterion, although widely used in the lowlands, has less relevance in the uplands where many of the communities are characterized by their relative species poverty.

The animal communities of the uplands of Britain are composed of species that are thought to be natural in the upland environment; few animal species have been introduced and invasive species are rare. Perhaps the greatest challenge facing conservationists is to re-create more natural animal communities in the uplands, with a richer variety of predators, together with the diversity of herbivores, detritivores, etc. The task is to manage an upland environment, presumably primarily for agriculture, with as full a complement of native animal species as is possible given the constraints of multiple land usage.

ACKNOWLEDGMENTS

The authors acknowledge the financial assistance of the Natural Environment Research Council, Nature Conservancy Council and North York Moors National Park Department by the award of an 'Agriculture and the Environment' special topic research grant (reference GST/02/195(AgE)).

REFERENCES

Barclay-Estrup, P. & Gimingham, C.H. (1969). The description and interpretation of cyclical processes in a heath community. I. Vegetational change in relation to the *Calluna* cycle. *Journal of Ecology*, **57**, 737–758.

Bibby, C.J. (1988). Impacts of agriculture on upland birds. *This volume.*

Brian, M.V., Mountford, M.D., Abbott, A. & Vincent, S. (1976). The changes in ant species distribution during ten years post-fire regeneration of a heath. *Journal of Animal Ecology*, **45**, 115–133.

Bristowe, W.S. (1958). *The World of Spiders*. Collins, London.

Brown, K.C. (1985). *Animals, Plants and Micro-organisms Introduced to the British Isles*. Report PECD7/8/60 to the Department of the Environment, London.

Butterfield, J. & Coulson, J.C. (1983). The carabid communities in peat and upland grasslands in northern England. *Holarctic Ecology*, **6**, 163–174.

Cameron, R.A.D. & Redfern, M. (1972). The terrestrial Mollusca of the Malham area. *Field Studies*, 3 589–602.

Chapman, S.B. & Webb, N.R. (1978). The productivity of *Calluna* heathland in southern England. *Production Ecology of some British Moors and Montane Grasslands* (Ed. by O.W. Heal & D.F. Perkins), pp. 247–262. Springer-Verlag, Berlin.

Coleman, D.C. (1985). Through a ped darkly: an ecological assessment of root–soil–microbial–faunal interactions. *Ecological Interactions in Soil* (Ed. by A.H. Fitter, D. Atkinson, D.J. Read & M.B. Usher), pp. 1–21. Blackwell Scientific Publications, Oxford.

Coulson, J.C. (1988). The structure and importance of invertebrate communities on peatlands and moorland, and effects of environmental and management changes. *This volume.*

Coulson, J.C. & Butterfield, J.E.L. (1985). The invertebrate communities of peat and upland grasslands in the North of England and some conservation implications. *Biological Conservation*, 34, 197–225.

Coulson, J.C. & Butterfield, J. (1986). The spider communities in peat and upland grasslands in northern England. *Holarctic Ecology*, 9, 229–239.

Coulson, J.C. & Whittaker, J.B. (1978). Ecology of moorland animals. *Production Ecology of British Moors and Mountain Grasslands* (Ed. by O.W. Heal & D.F. Perkins), pp. 52–93. Springer-Verlag, Berlin.

Coulson, J.C., Horobin, J.C., Butterfield, J. & Smith, G.R.J. (1976). The maintenance of annual life-cycles in two species of Tipulidae (Diptera); a field study relating development, temperature and altitude. *Journal of Animal Ecology*, 45, 215–232.

Coulson, J.C., Butterfield, J.E.L., Barratt, B.I.P. & Harrison, S. (1984). The spiders and harvestmen of some peat and upland grassland sites in Yorkshire. *Naturalist*, 109, 103–110.

Countryside Commission (1984). *A Better Future for the Uplands.* Countryside Commission, Cheltenham.

Curry, J.P. (1987). The invertebrate fauna of grassland and its influence on productivity. 1. The composition of the fauna. *Grass and Forage Science*, 42, 103–120.

Diemont, W.H. & Heil, G.W. (1984). Some long-term observations on cyclical and seral processes in Dutch heathlands. *Biological Conservation*, 30, 283–291.

Disney, R.H.L. (1986). Assessment using invertebrates: posing the problem. *Wildlife Conservation Evaluation* (Ed. by M.B. Usher), pp. 271–293. Chapman & Hall, London & New York.

Disney, R.H.L. (1987). The use of rapid sample surveys of insect fauna. *The Use of Invertebrates in Site Assessment for Conservation* (Ed. by M.L. Luff), pp. 19–28. Agricultural Environment Research Group, Newcastle upon Tyne.

Duffey, E. (1963). Ecological studies on the spider fauna of the Malham Tarn area. *Field Studies*, 1, 65–85.

Flint, J.H. (Ed.) (1963). The insects of the Malham Tarn area. *Proceedings of the Leeds Philosophical and Literary Society*, 9, 15–91.

Ford, E.D., Malcolm, D.C. & Atterson, J. (Ed.) (1979). *The Ecology of Even-aged Forest Plantations.* Institute of Terrestrial Ecology, Cambridge.

Gardner, S.M. & Usher, M.B. (in press). Insect abundance on burned and cut upland *Calluna* heath. *The Entomologist.*

Gimingham, C.H. (1972). *Ecology of Heathlands.* Chapman & Hall, London.

Gimingham, C.H. (1981). Conservation: European heathlands. *Ecosystems of the World, volume 9B* (Ed. by R.L. Specht), pp. 249–259. Elsevier, Amsterdam.

Gimingham, C.H. (1985). Age-related interactions between *Calluna vulgaris* and phytophagous insects. *Oikos*, 44, 12–16.

Gjaerevoll, O. (1988). Nature conservation in Norway. *This volume.*

Grant, S.A. (1968). Heather regeneration following burning: a survey. *Journal of the British Grassland Society*, 23, 26–33.

Grant, S.A. & Hunter, R.F. (1966). The effects of frequency and season of clipping on the morphology, productivity and chemical composition of *Calluna vulgaris* (L.) Hull. *New Phytologist*, 65, 125–133.

Grant, S.A., Lamb, W.I.C., Kerr, C.D. & Bolton, G.R. (1976). The utilization of blanket bog vegetation by grazing sheep. *Journal of Applied Ecology*, 13, 857–869.

Grant, S.A., Barthram, G.T., Lamb, W.I.C. & Milne, J.A. (1978). Effects of season and level of grazing on the utilization of heather by sheep. I. Responses of the sward. *Journal of the British Grassland Society*, **33**, 289–300.

Grant, S.A., Hamilton, W.J. & Souter, C. (1981). The responses of heather-dominated vegetation in north-east Scotland to grazing by red deer. *Journal of Ecology*, **69**, 189–204.

Green, R. (1978). Factors affecting diet of farmland skylarks, *Alauda arvensis*. *Journal of Animal Ecology*, **47**, 913–928.

Greenslade, P.J.M. (1968). Habitat and altitude distribution of Carabidae (Coleoptera) in Argyll, Scotland. *Transactions of the Royal Entomological Society of London*, **120**, 39–54.

Hadley, M. (1971). Aspects of the larval ecology and population dynamics of *Molophilus ater* Meigen (Diptera: Tipulidae) on Pennine moorland. *Journal of Animal Ecology*, **40**, 445–466.

Hale, W.G. (1966). A population study of moorland Collembola. *Pedobiologia*, **6**, 65–99.

Hewson, R. (1976). Grazing by mountain hares *Lepus timidus* L., red deer *Cervus elaphus* L. and red grouse *Lagopus l. scoticus* on heather moorland in north-east Scotland. *Journal of Applied Ecology*, **13**, 657–666.

Hewson, R. & Wilson, C.J. (1979). Home range and movements of Scottish blackface sheep in Lochaber, North-West Scotland. *Journal of Applied Ecology*, **16**, 743–751.

Hobbs, R.J. & Gimingham, C.H. (1984). Studies on fire in Scottish heathland communities. I. Fire characteristics. *Journal of Ecology*, **72**, 223–240.

Hobbs, R.J. & Gimingham, C.H. (1987). Vegetation, fire and herbivore interactions in heathland. *Advances in Ecological Research*, **16**, 87–173.

Holmes, P.F. (1960). The birds of Malham Moor. *Field Studies*, **1**, 49–60.

Hopkins, P.J. & Webb, N.R. (1984). The composition of the beetle and spider faunas of fragmented heathlands. *Journal of Applied Ecology*, **21**, 935–946.

Houston, K. (1971). Carabidae (Col.) from two areas of the North Pennines. *Entomologist's Monthly Magazine*, **107**, 1–4.

Hudson, P. (1986). *Red Grouse: the Biology and Management of a Wild Gamebird*. Game Conservancy Trust, Fordingbridge.

Hudson, P.J. (1988). Spatial variations, patterns and management options in upland bird communities. *This volume*.

Illius, A.W. & Gordon, I.J. (1987). The allometry of food intake in grazing ruminants. *Journal of Animal Ecology*, **56**, 989–999.

Lance, A.N. (1978). Territories and the food plant of individual red grouse. II. Territory size compared with an index of nutrient supply in heather. *Journal of Animal Ecology*, **47**, 307–313.

Leinaas, H.P. & Sømme, L. (1984). Adaptations in *Xenylla maritima* and *Anurophorus laricis* (Collembola) to lichen habitats on alpine rocks. *Oikos*, **43**, 197–206.

Merrett, P. (1976). Changes in the ground-living spider fauna after heathland fires in Dorset. *Bulletin of the British Arachnological Society*, **3**, 214–221.

Miller, B.J.F. (1974). *Studies of changes in the populations of invertebrates associated with cyclical processes in heathland*. Ph.D. thesis, University of Aberdeen.

Miller, G.R. (1968). Evidence for selective feeding on fertilized plots by red grouse, hares and rabbits. *Journal of Wildlife Management*, **32**, 849–853.

Miller, G.R. & Watson, A. (1978). Territories and the food plant of individual red grouse. I. Territory size, number of mates and brood size compared with the abundance, production and diversity of heather. *Journal of Animal Ecology*, **47**, 293–305.

Moore, N.W. (1962). The heaths of Dorset and their conservation. *Journal of Ecology*, **50**, 369–391.

Morris, M.G. (1978). Grassland management and invertebrate animals—a selective review. *Scientific Proceedings of the Royal Dublin Society, Series A*, **6**, 247–257.

Morris, M.G. & Lakhani, K.H. (1979). Responses of grassland invertebrates to management by cutting. I. Species diversity of Hemiptera. *Journal of Applied Ecology*, **16**, 77–98.

Moss, R. & Hewson, R. (1985). Effects on heather of heavy grazing by mountain hares. *Holarctic Ecology*, **8**, 280–284.

Moss, R., Welch, D. & Rothery, P. (1981). Effects of grazing by mountain hares and red deer on the production and chemical composition of heather. *Journal of Applied Ecology*, **18**, 487–496.

Murphy, D.H. (1955). Long-term changes in collembolan populations with special reference to moorland soils. *Soil Zoology* (Ed. by D.K.McE. Kevan), pp. 157–166. Butterworths, London.

Nature Conservancy Council (1986). *Nature Conservation and Afforestation in Britain.* Nature Conservancy Council, Peterborough.

Otto, C. & Svensson, B.S. (1982). Structure of communities of ground-living spiders along altitudinal gradients. *Holarctic Ecology,* **5,** 35–47.

Pearsall, W.H. (1950). *Mountains and Moorland.* Collins, London.

Pimm, S.L. (1982). *Food Webs.* Chapman & Hall, London.

Ratcliffe, D.A. (Ed.) (1977). *A Nature Conversation Review, Vol. 1.* Cambridge University Press, Cambridge.

Ratcliffe, D.A. (1986). The effects of afforestation on the wildlife of open habitats. *Trees and Wildlife in the Scottish Uplands* (Ed. by D. Jenkins), pp. 46–54. Institute of Terrestrial Ecology, Huntingdon.

Ratcliffe, D.A. & Thompson, D.B.A. (1988). The British uplands: their ecological character and international significance. *This volume.*

Rawes, M. & Welch, D. (1969). Upland productivity of vegetation and sheep at Moor House National Nature Reserve, Westmorland, England. *Oikos Supplement,* **11,** 1–72.

Savory, C.J. (1986). Utilization of different ages of heather on three Scottish moors by red grouse, mountain hares, sheep and red deer. *Holarctic Ecology,* **9,** 65–71.

Sømme, L. (1986). Tolerance to low temperatures and desiccation in insects from Andean Paramos. *Arctic and Alpine Research,* **18,** 253–257.

Spence, D. (1979). *Shetland's Living Landscape:* a Study in Island Plant Ecology. Thuleprint, Sandwick (Shetland).

Springett, J.A. (1970). The distribution and life histories of some moorland Enchytraeidae (Oligochaeta) in moorland soils. *Oikos,* **21,** 16–21.

Stewart, A.J.A. & Lance, A.N. (1983). Moor-draining: a review of impacts on land use. *Journal of Environmental Management,* **17,** 81–99.

Thompson, D.B.A., Galbraith, H. & Horsfield, D. (1987). Ecology and resources of Britain's mountain plateaux: land-use conflicts and impacts. *Agriculture and Conservation in the Hills and Uplands* (Ed. by M. Bell & R.G.H. Bunce), pp. 22–31. Institute of Terrestrial Ecology, Grange-over-Sands.

Thompson, D.B.A., Stroud, D.A. & Pienkowski, M.W. (1988). Afforestation and upland birds: consequences for population ecology. *This volume.*

Usher, M.B. (1980). An assessment of conservation values within a large Site of Special Scientific Interest in North Yorkshire. *Field Studies,* **5,** 323–348.

Usher, M.B. (1986). Wildlife conservation evaluation: attributes, criteria and values. *Wildlife Conservation Evaluation* (Ed. by M.B. Usher), pp. 3–44. Chapman & Hall, London & New York.

Usher, M.B. & Smart, L.M. (in press). Recolonisation of burnt and cut heathland in the North York Moors by arachnids. *The Naturalist.*

Watt, A.S. (1955). Bracken versus heather, a study in plant sociology. *Journal of Ecology,* **43,** 490–506.

Welch, D. (1984a). Studies in the grazing of heather moorland in north-east Scotland. I. Site descriptions and patterns of utilization. *Journal of Applied Ecology,* **21,** 179–195.

Welch, D. (1984b). Studies in the grazing of heather moorland in north-east Scotland. III. Floristics. *Journal of Applied Ecology,* **21,** 209–225.

Welch, D. (1984c). Studies in the grazing of heather moorland in north-east Scotland. IV. Seed dispersal and plant establishment in dung. *Journal of Applied Ecology,* **22,** 461–472.

Williamson, K. (1968). The bird communities in the Malham Tarn region of the Pennines. *Field Studies,* **2,** 651–668.

Wood, T.G. (1967a). Acari and Collembola of moorland soils from Yorkshire, England. I. Description of the sites and their populations. *Oikos,* **18,** 102–117.

Wood, T.G. (1967b). Acari and Collembola of moorland soils from Yorkshire, England. III. The micro-arthropod communities. *Oikos,* **18,** 277–292.

Woods, A. (1984). *Upland Landscape Change: a Review of Statistics.* Countryside Commission, Cheltenham.

Climate and the population dynamics of red deer in Scotland

S. D. ALBON AND T. H. CLUTTON-BROCK

Large Animal Research Group, Department of Zoology,
University of Cambridge, Cambridge CB2 3EJ

SUMMARY

1 Climate influences the population dynamics of deer both directly, via the energetic costs of thermoregulation, and indirectly, through changes in the distribution, quantity and quality of food plants.

2 The weather variables most closely associated with deer performance vary across Scotland.

3 In west coast populations year to year fluctuations in survival are negatively related to summer rainfall, via production of grasses and forbs, and positively related to winter temperature.

4 In central and eastern Scotland, heather is the main food, particularly in winter; annual heather production estimated from climatic variables is associated with winter weight loss in females but not with variation in mortality. Instead, winter mortality rises with increasing duration of snow lie.

5 In the short term, climatically induced changes in energy expenditure and food supply affect only the current year's birth- and death-rates. However, temperature in spring influences the birth weight of calves and appears to have permanent effects on the reproductive potential of a cohort.

INTRODUCTION

Red deer in Scotland are at the edge of the species range and on the open moors inhabit an environment characterized by lack of shelter and a low plane of nutrition (Mitchell, Staines & Welch 1977). In this marginal habitat density-independent processes may be more important than density-dependent processes (see Elliott 1987) and birth- and death-rates are likely to be sensitive to climatic variation. The direct effects of climate on deer performance have been demonstrated in terms of increased mortality associated with severe winter weather. For example, in central and eastern Scotland high mortality occurs in winters with long periods of snow lie (Watson 1971). In contrast, on the west coast island of Rhum (57°00′N, 6°20′W) long periods of snow lie are rare and high winter mortality is associated with cold, wet and windy conditions between January and March (Clutton-Brock & Albon 1982). This comparison emphasizes geographic variation in the importance of climatic factors along a gradient from the warm, moist oceanic climate in western Scotland to the colder, drier continental climate of central and eastern Scotland. The west–east climatic gradient across Scotland

93

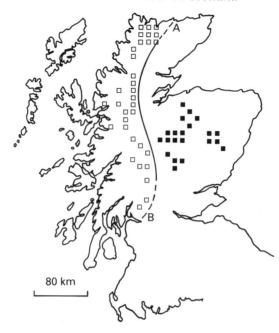

80 km

FIG. 1. The distribution of two types of heather moor in Scotland. *Trichophorum* and *Eriophorum* co-dominant with *Calluna* (□) and Arctostaphylo-Callunetum (■), separated by line AB, reflects the pronounced differences between the wet, mild oceanic climate of the west coast and the more continental dry, cool climate of the central and eastern regions (after McVean & Ratcliffe 1962).

also contributes to major differences in the vegetation communities (Fig. 1). Although diet selection has been studied in these contrasting habitats (Mitchell *et al.* 1977; Staines, Crisp & Parish 1982), the indirect effects of climate, operating through plant productivity, have been investigated only on Rhum. However, plant–atmosphere interactions have been studied extensively (see Grace, Ford & Jarvis 1981) and the case of heather provides an opportunity to estimate annual variation in primary production from meteorological measurements (Miller 1979).

Here we synthesize the information on annual variation in red deer performance, particularly birth- and death-rates, in relation to changes in weather and food supply for several different populations across Scotland. Since rainfall and temperature in different seasons and regions of Scotland exhibit oscillations of variable length (see Miller & Cooper 1976), we discuss the implications for the stability of red deer populations over decades or centuries.

SUMMER RAIN, GRASS GROWTH AND FECUNDITY

Although fecundity in unculled populations (Clutton-Brock, Guinness & Albon 1982a) and fertility in culled populations (Beddington 1973, Albon 1983) have been shown to be density-dependent, there have been few attempts to explain the residual variation. Beddington (1973) re-analysed Lowe's (1969) data, for fertility between 1958/9 and 1961/2 on Rhum, and showed that the residuals about the density-dependent relationship were correlated, negatively, with total rainfall between June and September.

More recent data from Rhum suggest that low rainfall at the time of peak primary production between May and July (three of the four driest months of the year) may depress deer performance through retarded grass production, while low rainfall in the autumn may enhance deer performance because it is associated with a large secondary peak in grass production. For example, the body weight of hinds in autumn, which influences the probability of fertility (Albon, Mitchell & Staines 1983), was related negatively to the maximum soil moisture deficit (Albon 1983). High soil moisture deficits are not uncommon in the Inner Hebrides (Green & Harding 1983) and can retard plant growth (Alcock & Lovett 1968; Munro & Davies 1973). This may explain why the time spent feeding on the preferred *Agrostis–Festuca* swards, which on Rhum overlie well-drained alluvial soils, was lower when early summer (May–July) was dry than when wet (Clutton-Brock *et al.* 1982b). However, fecundity was not related positively to rainfall in early summer, but instead was related negatively to rainfall in August and September. This negative association between fecundity and rainfall in late summer may have occurred because low September rainfall was associated with high net aerial primary production on the *Agrostis–Festuca* swards (see Iason, Duck & Clutton-Brock 1986 for methodology), which in turn influenced the live standing crop at the onset of the October rut (Fig. 2). Primary production at this time of the year is sometimes referred to as the 'autumn flush' and in sheep has important implications both for weight gain immediately prior to the rut and for fertility (Gunn, Doney & Russel 1969).

SEPTEMBER RAIN, WINTER FOOD AND MORTALITY

On Rhum, the standing crop of live grasses and forbs on the preferred *Agrostis–Festuca* grazings at the end of winter (April) was related to the live standing crop at the end of the previous summer (September). Consequently, we would predict a negative correlation between winter mortality and September live standing crop. Among males, mortality in yearlings and 8–10-year-olds declined by approximately 4 per cent for every 10 g m^{-2} increase in live standing crop (Fig. 3), after controlling for variation in mortality associated with differences in winter temperature. However, mortality among females was not correlated significantly with September live standing crop.

Since the live standing crop of grasses and forbs on the *Agrostis–Festuca* swards at the end of September was related positively to the net aerial primary production (NAPP) during that month and NAPP was related negatively to September rainfall, we would expect winter mortality to be correlated positively with September rainfall. Evidence from Rhum supported this prediction since

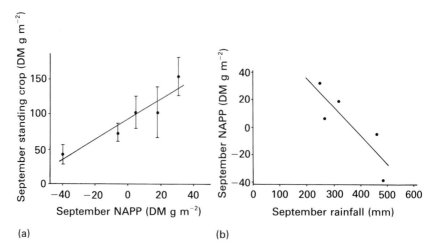

FIG. 2. (a) September standing crop (dry mass, g m^{-2}) of live grasses and forbs plotted against net aerial primary production (dry mass, g m^{-2}) in September measured on *Agrostis–Festuca* vegetation communities on Rhum between 1981 and 1985. Each point is the mean of six plots and bars show ±1 standard deviation. Regression line is $y = 92.5 + 1.417x$ ($r^2 = 0.880$). (b) September net aerial primary production (dry mass, g m^{-2}), measured on *Agrostis–Festuca* vegetation communities on Rhum, plotted against rainfall in the same month each year between 1981 and 1985. Regression line is $y = 71.2 - 0.192x$ ($r^2 = 0.727$).

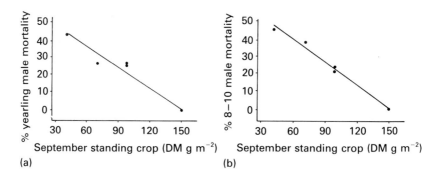

FIG. 3. Percentage mortality in winter among (a) yearling males and (b) 8–10-year-old males, plotted against the September standing crop (dry mass, g m^{-2}) of live grasses and herbs on *Agrostis–Festuca* vegetation communities on Rhum each year between 1981 and 1985. The mortality values are corrected for variation due to differences in winter (December–February) temperature. Regressions lines are, for yearlings, $y = 62.5 - 0.417x$ ($r^2 = 0.914$) and for 8–10-year-olds, $y = 66.9 - 0.441x$, ($r^2 = 0.967$).

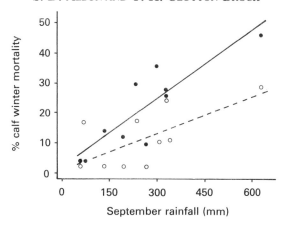

FIG. 4. Percentage winter mortality of both male calves (●) and female calves (○) plotted against September rainfall each year between 1971 and 1980. The mortality values are corrected for variation due to differences in winter (December–February) temperature. Regressions lines are, for males, $y = 2 \cdot 0 + 0 \cdot 0767x$ ($r^2 = 0 \cdot 680$) and, for females, $y = 0 \cdot 7 + 0 \cdot 0405x$ ($r^2 = 0 \cdot 423$).

between 1971 and 1980 mortality among males of all age classes rose significantly with increasing rainfall. Among male calves an increase of 100 mm of rain in September increased mortality by 8 per cent. Mortality in female calves also increased with increasing rainfall, although the magnitude of the increase was only about half that of males (Fig. 4).

Two other studies in western Scotland provide corroborative evidence supporting a positive relationship between winter mortality and rainfall in the previous summer. First, Beddington (1973), using total rainfall between June and September, found that mortality in deer older than 8 years of age was high after wet summers, particularly among males. The results for juveniles were less clear, and Beddington suggested that there may be an additive relationship of the form in which a poor winter following a wet summer accentuates mortality. Second, we analysed data on mortality collected on another Hebridean island, Jura, a century ago (Evans 1890). Over a 12-year span between 1878 and 1889 the mortality of females was correlated significantly and positively with rainfall in the previous August. Interestingly, the recent studies on Rhum found that, among females older than 4 years of age, mortality also rose with increasing August rainfall.

HEATHER GROWTH, WINTER WEIGHT LOSS AND DEER ABUNDANCE

In contrast to Rhum, red deer inhabiting the drier moors of central and eastern Scotland eat a winter diet in which the predominant forage is heather, *Calluna vulgaris* (Staines *et al.* 1982), especially the current year's growth (Moss, Welch & Rothery 1981). Annual variation in the production of new heather shoots at Kerloch Moor (56°50′N, 2°30′W) was closely related to temperature and rainfall between mid-April and August (Miller 1979). If the growing season was warm and dry, the production of new shoots was high compared with that in cool, wet summers. Assuming that it was valid to substitute meteorological data for Braemar (56°55′N, 3°30′W and 330 m above MSL), which during these summer months was similar to Kerloch, into the equation,

$$H = 29 \cdot 3T - 0 \cdot 168R + 56 \cdot 0$$

where H is the annual production of new shoots, T the mean daily temperature (°C) between mid-April and August, and R is the total rainfall (mm) from mid-April to August (Miller 1979), Albon (1983) estimated that the annual production of new heather shoots at Glenfeshie (57°N, 4°W) varied from $182 \cdot 6$ g m^{-2} to $218 \cdot 2$ g m^{-2} between 1966 and 1974.

Although offtake is generally a small proportion of the current year's growth, deer and other herbivores grazing heather select the shoots with highest concentrations of nitrogen and phosphorus (Moss *et al.* 1981). Thus, in years of high production there may be, on average, more high-quality shoots to select than in years of low production, even though at a particular site the average N content is inversely related to production across years (R. Moss, pers. comm.). As a result annual differences in heather production might reflect variation in potential nutrient intake, and we would predict significant correlations between aspects of deer performance and the variation in heather production. For example, early winter weight loss in hinds, calculated from animals shot in season between November and mid-February, was negatively correlated with the estimated annual heather production (Albon 1983), such that a reduction in production of 10 g m^{-2} increased daily weight loss by 20 g. In years when weight loss is rapid, hinds may be in danger of falling below the critical weight for survival (see Moen & Severinghaus 1981). As a result high mortality might be expected in winters following poor heather growth. However, in practice mortality searches at Glenfeshie did not show any relationship between mortality in any sex/age class and the estimated production of heather shoots in the previous summer (Albon 1983).

Although searches for dead animals at Glenfeshie did not reveal an association between mortality and the winter food supply, the population of hinds overwintering on the estate was correlated to annual estimated heather production (Fig. 5). Furthermore, the change in hind numbers between consecutive years was significantly correlated with the difference in the estimated heather production for the same pairs of years. The changes in numbers between consecutive years were often larger than could be attributed to changes in the magnitude of the cull, mortality or estimated recruitment (Mitchell, McCowan & Parish 1986). One possible explanation for the association between the spring population size and estimated heather production the previous summer is that hinds which tended to move on to the estate in summer stayed through the winter if the summer was warm and dry, when heather growth would have been good, but departed after cool, wet summers, when heather growth would have been poor. However, since we would expect summer weather to influence heather production in the adjacent

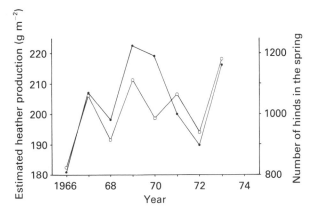

FIG. 5. Estimated annual heather production (○) (dry mass, g m⁻²) and the number of hinds counted in the following spring (●) at Glenfeshie, Inverness-shire, plotted each year from 1966/7 to 1973/4 ($r^2 = 0.62$).

glens it is difficult to understand why hinds should move to other localities unless they provided alternative winter forage.

WINTER SEVERITY AND MORTALITY

Severe winter weather may increase the risk of deer dying, either directly, through the increased energetic costs of thermoregulation (see Blaxter, Joyce & Wainman 1963), or indirectly, because of the dieback of forage (see Watson, Miller & Green 1966). In central and eastern Scotland several studies have shown that high mortality occurs in winters with long periods of snow lie (Watson 1971; Albon 1983; Mitchell 1984). At Glenfeshie, increasing snow lie, particularly in March and April, had a greater effect on mortality in calves than in hinds. Hind mortality was density-dependent, and only after accounting for variation in hind numbers was the relationship with days of snow lie significant. Mortality in calves increased at approximately 0·2 per cent per extra day's snow lie in March and April, compared with only 0·03 per cent in hinds (Fig. 6). Annual mortality among stags was negatively related to the mean weight in the previous autumn, which, in turn, was related to the duration of snow lie in the previous winter. Therefore, the effect of winter weather on stag mortality appeared to be lagged by a year.

In contrast, populations in western Scotland rarely experience long periods of snow lie. For example, on Rhum the average number of days with snow lie at sea level is less than 10 year⁻¹. High mortality was associated with low temperature combined with wet, windy weather in February, conditions that increase wind chill (Clutton-Brock & Albon 1982). Beddington (1973) suggested that there may be an additive relationship in which a poor winter following a poor summer

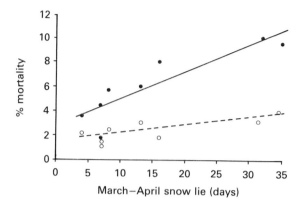

FIG. 6. Percentage winter mortality of both calves (●) and hinds (○) at Glenfeshie, Inverness-shire, plotted against the number of days of snow lie in March and April each year between 1967 and 1974. The mortality values for hinds are corrected for variation due to differences in density. Regression lines are, for calves, $y = 2.75 + 0.23x$ $(r^2 = 0.796)$ and, for hinds, $y = 1.66 + 0.033x$ $(r^2 = 0.503)$.

accentuates mortality, particularly among juveniles. More recent work on Rhum supports Beddington's suggestion by demonstrating increased mortality when wet summers were followed by cold winters (Fig. 7). Mortality in 5- to 12-year-old hinds increased by 2 per cent for each 1 °C decline in winter temperature. As at Glenfeshie, calves appeared to be particularly susceptible to very cold winter weather.

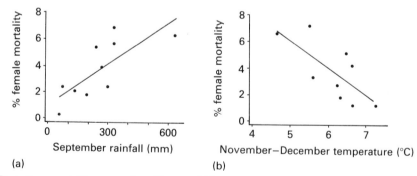

FIG. 7. Percentage mortality among 5- to 12-year-old females on Rhum each year between 1971 and 1980, plotted against (a) September rainfall, controlling for variation due to early winter (November–December) temperature and (b) November to December temperature, controlling for variation due to September rainfall. Multiple regression equation is $y = 13.375 + 0.0107$ (Sept. rain) $- 2.085$ (Nov–Dec temp.) $(r^2 = 0.591)$.

SPRING TEMPERATURE, COHORT QUALITY AND POPULATION FLUCTUATION

The examples of the influence of climate on deer performance described above all focus on short-term effects, usually the current or subsequent year's birth and death rates. However, recent work on Rhum suggests that the climatic environment into which a cohort is born may shape the fate of that cohort in the future, even if conditions ameliorate, and has important consequences for population fluctuation (Albon, Clutton-Brock & Guinness 1987).

The population inhabiting the North Block, Rhum, has increased steadily since the cessation of the cull in 1972 (Clutton-Brock *et al.* 1982a), and has been regulated by density-dependent changes in fecundity and mortality (Clutton-Brock, Major & Guinness 1985). However, each successive cohort recruited to the population has not necessarily had a lower reproductive success than the cohorts recruited in earlier years (Albon *et al.* 1987). Cohorts of females born after cold springs (April and May) tended to have low birth weights, presumably because poor grass growth retarded foetal development (Fig. 8a). As adults, cohorts born after cold springs were less likely to breed as 3-year-olds and had short lifespans (Figs. 8b and c). Furthermore, light-born cohorts gave birth to small offspring with low viability. The cumulative effect of these cohort differences in the components of reproductive success resulted in a sixfold difference in average fitness between individuals in the worst and best cohorts (Clutton-Brock, Albon & Guinness, in press).

The potential influence of cohort variation in reproductive performance on fluctuation in red deer numbers has been illustrated by a series of Leslie matrix models. Good, average and bad cohorts were recruited at random to a population under density-dependent control. When the proportion of good cohorts in the breeding population was high, numbers increased from the simple density-dependent asymptotic population by about 15 per cent (Fig. 9a). In contrast, when the proportion of good cohorts was small, numbers fell by about 15 per cent. Clearly, if April and May temperature followed some long-term change, then the amplitude of the fluctuations generated by cohort variation could be much larger. This can be illustrated by a simulation using the meteorological data for Stornoway from 1885 to present (meteorological data for Rhum not available before 1957). Between 1885 and 1910 60 per cent of springs were cold (falling in the lower third of the distribution), while from 1935 to 1959 only 16 per cent were cold. As a result the model predicted a low population in the first two decades of this century rising to a peak population around 1960, which was approximately 50 per cent greater than the lowest level (Fig. 9b). We do not suggest that this actually occurred, for many other factors affected deer numbers over this period, but the model shows that, if all else had been equal, these effects could have produced major changes in abundance.

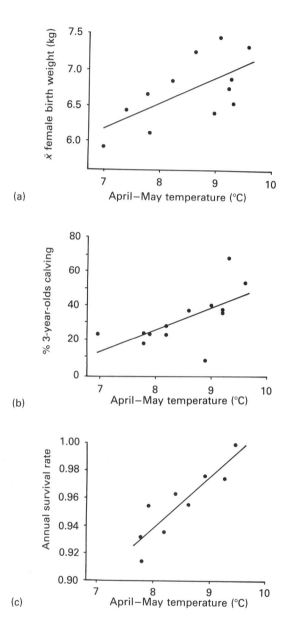

FIG. 8. Female cohort values of (a) mean birth weight (kg), (b) percentage calving as 3-year-olds and (c) mean annual survival rate (2 to 8 years old), plotted against April and May temperature (°C) in cohort's year of birth. Regression equations are (a) $y = 3.58 + 0.37x$ ($r^2 = 0.433$); (b) $y = 13.2x - 80$ ($r^2 = 0.408$); and (c) $y = 0.623 + 0.039x$ ($r^2 = 0.781$).

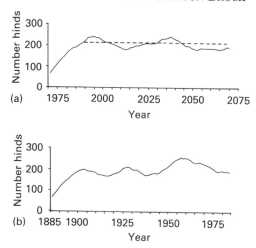

FIG. 9. Simulations of the changes in hind numbers in the North Block, Rhum, generated by a model incorporating both density dependence and cohort variation due to April and May temperature in the year of birth, (a) using a random sequence of cold, average and warm springs for 100 years beginning in 1971 (the dashed line K indicates the simple density-dependent asymptotic population), and (b) using the observed sequence of cold, average and warm springs for the 100 years from 1885 at Stornoway, Western Isles. In each case the starting population was 57 hinds.

DISCUSSION

The available evidence suggests that climatic variation may play an important role in the population dynamics of red deer in Scotland. Although the direct effects of climate, particularly the influence of winter weather on mortality, have been documented previously, recent research has begun to identify the nature and magnitude of some of the indirect effects of climate operating through variation in the food supply. The pronounced west–east gradient in climate and the associated vegetation communities results in deer performance being influenced by rainfall and grass production in the west and by temperature and heather production in the east.

During the summer, food is apparently abundant, with primary production exceeding offtake on all but the most preferred vegetation communities. However, seasonal variation in production may have important consequences for winter feeding. For example, in some age classes of males on Rhum, winter mortality was related to the standing crop of live grasses and forbs on the most heavily selected vegetation community at the end of the previous September. Year-to-year variation in September biomass was related closely to the actual production during September, which, in turn, was negatively related to September rainfall. Thus, winter mortality was influenced, in part, by the effect of rainfall six months earlier on the food supply.

Although, fertility on Rhum was not related to either primary production or the total live biomass in September between 1981 and 1985, it was negatively correlated to rainfall in September between 1971 and 1985. Therefore, this relationship may also represent an example of the effects of climate operating through variation in the food supply.

So far, we have argued that rainfall in late summer may affect fertility and mortality indirectly, through its influence on the food supply; however, it may also have a direct effect in terms of the energetic costs of thermoregulation (Beddington 1973) or in terms of feeding behaviour (J. Milne pers. comm.). In sheep, exposure to very wet conditions prior to mating delayed the onset of oestrus and depressed ovulation rates (Griffiths, Gunn & Doney 1970). In young growing animals climatic stress may inhibit rapid weight gain immediately prior to the winter and therefore the probability of surviving the winter. However, the direct energetic costs of wet summer weather are presumably lower than those associated with the wetter, windier and colder conditions of winter.

The high mortality associated with severe winter weather presumably reflects the direct increase in energetic costs of thermoregulation at low temperatures. Compared with hinds, calves and stags were more likely to die in severe winter weather. Stags enter the winter in poor condition, having lost up to 30 per cent of body weight and nearly all their fat reserves during the autumn rut, while calves, which are still growing, have very small fat reserves (Mitchell, McCowan & Nicholson 1976). Calves are also at a disadvantage because their small size elevates the relative expenditure of energy in maintaining body temperature. Furthermore, since calves are socially subordinate, the increased feeding competition when access to food is restricted will discriminate against them (Hall 1983).

A particularly striking case of the relative susceptibility of calves was recorded at Invermark, where in the winter of 1962–63, one of the worst this century, 64 per cent of calves died compared with only 17 per cent of hinds (Mitchell 1984). However, it may be misleading to give the impression that the severe winter of 1962–63 only affected the 1962 cohort, for the spring of 1963 was very cold and the average birth weight of calves was very low and neonatal mortality high (Mitchell 1984). By analogy with recent results from Rhum (Albon *et al.* 1987), we would predict that the members of the 1963 cohort that survived juvenile mortality never achieved compensatory growth, were late to mature, died relatively young and, during their short reproductive lives, experienced low fecundity and low offspring viability. As a result, the recruitment of the 1963 cohort to the breeding population may have introduced a lifelong lag of the effects of the severe winter of 1962–63.

The importance of variation in spring temperature is that through its effects on cohort quality it can generate population fluctuation. Even a random sequence of spring conditions could lead to substantial perturbations, and clearly if there were a long-term change in spring weather then the fluctuations in abundance might be large.

Some regular cycles in the climate at particular localities in Scotland have been described (see Miller & Cooper 1976). For example, in the rain-shadow of the Grampian mountains, rainfall in May exhibited 11·9 and 23-year cycles, while rainfall in June followed a 4·4-year cycle. Rainfall at this time of the year influenced the autumn body weight of hinds at Glenfeshie (Albon 1983) and also contributed to variation in heather growth at Kerloch (Miller 1979). Thus, we might expect to find evidence of cycles in deer abundance in east coast locations. Although a twofold variation in the numbers of red deer in Scotland since 1961 has been described (Clutton-Brock & Albon, in press) the period is too short to detect regular cycles. However, when the fluctuation was investigated for populations to the east of the vegetation boundary shown in Fig. 1 there was a significant correlation between temporal estimates of deer density and estimated annual heather production. In contrast, temporal estimates of deer density to the west of the vegetation boundary in Fig. 1 were less pronounced and not related to estimated annual heather production. Preliminary work suggests that the change in average density in west coast populations is related to September rainfall. However, since September rainfall varies with latitude, depending on the track of the prevailing westerlies, populations on Arran, Islay and Jura may not be in phase with those in Sutherland.

So far this discussion has been confined to the direct and indirect response of deer to climatic change and how this may vary across Scotland. However, man has a greater potential for ecological change in the Highlands. Changes in management policies, such as the size of the cull, or changes in agricultural practices, especially the number of sheep, which overlap with deer in their use of habitat (Osborne 1984), may be at least as important as climatically induced change. For example, the increase in deer density between 1968 and 1977 coincided with a 20 per cent decline in the number of ewes (Clutton-Brock & Albon, in press). Further research is needed to predict the effects of both change in land use and change in climate on the population dynamics of red deer in the Highlands.

ACKNOWLEDGMENTS

We are especially grateful to Brian Mitchell for access to unpublished data for Glenfeshie and to Steven Mithen for doing the simulations in Fig. 9. Also, we wish to thank our colleagues, Fiona Guinness, Callan Duck, Glenn Iason, Jerry Kinsley, Richard Clarke and David Green, for assistance with monitoring the deer and vegetation on Rhum. The work on Rhum has been supported by NERC, SERC, the Royal Society, and the Leverhulme Trust. We are grateful to Gordon Miller and Adam Watson for helpful discussion and wish to thank Iain Gordon, Sheila Grant, Nigel Leader-Williams, Robert Moss and Brian Staines for constructive comments on earlier drafts of this paper.

REFERENCES

Albon, S.D. (1983). *Ecological aspects of growth, reproduction and mortality in female red deer.* Ph.D. thesis, University of East Anglia.

Albon, S.D., Mitchell, B. & Staines, B.W. (1983). Fertility and body weight in female red deer: a density-dependent relationship. *Journal of Animal Ecology*, **56**, 969–980.

Albon, S.D., Clutton-Brock, T.H. & Guinness, F.E. (1987). Early development and population dynamics in red deer. II. Density-independent effects and cohort variation. *Journal of Animal Ecology*, **56**, 69–81.

Alcock, M.B. & Lovett, J.V. (1968). Analysis of environmental influence on productivity. *Occasional Symposium of the British Grassland Society*, **4**, 20–29.

Beddington, J.R. (1973). *The exploitation of red deer* (Cervus elaphus L.) *in Scotland.* Ph.D. thesis, University of Edinburgh.

Blaxter, K.L., Joyce, J.P. & Wainman, F.W. (1963). Effect of air velocity on the heat losses of sheep and cattle. *Nature*, 198, 1115–1116.

Clutton-Brock, T.H. & Albon, S.D. (1982). Winter mortality in red deer (*Cervus elaphus*). *Journal of Zoology*, 198, 515–519.

Clutton-Brock, T.H. & Albon, S.D. (in press). *Red Deer in the Highlands: the Ecology of a Marginal Population.* Blackwell Scientific Publications, Oxford.

Clutton-Brock, T.H., Guinness, F.E. & Albon, S.D. (1982a). *Red Deer: Behaviour and Ecology of Two Sexes.* Chicago University Press, Chicago.

Clutton-Brock, T.H., Iason, G.R., Albon, S.D. & Guinness, F.E. (1982b). Effects of lactation on feeding behaviour and habitat use in wild red deer hinds. *Journal of Zoology*, 198, 227–236.

Clutton-Brock, T.H., Major, M. & Guinness, F.E. (1985). Population regulation in male and female red deer. *Journal of Animal Ecology*, 54, 831–846.

Clutton-Brock, T.H., Albon, S.D. & Guinness, F.E. (in press). Reproductive success in red deer. *Reproductive Success: Studies of Individual Variation in Contrasting Breeding Systems* (Ed. by T.H. Clutton-Brock). Chicago University Press, Chicago.

Elliott, J.M. (1987). Population regulation in contrasting populations of trout *Salmo trutta* in two Lake District streams. *Journal of Animal Ecology*, 56, 83–98.

Evans, H. (1890). *Some Account of Jura Red Deer.* Carter, Derby.

Grace, J., Ford, E.D. & Jarvis, P.G. (1981). *Plants and Their Atmospheric Environment.* Blackwell Scientific Publications, Oxford.

Green, F.H.W. & Harding, R.J. (1983). Climate of the Inner Hebrides. *The Natural Environment of the Inner Hebrides* (Ed. by J.M. Boyd & D.R. Jones). *Proceedings of the Royal Society of Edinburgh*, 83B, 121–140.

Griffiths, J.G., Gunn, R.G. & Doney, J.M. (1970). Fertility in Scottish Blackface ewes as influenced by climatic stress. *Journal of Agricultural Science*, 75, 485–488.

Gunn, R.G., Doney, J.M. & Russel, A.J.F. (1969). Fertility in Scottish Blackface ewes as influenced by nutrition and body condition at mating. *Journal of Agricultural Science*, 73, 289–294.

Hall, M.J. (1983). Social organisation in an enclosed group of red deer (*Cervus elaphus* L.) on Rhum. I. The dominance hierarchy. *Zeitschrift für Tierpsychologie*, 61, 250–262.

Iason, G.R., Duck, C.D. & Clutton-Brock, T.H. (1986). Grazing and reproductive success of red deer: the effect of local enrichment by gull colonies. *Journal of Animal Ecology*, 55, 507–516.

Lowe, V.P.W. (1969). Population dynamics of the red deer (*Cervus elaphus* L.) on Rhum. *Journal of Animal Ecology*, 38, 425–457.

McVean, D.N. & Ratcliffe, D.A. (1962). *Plant Communities of the Scottish Highlands*, Nature Conservancy Monograph No.1. HMSO, London.

Miller, G.R. (1979). Quantity and quality of the annual production of shrubs and flowers by *Calluna vulgaris* in North-east Scotland. *Journal of Ecology*, 67, 109–129.

Miller, H.G. & Cooper, J.M. (1976). Tree growth and climatic cycles in the rain shadow of the Grampian mountains. *Nature*, 260, 697–698.

Mitchell, B. (1984). Effects of the severe winter of 1962/63 on red deer hinds and calves in North-east Scotland. *Deer*, 6, 81–84.

Mitchell, B., McCowan, D. & Nicholson, I.A. (1976). Annual cycles of body weight and condition in Scottish red deer (*Cervus elaphus*). *Journal of Zoology,* **180**, 107–127.

Mitchell, B., Staines, B.W. & Welch, D. (1977). *Ecology of Red Deer: a Research Review Relevant to Their Management.* Institute of Terrestrial Ecology, Cambridge.

Mitchell, B., McCowan, D. & Parish, T. (1986). Performance and population dynamics in relation to management of red deer *Cervus elaphus* at Glenfeshie, Inverness-shire, Scotland. *Biological Conservation,* **37**, 237–267.

Moen, A.N. & Severinghans, C.W. (1981). The annual weight cycle and survival of white-tailed deer in New York. *New York Fish and Game Journal,* **28**, 162–177.

Moss, R., Welch, D. & Rothery, P. (1981). Effects of grazing by mountain hares and red deer on the production and chemical composition of heather. *Journal of Applied Ecology,* **18**, 487–496.

Munro, J.M.M. & Davies, D.A. (1973). Potential pasture production in the uplands of Wales. 2: Climatic limitation on production. *Journal of the British Grassland Society,* **28**, 161–169.

Osborne, B.C. (1984). Habitat uses by red deer (*Cervus elaphus* L.) and hill sheep in the West Highlands. *Journal of Applied Ecology,* **21**, 497–506.

Staines, B.W., Crisp, J.M. & Parish, T. (1982). Differences in the quality of food eaten by red deer (*Cervus elaphus*) stags and hinds in winter. *Journal of Applied Ecology,* **19**, 65–78.

Watson, A. (1971). Climate, antler-shedding and the performance of red deer in North-east Scotland. *Journal of Applied Ecology,* **8**, 53–68.

Watson, A., Miller, G.R. & Green, F.H.W. (1966). Winter browning of heather (*Calluna vulgaris)* and other moorland plants. *Transactions Botanical Society of Edinburgh,* **40**, 195–203.

Improved automated classification of upland environments utilizing high-resolution satellite data

A. R. JONES AND B. K. WYATT

NERC Remote Sensing Applications Centre, Institute of Terrestrial Ecology,
Penrhos Road, Bangor, Gwynedd LL57 2LQ

SUMMARY

1 High-resolution multispectral sensors aboard the SPOT-1 satellite have been designed to detect radiation at wavelengths which are particularly sensitive to vegetation. Analysis of digital imagery from SPOT is therefore a potentially effective technique for the mapping and monitoring of semi-natural upland vegetation communities.

2 Despite good discrimination of broad cover types, overall levels of classification accuracy were poor. This was primarily due to the complex heterogeneity of upland plant communities and a high interband correlation in the HRV data.

3 To improve the discriminatory potential of the satellite data, digital terrain information has been utilized as additional inputs into automated classification procedures.

4 This paper gives examples derived from a test site in southern Snowdonia, North Wales.

INTRODUCTION

The upland environment, defined here as land with a mean altitude over 122 m (Ball *et al.* 1982), is an important component of the British landscape. Although upland ecology is dominated by physical environment factors (e.g. climate, geology and topography), the situation is not static but changes in response to past, present and new land-use patterns. These changes include loss of moorland due to grazing pressures and management regimes, expansion and reduction of agriculture (e.g. improvement and reversion of pastures) and expansion of forestry, not to mention an increase in pressures from leisure activities (see Ratcliffe & Thompson 1988). To improve the understanding of these changes and to attempt to predict future patterns, accurate land cover maps need to be produced easily and cheaply on a systematic basis.

However, a number of factors make the British uplands a difficult environment in which to carry out extensive ecological survey. There is extreme diversity and complexity in the species composition of the natural vegetation communities encountered. Boundaries between these communities are diffuse and the land cover is spatially variable. Physical inaccessibility and the rugged nature of the

109

terrain make conventional ground-based survey difficult. Consequently, remote sensing offers a seemingly ideal tool for the mapping and monitoring of these relatively inaccessible areas (Weaver 1984; Frank & Thorn 1985; Nellis 1986). However, the uplands present the user of remotely sensed data with further problems. The major one is that spectral responses observed from a surface are influenced by physiological factors (e.g. health of plants), ecological factors (e.g. canopy composition), atmospheric conditions and the orientation of the surface with respect to the light source and sensor position (Hutchinson 1982; Hall-Köynes 1987; McMorrow & Hume 1986; Stohr & West 1985). Numerous techniques have been proposed to reduce the effect on spectral information. Band ratioing (Holben & Justice 1981), the separation of training sets into subclasses determined by their orientation (Williams 1988), and use of a digital elevation model (DEM) (Justice, Wharton & Holben 1981) have all been tried with varying degrees of success. This paper illustrates the improvements in land cover classification that can be obtained from remotely sensed imagery when the effects of terrain on the data are suppressed using digital terrain information.

STUDY AREA

The work was carried out in a 10×10 km study area centred on the coniferous forest of Coed-y-Brenin, situated near the town of Dolgellau, in southern Snowdonia, North Wales (Fig. 1). The study site has varied topography with deep glaciated valleys, dissected plateau and grassy mountain slopes with numerous rock outcrops. Local relief is around 900 m with extensive areas of steeply sloping ground (>30°). The climate is generally mild, due to maritime influences, with high precipitation (>3000 mm year^{-1}). Land cover varies from coniferous plantations, deciduous woodlands, wetlands or peatlands (e.g. with *Juncus* spp. and blanket bog vegetation), numerous mountain grasses (*Nardus* spp., *Agrostis* spp.) to extensive areas of *Calluna vulgaris*-dominated moorland and bracken (*Pteridium aquilinum*)-infested pastureland.

DATA SOURCES

Two types of data were utilized in this study: SPOT HRV multispectral imagery and a digital elevation model (DEM). The SPOT-1 satellite, launched in early 1986, carries two identical high-resolution visible (HRV) sensors, which differ quite markedly from the sensors on board the Landsat series of Earth observation satellites (Jones 1987). These include new sensing technology (charge coupled devices as opposed to line scanners), finer spatial resolution (20 m and 10 m pixels) and the ability to provide stereoscopic imagery. However, the HRV sensors have only three multispectral channels (green, red and near infra-red), compared with the seven data channels from the earlier Landsat Thematic Mapper. The work in this paper utilized a SPOT-1 HRV multispectral image (K;J: 22–243) from 17 October 1986. The raw digital data were converted to

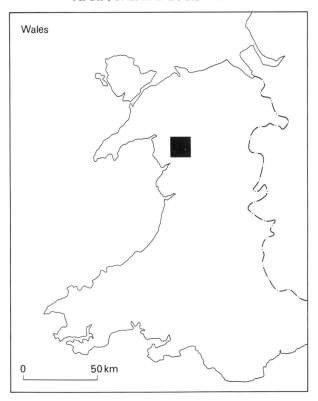

FIG. 1. Location of Coed-y-Brenin study area (■).

radiance values using calibration information supplied with the image. The imagery was then geometrically registered to the British National Grid with a mean error of less than 0·2 pixels (< 5 m). A subscene corresponding to a 1:25,000 Ordnance Survey (OS) sheet (SH 72) was then extracted (Fig. 2).

A DEM of the study area was produced by automatically scanning the contour plate of the same 1:25,000 OS sheet. The resulting data were edited manually on a Sysscan digital mapping system at ITE Bangor in order to remove superfluous linear information, interpolate contours in areas of cliff masks, and remove all but the main 100-foot (c. 30 m) contours. These vector contours were processed using a terrain-modelling software package (PANACEA) to generate a DEM with altitude values at horizontal intervals of 19·53 m. The DEM was used to produce a grid-based elevation image. As both the elevation model and satellite imagery were now on the same OS base, registration of the data was not a problem (Fig. 3.)

Visual analysis of the HRV data revealed low contrast in the raw imagery with most land cover discrimination resulting from the near infra-red data. Statistical examination of the data confirmed the relatively narrow range of digital number values (Table 1). A high level of correlation between the two visible bands

TABLE 1. Statistical analysis of HRV data. DN denotes digital numbers

Statistical parameters	Band number			Correlation matrix				Eigenvalues (%)
	1	2	3					
Minimum DN	15	7	8	1	1·000			93·9
Maximum DN	50	27	94	2	0·898	1·000		5.6
Mean DN	21	13	37	3	0·708	0·559	1·000	0·5
Standard deviation	3	3	12					

($r = 0.89$, for 512 by 512 data sets with 3 spectral bands) reduced the discriminatory power of the data. This was confirmed by rather poor false-colour composites and by principal component analysis. Inspection of the eigenvalues showed that 94 per cent of scene variance was explained by the first principal component (PC) with only 0·5 per cent by PC3.

Simple image enhancement techniques applied to the data revealed that a discrimination of broad cover types was possible. Coniferous forest, areas of clear felling, moorland, improved and unimproved grasslands were easily identifiable. However, it was apparent that areas with homogeneous vegetation types displayed different intensity levels on slopes of differing aspects, thus necessitating a correction to suppress the effects of topography. Furthermore, some slight blurring occurs in part of the image due to radiometric noise of the sensor.

Following the production of the DEM, the next stage was to develop procedures for radiometrically preprocessing multispectral data. This minimized variation in the reflectance signal due to the effects of topography, which are particularly marked on south-facing slopes. A program was written to perform radiometric correction of the multispectral imagery assuming a simple Lambertian model:

$$L = Ln[\cos(i)]$$

where L is the measured radiance, Ln is the normalized radiance and i is the angle between the surface normal and the direction of the sun. This model assumes that the intensity of reflected radiation is independent of the direction of view, and corrects for changes in irradiation due to topography. Irradiation is assumed to be proportional to the cosine of the local angle of incidence. Solar azimuth and elevation are recorded in the calibration data associated with the satellite data. The local angle of solar irradiance was then derived by computing partial derivatives of the surface of the elevation model. Visual analysis of the image indicated that some basic assumptions in this model were inadequate because it resulted in over-correction in many places. A non-Lambertian model has recently been implemented; preliminary results suggest that the revised model is a better representation of the actual behaviour of natural vegetation canopies.

IMAGE CLASSIFICATION

It is possible to classify land cover characteristics from a remotely sensed image utilizing their varying spectral response. Two approaches of classification were applied to the data: supervised and unsupervised. In the supervised method, classes are defined *a priori* by the use of known ground reference data in small regions to classify the entire image. In an unsupervised classification, a clustering algorithm is applied to the data and classes are defined *a posteriori* as a result of inspection and interpretation against ground reference points (pp. 208–201 in Curran 1985; pp. 86–94 in Townshend 1981).

An important requirement of supervised classification is the need to draw up, in advance of classification, a list of ground cover types which are spectrally separable. Within the image to be classified, there must be representative examples of each class which can be selected as ground reference sites. For upland vegetation, both these points can prove difficult and depend a great deal on the nature and availability of the reference data.

Unsupervised classification has been shown to be an effective technique for discriminating upland land cover units (McMorrow & Hume 1986), especially where it is difficult to define suitable homogeneous training sites due to size, topography and variability of the vegetation cover. It is particularly applicable in situations where detectable within- and between-class variation is not adequately described by the available documentary ground reference data (e.g. maps, site descriptions). An important criterion which determines the suitability of imagery for unsupervised classification is that spectral classes will relate well to vegetation classes or types. Since SPOT HRV wavelengths were specifically selected for their vegetation investigation potential (Jones 1987), HRV imagery is expected to perform well in this respect.

RESULTS

We present an initial examination of the data; fuller conclusions will follow. Both the raw and terrain-corrected images showed much discrimination of cover types. Ten classes were defined to evaluate the effectiveness of the various data sets for discriminating land cover (Table 2). Ground reference data existed in the form of a land cover map compiled by the Institute of Terrestrial Ecology (ITE), which was augmented by colour aerial photography, and was almost contemporary with satellite imagery and airborne multispectral scanner data. An up-to-date land cover map is currently being assembled from these various sources (coupled with several field visits, already undertaken).

Transformed pairwise divergence, a measure of class separability (Swain & Davis 1978), indicated that overall the classes were readily separated (Table 2). There was some confusion between peatlands, bracken and *Nardus* in all three data sets (raw, Lambertian and non-Lambertian). These problems may be due in

TABLE 2. Transformed pairwise divergence matrix for raw data. Values of 100 indicate complete separation

Land classes	1	2	3	4	5	6	7	8	9	10
1 Conifers	0									
2 Felled	100	0								
3 Deciduous	64	100	0							
4 Moor	95	100	97	0						
5 Peatlands	99	100	100	100	0					
6 Unimproved	99	99	99	97	97	0				
7 Improved	99	100	100	100	99	100	0			
8 *Nardus*	100	99	100	100	48	97	99	0		
9 Bracken	99	100	100	99	31	87	100	91	0	
10 Water	100	100	100	100	100	100	100	100	100	0

part to the complex heterogeneity of upland plant communities. Visual inspection of the results from supervised classification provided a number of interesting results (Fig. 4). All three data sets generated classifications which subjectively conformed well with the ground reference data. They all appear to overestimate the extent of the peatland class, however. The classification of the Lambertian-corrected data overestimated the extent of moorland. Overall, the products of classification of Lambertian-corrected data appeared less satisfactory than either the uncorrected imagery or the corrected imagery using the non-Lambertian model.

Quantitative evaluation of classification results indicated the overall accuracy to be 60 per cent, 52 per cent and 67 per cent for raw, Lambertian and non-Lambertian data sets, respectively. The non-Lambertian corrected imagery showed highest accuracy (Fig. 5). Absolute classification accuracy is generally poor. To some extent this was due to the overestimation of peatlands and moorlands attributable to inadequate training data (see above). Examination of classification accuracy for each land class demonstrates that the classification performs worst for the natural and semi-natural vegetation classes (Table 3), which tend to be intrinsically variable in nature. This suggests that classification of upland vegetation based solely upon spectral information will never be entirely satisfactory, and that an alternative approach is needed.

The results presented in Table 3 also indicate that, for certain cover types, the Lambertian reflectance model performs better than the non-Lambertian model. If further work confirms that this is true we must conclude that models for the radiometric correction of multispectral data for terrain effects are not independent of the nature of the ground cover. In these circumstances, radiometric correction of imagery would require an iterative approach, based upon a knowledge of the response of different individual canopies.

Unsupervised classifications were also carried out on the same data sets, with broadly similar results. The most satisfactory clusters were produced from non-Lambertian data. Most of the clusters produced using the Lambertian data were

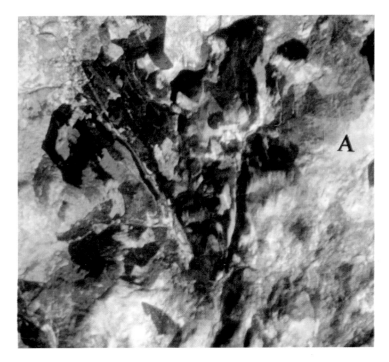

FIG. 2. A SPOT HRV false-colour multispectral image of the study area. The dark red regions are coniferous woodland, light red regions are improved pasture, light blue regions are mountain grasses with bracken and rock outcrops (A, mountain of Rhobell Fawr). The scene is 10 × 10 km and registered to the national grid. Image copyright CNES 1986.

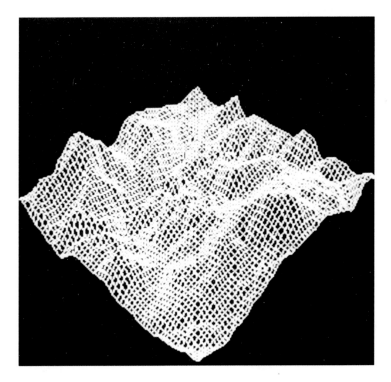

FIG. 3. A grid-perspective view of the digital terrain model. Viewpoint is slightly elevated away from the south-east.

FIG. 4. A supervised classification of Fig. 2. Clear felling (blue) and moorland (purple) areas are clearly visible. Area A (Rhobell Fawr) is a rounded mountain with a relatively uniform cover of *Nardus*-dominated grassland (yellow). Due to higher illumination of south-facing slopes, only the southern side of the mountain has been classified as *Nardus* grassland when using the uncorrected data. Image copyright CNES 1986.

FIG. 5. A supervised classification of Fig. 2 using the non-Lambertian correction (see comparison with Fig. 4). The suppression of topographic noise has resulted in a more accurate classification of *Nardus* grasslands on both the northern and southern sides of the mountain (A). Image copyright CNES 1986.

TABLE 3. Individual land class accuracy of the raw, Lambertian and
non-Lambertian data sets in comparison with the land cover map (%)

Land class	Raw	Data sets Lambertian	Non-Lambertian
Conifers	100	42	83
Felled	50	50	100
Deciduous	67	67	33
Moor	34	34	90
Peatlands	30	30	67
Unimproved	50	63	50
Improved	50	75	100
Nardus	75	100	80
Bracken	60	60	70
Water	100	100	100

meaningless. Topographic information (elevation, slope and aspect), derived from the DEM, was added to spectral data and used as additional inputs into unsupervised classification. The addition of the elevation data produced a significant improvement in the discrimination of land cover types, with the separation of high-altitude mountain grasses, improved pasture and lowland grasses being very pronounced. Use of slope and aspect data in the clustering did not result in any significant improvement.

DISCUSSION

The topographic corrections discussed above assume that the procedures used are independent of (i) wavelength of radiation and (ii) the nature of the vegetation canopy. The first of these assumptions is almost certainly invalid (Justice *et al.* 1981); our evidence suggests that behaviour of reflected radiation is also determined by (ii). Further refinement of the correction procedures is planned in two main areas. First, the computer program will be modified to permit use of different models of reflectance in each wavelength. Second the behaviour of different reflection models will be developed for each of these canopies. The radiometric correction would then be applied iteratively, the image being classified after each iteration to provide new ground cover information until stability is achieved.

Classifications based on spectral data alone ignore much existing ecological knowledge about the distribution of vegetation. A future step in this project is to develop procedures which can exploit our understanding of, for example, relationships between topographic variables and the spatial distribution of vegetation communities encountered within upland study areas. This knowledge will then be built into a series of 'decision-rule' classifiers which will utilize prior probabilities derived for various land cover units based upon their topographic environment. At each stage in the refinement of the classification process, the results will be systematically evaluated against ground reference data in order to identify significant factors and to quantify the improvements that have been achieved.

This work illustrates the value of digital terrain data in the processing and analysis of remotely sensed imagery in areas of high relief. The isolation of topographic effects coupled with the use of information derived from digital elevation models will improve our analysis. This and the application of improved classification algorithms will substantially increase the discrimination and accuracy of satellite data for land cover mapping and enhance the number of useful application for these data.

ACKNOWLEDGMENTS

This paper describes work carried out as part of SPOT PEPS (Projet Évaluation Préliminaire du SPOT) Project No. 226, awarded to the NERC Remote Sensing Application Centre at ITE Bangor, the NERC Unit for Thematic Information Systems (NUTIS) at Reading University and representatives of the British National Space Centre. We should like to express our gratitude to Prof. John Townshend and Dr J. Settle (NUTIS) and Dr S. Briggs (BNSC) for their comments during the project. We thank Dr G. Robinson (NUTIS), Nigel Brown and Dave Norris (ITE) for their assistance in the production of the DEM.

REFERENCES

Ball, D., Dale, J., Sheail, J. & Heal, O. (1982). *Vegetation Change in Upland Landscape.* ITE, Merlewood.

Curran, P. (1985). *Principles of Remote Sensing.* Longman, London.

Frank, T. & Thorn, C. (1985). Stratifying alpine tundra for geomorphic studies using digitized aerial imagery. *Alpine and Arctic Research,* 17, 179–188.

Hall-Köynes, K. (1987). The topographic effect on Landsat data in gently undulating terrain in southern Sweden. *International Journal of Remote Sensing,* 8, 157–168.

Holben, B. & Justice, C. (1981). An examination of spectral band ratioing to reduce the topographic effect on remotely sensed data. *International Journal of Remote Sensing,* 2, 115.

Hutchinson, C. (1982). Techniques for combining Landsat and ancillary data for digital classification improvement. *Photogrammetric Engineering and Remote Sensing,* 48, 123–130.

Jones, A. (1987). *SPOT PEPS Technical Meeting Report No. 1.* NERC Remote Sensing Application Centre, NERC/ITE, Bangor.

Justice, C., Wharton, C. & Holben, B. (1981). Application of digital terrain data to quantify and reduce the topographic effect on Landsat data. *International Journal of Remote Sensing,* 3, 213–230.

McMorrow, J. & Hume, E. (1986). Problems of applying multispectral classification to upland vegetation. *Proceedings of a Symposium held by Commission IV of the International Society for Photogrammetry and Remote Sensing (ISPRS) and the Remote Sensing Society (RSS). Edinburgh, 1986,* 610–620.

Nellis, M. (1986). Remote sensing for monitoring rangeland management strategies in the Kansas Flint Hills. *Proceedings of a Symposium held by Commission IV ISPRS AND RSS, Edinburgh, 1986,* 370–375.

Ratcliffe, D.A. & Thompson, D.B.A. (1988). The British uplands: their ecological character and international significance. *This volume.*

Stohr, C. & West, T. (1985). Terrain and look angle effects upon multispectral scanner response. *Photogrammetric Engineering and Remote Sensing,* **51**, 229.

Swain, P. & Davis, S. (1978). *Remote Sensing: the Quantitative Approach.* McGraw-Hill, New York.

Townshend, J. (1981). *Terrain Analysis and Remote Sensing.* Allen & Unwin, London.

Weaver, R. (1984). Integration of remote sensing data for moorland mapping. *Proceedings of the 10th International Conference of the Remote Sensing Society, Reading, 1984,* 194.

Williams, J. (1988). Improving the mapping of upland vegetation from TM data using pre- and post-classification image filtration. *Proceedings of Workshop on Remote Sensing of Grasslands, Bangor, 1987.*

Vegetation survey in upland Wales: reconciling ground and satellite surveys

J. H. WILLIAMS

Institute of Terrestrial Ecology, Penrhos Road, Bangor, Gwynedd LL57 2LQ

INTRODUCTION

Changes in the nature and distribution of upland vegetation resulting from pasture improvement, afforestation and bracken encroachment have increased the need for monitoring. The Nature Conservancy Council, for example, uses upland vegetation maps when considering the designation of new protected areas and during the preparation of management agreements. Furthermore, county planning authorities now have a mandatory responsibility for the maintenance of up-to-date maps of moor and heath areas within the National Parks as a result of the 1981 Wildlife and Countryside Act.

For most of these applications there is a requirement for the identification of plant communities rather than individual plant species. Ratcliffe & Birks's (1980) classification, used by the Nature Conservancy Council, is based on vegetation type, physiognomy and dominance, factors which have been shown to relate to spectral response (Justice & Townshend 1982). Vegetation mapping methods using aerial photographs and field survey (Fuller 1981) have tradition-ally provided the basis for monitoring the uplands (Budd 1987). Aerial photograph interpretation has the advantages of being relatively simple, rapid, reasonably accurate, and cheap (Hume, McMorrow & Southey 1986). However, these surveys are only undertaken at 10- to 15-year intervals and they are usually based on sampling methods. The opportunity therefore arises for satellite sensor data to 'fill in the gaps' in time and in aerial coverage because they can provide a repeatable and complete census. Satellite data have already been used for applications in the UK uplands (Weaver 1984, 1987; Jewell & Brown 1987; Jones *et al.* 1987; Ward, Weaver & Brown 1987; Wyatt 1987; Williams, Brown & Norris, in press). The results of many of these studies have been encouraging, but most workers have found that the upland environment is extremely complex, demanding a correspondingly sophisticated methodology.

Traditional field survey methods and satellite survey techniques use quite different criteria for classifying upland terrain. While there are difficulties in maintaining consistent performance between two field surveyors, there are even greater difficulties in trying to obtain comparable classifications using two quite different methodologies (McMorrow & Hume 1986).

Satellite sensor data are influenced by a wide range of physical and environmental variables (Curran 1985) not necessarily considered by traditional field survey methods. The result of this is that the products of satellite surveys of

119

upland terrain usually show a greater degree of variability than do vegetation maps derived from field survey. A method for simplifying the satellite survey products is needed to make the two types of map more comparable. This often has the additional benefit of making the satellite survey products easier to interpret. The problem that remains is to find an optimum technique for simplifying the satellite survey products while minimizing any adverse effects on the accuracy of mapping each of the vegetation types. Image filtering techniques have been used for simplifying satellite survey products, but there are several different filter types and each one can be applied with a range of 'strengths' or kernel sizes.

The need to identify plant communities, rather than individual plant species, affects the choice of spatial resolution and the methods used in image classification. Landsat Thematic Mapper data with a spatial resolution of 30 × 30 m were thought to be suitable for mapping the extensive vegetation communities in the Welsh uplands. If more detailed mapping is required, then aircraft scanner data (Williams 1984) might prove more suitable. The filtration methods used in this study further reduce the influence of the individual components to obtain a 'community spectral response' rather than a 'species spectral response'.

It is difficult to know where to find the most realistic model of the vegetation cover for an area. Upland vegetation is fairly stable but its appearance and composition can alter. Grazing, the use of fire for pasture management, drought and erosion can alter the uplands from year to year. Field surveys usually record the vegetation at a limited number of sample sites, and then only at given points in time. A point of reference, however, has to be chosen and a field survey conducted close to the date of acquiring satellite data can provide a reasonable basis for a comparison study.

MATERIALS AND METHODS

Study site

The vegetation of the Glyderau mountains in Snowdonia was mapped by the Nature Conservancy Council (1986) during the summer of 1984 using black-and-white aerial photographs and field survey. A cloud-free scene of the study area was acquired by the Landsat-5 Thematic Mapper (TM) on 22 July 1984. The study area lies south of the A5 road between Llyn Ogwen and Capel Curig and covers 1384 ha of rough grazing land. The land rises from 250 m to nearly 1000 m on Glyder Fach. Eleven classes from Ratcliffe & Birks's (1980) upland vegetation classification were found: *Agrostis–Festuca*, *Nardus–Juncus* and *Molina* grasslands; *Calluna*, *Vaccinium* and *Rhacomitrium* heaths; bracken (*Pteridium acquilinum*); soligenous flushes; blanket bog; crags; and scree.

Treatment of the data

The NCC vegetation map was vector-digitized, converted to raster format and transferred to an International Imaging Systems (I²S) image processing system. Using control points and a nearest-neighbour resampling procedure, the TM data were registered to the map data. Image filters can be applied both before and after classification of satellite sensor data. Mean and median filters are appropriate for pre-classification filtering, with median filters usually being preferred because of their boundary-preserving properties (Cushnie & Atkinson 1985).

After preparing four images for classification (Table 1), training data were extracted for each vegetation class using the NCC vegetation map for reference. This placed an implicit trust in the ground survey. A fast maximum likelihood classifier (Briggs & Settle 1985) was used to classify each of the four TM images. The four resulting classified images were then processed with a post-classification image filter. Majority, or logic, filters are appropriate for this process because the class values are arbitrary rather than quantized (Townsend 1986). With four pre-classification filters (none and 3×3, 5×5, 7×7 kernel) and four post-classification filters (none and 3×3, 5×5, 7×7 kernel), there were 16 classified TM images. Correspondence was then measured between the digitized NCC vegetation map and the 16 classified TM images.

RESULTS

Small areas of homogeneous vegetation, such as bracken, with relatively distinct spectral characteristics, did not appear to benefit from image filtration. Post-classification filtration only increased correspondence by a small amount. Large areas of homogeneous vegetation, including *Nardus–Juncus* grassland, *Agrostis–Festuca* grassland, *Calluna* heath and *Molinia* grassland, were found to have greater correspondence after image filtration. Some of the increases resulting from pre-classification filtration were large (*Agrostis–Festuca* grassland), with post-classification filtration accounting for smaller increases.

Large areas of heterogeneous vegetation, including *Vaccinium* heath and *Rhacomitrium* heath, did not show any major increases in correspondence after pre-classification image filtration, but there were some small increases after post-classification filtration. Small areas of heterogeneous vegetation, such as blanket bog and, in particular, soligenous flushes, showed increases in correspondence after both pre- and post-classification image filtration.

The best overall correspondence with ground survey was obtained after use of the 'strongest' version (7×7 kernel) of each filter type (Table 1). The intermediate treatments, however, produced a better correspondence for some classes.

DISCUSSION

Having made these comparisons between the ground and satellite survey products it was possible to make some general comments about optimizing the techniques for reconciling the two survey methods. Pre-classification filtration only 'diluted' the appearance of bracken with its distinct spectral characteristics, leading to confusion with neighbouring, or spectrally similar, classes. There was only a small increase in correspondence after post-classification filtering because there were relatively few gaps in the classification of these small but distinct

TABLE 1. The correspondence matrix leading diagonals obtained by comparing the ground survey map with each of the sixteen Thematic Mapper vegetation classifications. The class codes are 2, *Nardus–Juncus*; 3, *Agrostis–Festuca*; 4, Crags; 5, *Vaccinium*; 6, *Rhacomitrium*; 7, soligenous flushes; 8, blanket bog; 9, *Calluna*; 10, *Molinia*; 11, bracken; and 12, scree. Brackets indicate maximum correspondence for each class. There were improvements in correspondence of between 10 and 156 per cent as a result of filtering rather than not filtering

	Filter used															
PRE	0	0	0	0	3×3	3×3	3×3	3×3	5×5	5×5	5×5	5×5	7×7	7×7	7×7	7×7
POST	0	3×3	5×5	7×7	0	3×3	5×5	7×7	0	3×3	5×5	7×7	0	3×3	5×5	7×7
Class																
2	32·1	32·4	33·3	33·6	28·9	29·3	29·9	30·3	30·5	31·2	32·5	33·0	33·6	34·5	35·5	(36·1)
3	24·9	25·3	24·6	24·6	39·3	41·1	42·8	43·7	38·4	39·3	39·9	40·7	57·8	61·7	63·2	(63·8)
4	45·7	49·4	54·1	53·9	43·4	46·1	48·2	49·5	51·6	52·8	54·4	53·7	52·9	55·4	55·8	(57·1)
5	58·4	62·6	67·2	(69·7)	55·1	58·7	61·8	63·9	57·8	60·5	62·7	65·0	62·9	65·4	67·6	69·1
6	75·6	78·9	81·7	83·8	71·1	72·8	75·4	77·9	71·9	74·5	73·4	74·2	78·3	80·8	83·1	(86·0)
7	38·5	39·6	40·1	41·0	35·3	37·1	37·8	37·1	42·2	43·6	45·3	46·7	47·3	49·2	50·5	(52·3)
8	68·1	72·9	77·6	(79·6)	64·5	67·1	69·2	71·1	68·7	70·2	71·9	73·2	71·2	72·7	73·9	74·9
9	33·7	34·4	35·2	36·0	27·1	27·7	27·7	28·1	37·6	39·0	40·4	(41·3)	37·6	38·8	39·5	40·3
10	63·9	66·6	68·8	70·3	47·8	50·8	52·8	54·4	55·3	57·1	59·6	61·5	68·3	68·8	69·9	(70·5)
11	42·7	44·9	(47·2)	47·0	41·8	44·5	45·8	45·7	41·9	44·0	45·9	46·0	39·2	40·1	41·6	42·2
12	41·8	45·1	51·3	51·1	38·0	43·4	49·1	52·8	40·0	43·4	48·1	52·7	45·3	47·6	50·5	(53·0)

areas. It should be noted that the bracken canopies can sometimes be spectrally confused with the better grasslands, but during the drought summer of 1984 many of these grasslands were very dry and brown and not in their usual lush green state; bracken, however, retained its distinct green and dense canopy.

Amongst some of the large areas of homogeneous vegetation were many small rocks, boulders, patches of exposed soil and other non-vegetated debris. Pre-classification filtering helped to reduce the effect that these small objects had on the spectral characterization of the classes; this, in turn, produced a classification which was closer to the field survey, increasing correspondence. Field surveys tend to concentrate on the botanical nature of the terrain; median filtration will help to reduce the influence of minor non-vegetated components in the terrain, 'concentrating' the attention of the machine classifier on the vegetation.

Post-classification image-processing with a majority-logic filter increased correspondence between the two survey products in all cases. This type of filter produced a 'smoothing' or generalization of the image with the 7×7 kernel producing a more marked effect than the 3×3 or 5×5 filters. Some heterogeneous vegetation classes had components that were spectrally similar to other classes, or components in other classes. The isolated pixels representing these components were, in some cases, classified into classes other than the one representing the majority of their neighbours. The effect of the majority-logic filter was to re-classify these isolated differences if a majority of one class existed within the kernel. An increase in correspondence always resulted because the classifications were being compared with a simplified field survey map.

Image-filtering provides a technique for simplifying raster-format vegetation maps. In general, a fairly 'strong' or large kernel-size filter increased the correspondence between the vegetation map derived from ground survey and those produced by satellite survey. However, correspondence was reduced for at least one important vegetation type (bracken), and a pilot study has to be strongly recommended if this technique is to be used on sites where the characteristics of the ground cover are different from those found in this study.

ACKNOWLEDGMENTS

The author is grateful for support from the Natural Environment Research Council under research studentship GT4/84/TLS/61. The Nature Conservancy Council are thanked for making the Glyderau vegetation map and ancillary data available for this study. The author is grateful to the Digital Cartography Service at the Institute of Terrestrial Ecology, Bangor, for digitizing the vegetation map.

REFERENCES

Briggs, S.A. & Settle, J.J. (1985). A fast Maximum Likelihood classifier. *Remote Sensing: Data Acquisition, Management and Application* (Ed. by T. Allan, A. Else & N. Hutchings), pp. 249–255. Remote Sensing Society, University of Reading.

Budd, J.T.C. (1987). Remote sensing applied to the work of the NCC in upland areas. *The Ecology and Management of Upland Habitats: the Role of Remote Sensing* (Ed. by R. Weaver & S. Ward), pp. B1–B7. Remote Sensing Special Publication No. 2, Department of Geography, University of Aberdeen.

Curran, P.J. (1985). *Principles of Remote Sensing.* Longman, London.

Cushnie, J.L. & Atkinson, P. (1985). Effect of spatial filtering on scene noise and boundary detail in Thematic Mapper imagery. *Photogrammetric Engineering and Remote Sensing,* 51, 1483–1493.

Fuller, R.M. (1981). Aerial photographs as records of changing vegetation patterns. *Ecological Mapping from Ground, Air and Space* (Ed. by R.M. Fuller), pp. 57–68. Institute of Terrestrial Ecology, Huntingdon.

Hume, E., McMorrow, J. & Southey, J. (1986). Mapping semi-natural grassland communities from panchromatic aerial photographs and digital images at SPOT wavelengths. *Mapping from Modern Imagery* (Ed. by J. Farrow & B. Wright), pp. 386–395. Remote Sensing Society, University of Nottingham.

Jewell, N. & Brown, R.W. (1987). The use of Landsat TM data for vegetation mapping in the North York Moors National Park. *The Ecology and Management of Upland Habitats: the Role of Remote Sensing* (Ed. by R. Weaver & S. Ward), pp. E1–E8. Remote Sensing Special Publication No. 2, Department of Geography, University of Aberdeen.

Jones, A., Wyatt, B., Settle, J. & Robinson, G. (1987). The use of a DTM for topographic correction and classification of SPOT HRV data for ecological mapping in upland environments. *Advances in Digital Image Processing* (Ed. by P.M. Mather), pp. 488–497. Remote Sensing Society, University of Nottingham.

Justice, C. & Townshend, J.R.G. (1982). A comparison of unsupervised classification procedures on Landsat MSS data for an area of complex surface conditions in Basilicata, Southern Italy. *Remote Sensing of Environment,* 12, 407–420.

McMorrow, J. & Hume, E. (1986). Problems of applying multispectral classification to upland vegetation. *Mapping from Modern Imagery* (Ed. by J. Farrow & B. Wright), pp. 610–620. Remote Sensing Society, University of Nottingham.

Nature Conservancy Council (1986). *Upland Vegetation Survey — Eryri.* Nature Conservancy Council, Bangor.

Ratcliffe, D.A. & Birks, H.J.B. (1980). *A Classification of Upland Vegetation Types in Britain.* Nature Conservancy Council, Bangor.

Townsend, F.E. (1986). The enhancement of computer classifications by logical smoothing. *Photogrammetric Engineering and Remote Sensing,* 52, 213–221.

Ward, S.A., Weaver, R.E. & Brown, R.W. (1987). Monitoring heather burning in the North York Moors National Park using multi-temporal Thematic Mapper data. *The Ecology and Management of Upland Habitats: the Role of Remote Sensing* (Ed. by R. Weaver & S. Ward), pp. F1–F4. Remote Sensing Special Publication No. 2, Department of Geography, University of Aberdeen.

Weaver, R.E. (1984). Integration of remote sensing data for moorland mapping. *Satellite Remote Sensing—Review and Preview* (Ed. by J. Hardy & J. Brookling), pp. 191–220. Remote Sensing Society, University of Reading.

Weaver, R.E. (1987). Using multi-spectral scanner data to study vegetation succession in upland Scotland. *The Ecology and Management of Upland Habitats: the Role of Remote Sensing* (Ed. by R. Weaver & S. Ward), pp. H1–H11. Remote Sensing Special Publication No. 2, Department of Geography, University of Aberdeen.

Williams, D.F. (1984). Overview of the NERC Airborne Thematic Mapper Campaign of September 1982. *International Journal of Remote Sensing,* 5, 631–634.

Williams, J.H., Brown, N.J. & Norris, D. (in press). An experimental GIS for upland management information in Snowdonia. *Applications of Remote Sensing* (Ed. by R. Stone, A. Southgate & C. McVean). Department of Geography, University of Durham.

Wyatt, B.K. (1987). Remote sensing applications in upland ecology : a review of research at ITE. *The Ecology and Management of Upland Habitats: the Role of Remote Sensing* (Ed. by R. Weaver & S. Ward), pp. C1–C9. Remote Sensing Special Publication No. 2, Department of Geography, University of Aberdeen.

Effects of latitude, oceanicity and geology on alpine and sub-alpine soils in Scotland

G. HUDSON

The Macaulay Land Use Research Institute, Craigiebuckler, Aberdeen AB9 2QJ

SUMMARY

1 Alpine and sub-alpine soils in Scotland occupy the summits and upper slopes of mountains and high hills. The soils and parent materials are described, together with the climatic conditions which influence their development.

2 The altitude of the sub-alpine or alpine soil map unit boundary was recorded for 33 hills and mountains along a south-to-north transect and for 19 hills and mountains along a west-to-east transect. The average altitudes were compared to distance north or east of an arbitrary origin and third-order polynomial functions were fitted to the data. There is a wide scatter of individual data points on either side of the best-fitting polynomials.

3 Aspect and geology do not appear to influence the elevation of sub-alpine or alpine soils, but on steep slopes these soils occur at lower altitudes. The boundary rises from west to east as oceanicity reduces.

4 The influence of snow conditions is important in sub-alpine and alpine soil development and local variations in amount and persistence of snow cover may account for 'noise' in the data.

INTRODUCTION

Alpine and sub-alpine soils occupy the summits and upper slopes of mountains and high hills throughout Scotland where the climate is cold (Fig. 1). The alpine soils are confined to ridge crests and summits in the west and north of the country, and occur on the extensive high plateaux of the Cairngorm and Monadhliath massifs in the Grampian Mountains. Sub-alpine soils occupy plateaux, summits or upper slopes of hills and mountains, at a lower elevation than the alpine soils.

Sub-alpine and alpine soils and their associated vegetation are fragile, and potential changes in land use need careful evaluation. The regenerative capacity of plants in cold conditions is poor and areas damaged by heavy grazing or recreational pressures will be slow to recover. The bare patches of soil resulting from such damage are susceptible to erosion by water during periods of snowmelt or heavy rain, or to erosion by wind in dry spring or summer weather. The study of the distribution of these soils is an important first step in monitoring the effects of such changes.

The relationship of the distribution of sub-alpine and alpine soils to latitude

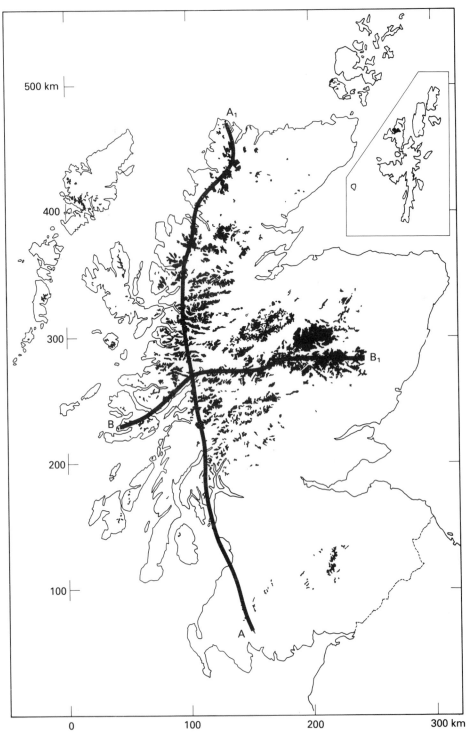

FIG. 1. Distribution of alpine and sub-alpine soils. The map polygons were sampled along transects A–A₁ and B–B₁. The localities at the ends of the transects are A, Cairnsmore of Fleet; A₁, Sgribhis Bheinn; B, Bearraich; and B₁, Mount Keen.

and longitude is known in broad terms. These soils generally occur at low elevations on mountains and hills in the west and north of Scotland and at high elevations in the south and east. Thus, for a hill of comparable size and shape, the area occupied by sub-alpine or alpine soils is greater in the north or west of the country. This paper presents empirically derived mathematical models of the variation in the lower boundary of sub-alpine soils along two transects and considers some environmental factors which influence this altitude.

In the global context of cold climate regions, two major climatic types, the maritime type and the continental type, can be recognized (Reiger 1983). The cold upland areas in Scotland have a high precipitation with fairly uniform distribution between the summer and winter months, although in the east precipitation is slightly higher in the summer, and the climate is thus closely allied to the maritime type. The maritime influence is greater in the west of the country and around the north coast.

The level of oceanicity in Scotland has been classified into three subsectors; hyperoceanic, euoceanic and hemioceanic, based on thermal criteria (accumulated day-degrees of frost at lower altitudes), and mapped at the 1:625,000 scale (Birse 1971). Alpine and sub-alpine soils occur in all of these subsectors, but are generally confined to the upper and lower oroarctic and orohemiarctic thermal subzones of Birse (1971), where accumulated temperatures (day-degrees above 5·6 °C) are less than 675 day-degrees C year^{-1}.

Cryic deposits , the parent materials of alpine soils, are formed by seasonally active freeze–thaw processes which fracture rocks and stones into smaller pieces and loosen the drifts, which may move under the influence of gravity to form terracettes or stone stripes, or may form stone polygons *in situ*. Occurring at lower elevations, sub-alpine soils are developed on colluvial, morainic or shallow locally derived drifts, subject to minimal freeze–thaw activity under present climatic conditions.

Alpine and sub-alpine soils have low rates of chemical weathering and organic matter production and, in the oroarctic thermal subzones, freeze–thaw cycles mix the surface horizons. Cryoturbation of the soil profile in alpine soils is most active in the upper layers, mixing mineral and organic matter annually, to form a dark-coloured humus-enriched horizon (the 'Alpine A' of Romans, Stevens & Robertson 1966). Organic matter breakdown and transport in the mineral horizons is also slow, and accumulation of organic material, in the form of colloidal humus deposits around stones, humus staining and thick Bh horizons, is commonly observed. A morphological description of an alpine podzol from the Cairngorm Mountains is given in Table 1. The organic matter content in this profile is 14·1 per cent in the Ah horizon, 4·4 per cent in the A horizon and 12·3 per cent in the Bh horizon, falling to 0·8 per cent in the B horizon.

Sub-alpine soils occur at lower elevations than alpine soils, given a similar environment, and cryoturbation of their upper horizons is weak due to a low frequency of freeze–thaw cycles or intensity of freezing conditions. The horizons are only slightly mixed, with mineral grains incorporated into the organic horizons;

TABLE 1. Morphological description of an alpine podzol from the Allt a'Mharcaidh catchment, Badenoch and Strathspey District

Profile name: Mharcaidh
National Grid: NH 884500
Slope: 19°, straight, simple
Altitude: 762 m   Aspect/bearing: NE, 51°
Rock outcrops: none   Boulders: slightly bouldery
Vegetation: Loiseleurio-*Alectorietum nigricantis* (alpine azalea–lichen heath) (Robertson 1984)
Flushing: no flushing
Soil drainage: free   Site drainage: normal
Association: Countesswells   Series: Rinnes
Parent material: cryogenic   Rock type: granite
Major soil subgroup: alpine podzol
Climate: hemioceanic extremely humid lower oroarctic

Profile description:

Ah	0–8 cm	black 10YR2/1 humose loamy sand; no mottles; weak fine subangular blocky structure; moist; very friable; many very fine fibrous and common fine fibrous roots; common very small angular granite and few medium subangular granite stones; clear wavy boundary
A	8–19 cm	dark brown 10YR4/3 loamy sand; no mottles; weak medium subangular blocky structure; moist; very friable; many very fine fibrous and few fine woody roots; common very small angular granite and few medium subangular granite stones; clear wavy boundary
Bh	19–36 cm	black 5YR2/1 humose loamy sand; no mottles; weak fine subangular blocky structure; moist; very friable; common very fine fibrous and few fine fibrous roots; common medium subangular granite and many very small subangular granite stones; clear wavy boundary
B	36–67 cm	dark yellowish brown 10YR4/4 loamy fine sand; no mottles; weak medium subangular blocky structure; moist; very friable; few very fine fibrous roots; many small subangular granite and common medium subangular granite stones; gradual wavy boundary
C	67–97 cm	yellowish brown 10YR5/4 coarse loamy sand; no mottles; weak fine subangular blocky structure; moist; very friable; no roots; many small subangular granite and common medium subangular granite stones

these soils generally have thick Bh horizons. A morphological description of a sub-alpine podzol from Aonach Mor is given in Table 2. The organic matter content is 7·2 per cent in the Eh horizon, 6·5 per cent in the Bh horizon and 2·6 per cent in the B horizon, generally lower in the Bh than in the alpine podzols at Aonach Mor.

A soil map can be regarded as a structural model of the spatial differences among soils at the landscape level, with the introduction of mapping units as simplifying concepts (Dijkerman 1974). The 1:250,000 scale soil maps of Scotland (Soil Survey of Scotland 1982) are models of soil landscape systems. The total number of soil mapping units for Scotland at this scale is 576 (excluding rock), of which 42 comprise alpine soils or sub-alpine soils as the main soil types, sometimes co-dominant or occurring with peat, rankers or lithosols. These soils cover 5159 km^2 or 6·7 per cent of the land area of Scotland, and have been charac-

TABLE 2. Morphological description of a sub-alpine podzol from Aonach Mor, Lochaber District

Profile name: Aonach Mor Z
National Grid: NN 186756
Slope: 9°, straight, simple
Altitude: 635 m   Aspect/bearing: N, 2°
Rock outcrops: none   Boulders: slightly bouldery
Vegetation: Junco-squarrosi-*Festucetum tenuifoliae* (flying bent grassland)
Flushing: yes
Soil drainage: poorly drained above iron pan   Site drainage: normal
Association: Countesswells   Series: Saighdeir
Parent material: morainic   Rock type: granodiorite
Major soil subgroup: sub-alpine podzol
Climate: euoceanic perhumid orohemiarctic

Profile description:

0	0–19 cm	black 2·5Y2/0 no mineral content; amorphous; wet; very weak coarse angular blocky structure; many very fine fibrous roots; clear smooth boundary
Eh	19–30 cm	dark grey 10YR4/1 coarse loamy sand; no mottles; massive structure; moist; friable; many very fine fibrous roots; many large subrounded granodiorite stones; clear smooth boundary
Bh	30–42 cm	very dark grey 5YR3/1 humose loamy sand; no mottles; massive structure; wet; slightly plastic and not sticky; common very fine fibrous roots; many large subrounded granodiorite stones; clear smooth boundary
B	42–53 cm	olive grey 5Y4/2 loamy sand; no mottles; massive structure; very moist; friable; few very fine fibrous roots; many large subrounded granodiorite stones.
Bf	53	weak thin iron pan
BC	53–75 cm	light brownish grey 2·5Y6/3 loamy sand; no mottles; weak fine platy structure; moist; friable; no roots; common medium subrounded granodiorite stones

terized into three soil landscape systems having different soil types, landforms and plant communities.

METHODS

The soil boundaries on the 1:250,000 scale soil maps have been vector-digitized, allowing selective manipulation of the computer file to extract and print line segments which delineate defined soil, landform or vegetation features. Map units consisting predominantly of sub-alpine or alpine soils were selected from the 1:250,000 map legend (Soil Survey of Scotland 1984) and the corresponding map polygons occurring throughout Scotland were printed from the digitized data (Fig. 1). Two sampling transects were drawn on this map, south to north (transect A–A₁) from Cairnsmore of Fleet (NX 501671) to Sgribhis Bheinn (NC 319714) and west to east (transect B–B₁) from Bearraich (NM 417274) to Mount Keen (NO 409869), crossing as many mountain soil landscape areas as possible in order to sample a reasonable density of map polygons. The largest map polygon, from the variable number of polygons of differing sizes in each 10 km × 10 km

TABLE 3. Variables recorded at sampling points along polygon boundaries (see text for discussion of selection of sampling points)

Grid reference
Altitude, in m(A)
Slope, in degrees (to the nearest 5°)
Bearing, in degrees (to the nearest 10°)
Distance north (N) of a west-to-east baseline through the origin,
    NX 000000, in km proportional to latitude
Distance east (E) of a north-to-south baseline through the origin,
    NX 000000, in km proportional to longitude
Rainfall, in mm (Meteorological Office 1977)
Bioclimatic subregion above the boundary (Birse 1971)
1:250,000 map unit number (Soil Survey of Scotland 1982)
Geology, interpreted from map unit number

grid square crossed by the transects, was chosen and several variables were recorded at sample sites along its boundary. The map polygons were classed as 'small' if they were less than about 5 km² in extent. The map unit boundary was sampled at eight points around small polygons (situated along the principal compass points from their centre) and at roughly 2-kilometre intervals around the larger polygons. At each sampling point, ten variables, listed in Table 3, were recorded from 1:50,000 or 1:625,000 scale maps.

RESULTS

On transect A–A$_1$, data were gathered from 514 sample sites on 33 map polygons, of which 10 were classed as 'small'. Transect B–B$_1$ had 302 sample sites on 19 map polygons, of which 5 were classed as 'small'. The average elevation of the sub-alpine or alpine soil boundary was calculated for each map polygon and used to assess between-site variance of altitude in relation to latitude, longitude, oceanicity and geology. Additionally, within-site variance of altitude with slope, aspect and geology was examined.

Variation of elevation with latitude

The average elevation of the sub-alpine or alpine soil boundary on transect A–A$_1$ was plotted against the distance north of the baseline through the origin. The elevation decreases with increasing northerly latitude and a third-order polynomial function was fitted to the data by multiple linear regression (Fig. 2). The fitted curve is

$$A = 874.79 - 3.31N + 0.012N^2 - 0.000013N^3$$

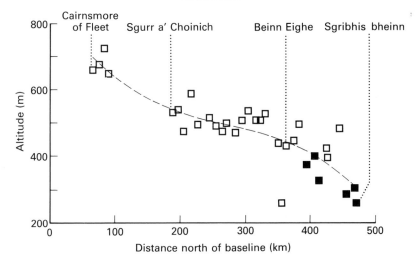

FIG. 2. Plot of elevation against distance north for transect A–A$_1$. The squares indicate site averages for each value and the curve is the best-fitting third-order polynomial. The symbol □ indicates polygons in the euoceanic subsector and the symbol ■ indicates polygons in the hyperoceanic subsector.

where the symbols are listed in Table 3 and the standard errors of the coefficients for N, N^2 and N^3 are 1·42, 0·0059 and 0·0000072 respectively. The results of the regression analysis are typical of a curve with a dome-shaped progression of slope. The conceptual complexity of the third-order polynomial means that it can only be used as an empirical approximating function to the reduction of elevation with distance north. The multiple correlation coefficient (R) of 0·88 (with 29 d.f.) indicates a good fit of the curve to the data. The slope of the curve is at a maximum along two sections of the transect: those between Cairnsmore of Fleet (NX 501671) and Sgurr a' Choinnich (NX 160957) and between Beinn Eighe (NG 951611) and Sgribhis Bheinn (NC 319714), indicating that the rate of change of the boundary elevation with distance north is at a maximum.

Variation of elevation with longitude

The average elevation of the sub-alpine or alpine boundary of 19 polygons on transect B–B$_1$ was plotted against distance east of the baseline through the origin. The elevation drifts higher with increasing easterly longitude and a third-order polynomial curve was fitted to the data by multiple linear regression (Fig. 3). The fitted curve is

$$A = 478·49 - 2·79E + 0·048E^2 - 0·00014E^3$$

where the symbols are listed in Table 3 and standard errors of coefficients for E, E^2 and E^3 are 4·33, 0·033 and 0·000078 respectively. The results of the regression

Alpine and sub-alpine soils

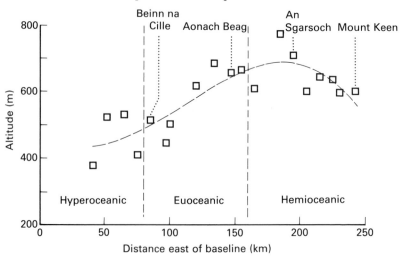

FIG. 3. Plot of elevation against distance east for transect B–B₁. The squares indicate site averages for each value and the curve is the best-fitting third-order polynomial. The oceanicity subsectors of Birse (1971) are also shown.

analysis are typical of a curve with a dome-shaped progression of slope with distance east. The multiple correlation coefficient (R) is 0·85 (with 15 d.f.), indicating a good fit. The slope of the curve rises steeply from Beinn na Cille (NM 854542), Kingairloch to Aonach Beag (NN 459741), Ben Alder, where the elevation of the boundary is increasing, and falls steeply from An Sgarsoch (NN 933837) to Mount Keen (NO 409869), where the boundary elevation is decreasing.

Variation of elevation in relation to oceanicity

The degree of oceanicity lessens from west to east along transect B–B₁ and the average elevation of the sub-alpine or alpine soil boundary shows the greatest rate of change, increasing across the euoceanic subsector from west to east (Fig. 3). There is also an apparent marked reduction as the eastern end of the transect approaches Mount Keen in the hemioceanic subsector. Transect A–A₁ mainly occurs in the euoceanic subsector except for two groups of three polygons in the hyperoceanic subsector, which appear to have lower elevations than expected (Fig. 2).

Within-site variation of elevation in relation to slope, aspect and geology

For transect B–B₁, using an analysis of covariance approach, the average influence of aspect and slope was investigated. The analysis indicates there is no effect due to aspect but there is an effect of slope equal to −3·6 m (±1·03) per degree of slope. The elevation appears to be some 70 lower where the sub-alpine or alpine soils are developed on granite compared to schists or quartzites.

DISCUSSION

The elevation of the lower limit of sub-alpine or alpine soils varies throughout Scotland and has been expressed using conceptual or verbal models. The boundary elevation is influenced by a range of environmental factors including temperature, exposure, precipitation (the frequency of which is influenced by Scotland's position in a maritime climate zone, tending to continental in the Cairngorms), geology and oceanicity. The accumulated temperature for any given altitude also changes from place to place throughout Scotland, generally increasing in southerly latitudes and, possibly, on south-facing slopes, although the elevation of the boundary does not appear to be influenced by aspect.

The identification of causative elements in soil boundary conditions has considerable value in land-use assessment. Elements which are important in establishing ecological land types, for example the sub-alpine and alpine soil zones on Scottish hills, also affect plant growth. Thus, the lower edge of the sub-alpine soil zone to all intents and purposes defines the limit of tree growth. A further investigation of climatic gradients below the boundary may be of value in establishing commercial planting limits.

Empirically derived mathematical functions show that the elevation of the sub-alpine soil boundary is related to distance north or east of two baselines. The data are 'noisy' with a wide scatter to either side of the line of best fit because other variables that also influence sub-alpine soil formation change along the transects, and further work is necessary to enlarge the data set and determine the contribution of these variables.

At the elevation of the boundary, amounts of annual rainfall, including snow, change along each transect at different rates. Bearraich on the west end of transect B–B$_1$ has an annual total of around 1900 mm and the totals rise rapidly along the transect to Ben Nevis, where the annual total is about 2900 mm. There is a rapid fall over the next 90 km east, until An Sgarsoch (NN 933837), from where the annual total of 1500 mm is sustained to the end of the transect. The proportion of the annual total falling as snow is likely to be significant as the depth of snow, the length of snow lie and degree of drifting affect soil temperatures. Deep snow falling early in the winter, and remaining on the ground, effectively insulates the soil from a further lowering of temperature, and, if snow falls late in the winter on bare, frozen soil, it will reduce the rate of thawing in warmer weather.

The elevation of the boundary appears to be influenced by slope, which was recorded above the boundary. The slope varies between sites and could be an important factor contributing to the data scatter about the line of best fit. The reduction in elevation of the boundary with steepening slope may be related to the occurrence of peat at high altitude on gentle slopes, particularly on the middle and eastern section of transect B–B$_1$. Where this high-level peat occurs in the orohemiarctic thermal subzone, it occupies land which would otherwise carry sub-alpine soils; as slopes increase, the wet hydrologic conditions conducive to peat formation become drier, and sub-alpine soils occupy their full range of physiographic position. On steep land, particularly where extreme rockiness prevails, the soil boundary,interpreted on air photographs, is sometimes located at a lower altitude to include crags and scree; such is the case at Liathath (on transect A–A$_1$), where the polygon has a much lower average elevation than adjacent polygons.

Oceanicity appears to influence the boundary elevation, which is low in the hyperoceanic subsector, rises rapidly from west to east across the euoceanic subsector and is high in the hemioceanic subsector. Exposure to wind is also high in the western hyperoceanic subsector of Scotland and this factor appears to be significant in reducing the altitude at which the orohemiarctic thermal subzone of Birse (1971) occurs. The elevation at which high wind speeds occur probably rises across the euoceanic subsector due to the underlying landmass deflecting the moving air upwards when the wind is in the westerly quarter. The drop in elevation of the boundary at the eastern end of the transect B–B$_1$, near Mount Keen, may result from exposure to east winds, combined with low accumulated day-degrees of frost.

The polynomial expressions derived for two transects provide a mathematical model of the sub-alpine or alpine soil boundary which occurs throughout Scotland. By extending the data set, derivation of a model to cover the whole country would improve the understanding of the relative contribution of factors which influence the genesis and distribution of sub-alpine and alpine soils.

ACKNOWLEDGMENTS

I would like to thank several people who have contributed to the research described here: Professor J. S. Bibby for guiding the direction of the work, A. Lilly, D. W. Futty, C. J. Bown and M. J. Adams for discussion and L. Robertson for data handling. The analysis of covariance was performed by Dr M. Franklin, and A. D. Moir prepared the digitized map and produced the illustrations.

The 1:250,000 soil maps, on which this work is based, were compiled by staff of the Department of Soil Survey, Macaulay Land Use Research Institute, Aberdeen.

REFERENCES

Birse, E.L. (1971). *Assessment of Climatic Conditions in Scotland. 3. The Bioclimatic Sub-regions.* Macaulay Institute for Soil Research, Aberdeen.

Dijkerman, J.S. (1974). Pedology as a science: the role of data, models and theories in the study of natural soil systems. *Geoderma,* **11**, 73–93.

Meteorological Office (1977). *Average Annual Rainfall: Northern Britain.* Meteorological Office, London.

Reiger, S. (1983). *The Genesis and Classification of Cold Soils.* Academic Press, London.

Robertson, J.S. (1984). *A Key to the Common Plant Communities of Scotland.* Macaulay Institute for Soil Research, Aberdeen.

Romans, J.C.C., Stevens, J.H. & Robertson, L. (1966). Alpine soils of North-East Scotland. *Journal of Soil Science,* **17**, 184–199.

Soil Survey of Scotland (1982). *1:250 000 scale Soil Maps, sheets 1–7.* Ordnance Survey, Southampton.

Soil Survey of Scotland (1984). *Organization and Methods: Soil and Land Capability for Agriculture 1:250 000 Survey.* Macaulay Institute for Soil Research, Aberdeen.

Climate and microclimate of the uplands

J. GRACE AND M. H. UNSWORTH*

Department of Forestry and Natural Resources, University of Edinburgh, Edinburgh EH9 3JU and
*Department of Physiology and Environmental Science, University of Nottingham,
School of Agriculture, Sutton Bonington, Loughborough LE12 5RD

SUMMARY

1 Climate in the British uplands is characterized by substantial decreases with altitude in air temperature and radiant energy, and increases in wind speed and rainfall.

2 Temperatures of plants and soils depend strongly on shelter, surface structure and water availability.

3 Surface temperatures of exposed trees and shrubs are coupled closely to air temperature, and this coupling also makes them effective collectors of wind-driven cloud, and of gaseous and particulate pollutants, leading to potentially large inputs.

4 Local topography results in substantial variations in plant and soil temperatures, especially when the sun is shining. Plant responses to temperature vary between species, and are not well characterized for the low temperatures prevailing for much of the time.

5 Reproductive success may be more sensitive to temperature than is vegetative growth.

INTRODUCTION

The main characteristics of the climate of upland Britain have been well documented in two texts (Manley 1952; Chandler & Gregory 1976), but we have selected several aspects which need to be re-emphasized in the context of this symposium. In addition, we outline new work on the deposition of atmospheric pollutants on upland vegetation, and several recent studies relevent to the microclimatolgy and energy balance of upland surfaces. Finally, we briefly examine the response of plants to the upland climate.

ALTITUDINAL VARIATION

The rates of change of the main climatological variables with altitude above sea level in Britain are fairly well established, although there is much variation from place to place and especially from east to west (Table 1).

Solar radiation and hours of sunshine decline appreciably over the first kilometre above sea level, as the persistence of cloud increases. However, sites above 1000 m are frequently above the clouds, and an increase in solar radiation with altitude is observed above this level. On the highest peaks in Britain, the

TABLE 1. Altitudinal gradients in some climatological variables in the British Isles.
Sources: Manley (1969), Chandler & Gregory (1976), Harding (1979a, b) and Ballantyne (1983)

	(+ or −)	Change per km	Typical sea level value
Solar radiation	(−)	2·5–3·0 MJ m^{-2} day^{-1}	10 MJ m^{-2} day^{-1}
Sunshine	(−)	1·3 h day^{-1}	3·4 h day^{-1}
Air temperature			
Mean	(−)	6–9 °C	10 °C
Maximum	(−)	7–9 °C	13 °C
Minimum	(−)	5–7 °C	7 °C
Wind speed	(+)	6–7 m s^{-1}	5 m s^{-1}
Precipitation	(+)	1000–3000 mm	1000 mm
Number of days with snow cover per year	(+)	100–200 days	10 days

irradiance on a clear day may exceed that at lowland stations by about 30 per cent as a result of the diminished atmospheric pathlength (Harding 1979a).

Temperature declines with altitude, but lapse rates of mean temperature vary with site and season, and may exceed the 6 °C km^{-1} which is often assumed (Harding 1979b; Green & Harding 1980). Maximum temperatures tend to decrease with height more rapidly than the mean temperature does, and minimum temperatures less rapidly.

Precipitation increases rapidly with altitude, more so in the west of the country than the east (Ballantyne 1983). The uplands are particularly wet, not only for this reason but also because the low radiation and high humidity of the air result in diminished rates of evaporation and transpiration.

Wind speeds are especially high, western and upland Britain being one of the windiest parts of the world. For example, the daily mean wind speed exceeds 10 m s^{-1} (Force 6) on one day in three at Great Dun Fell, Cumbria (857 m), and gusts of twice the daily mean are common at most upland sites, emphasizing the futility of quoting mean values for biological systems which often respond non-linearly. The high wind speeds and the associated high incidence of storms place restrictions on land use, especially on forestry. As well as obvious mechanical effects of the force of the wind on structures, wind influences exchanges of heat, water, pollutant gases and aerosols between the surface and the atmosphere, in a manner which may have profound consequences for vegetation, soils and their associated microclimates.

ENERGY BALANCE

Surfaces absorb radiant energy from the sun at a rate which depends on the sky–surface geometry and on the short-wave reflectance of the surface. In the

British uplands, the influence of slope and aspect is less important than in high mountain areas of the world because only a small part of the incident energy is received as the direct solar beam. At the relatively low altitude of a Derbyshire Dale the diffuse fraction of solar radiation in midsummer was as high as 0·85 (Rorison, Sutton & Hunt 1986). Many upland sites are likely to be cloudier than this, with a correspondingly higher diffuse fraction. Consequently, differences in the energy supply to slopes of differing aspect, and topoclimates in general, may be less significant in the British uplands than they are elsewhere, for example in the European Alps.

The short-wave reflectances of some of the main types of vegetation have been established: heather moorland and spruce plantation hardly differ from each other, both reflecting 9–17 per cent of the incident radiation, whilst short grass is known to reflect 20–35 per cent (Jarvis, James & Landsberg 1976; Miranda, Jarvis & Grace 1984). The native pinewood of former times may have had a rather low reflectance, judging from values obtained in pine plantations (Stewart 1971).

Radiant energy absorbed by vegetation is dissipated as sensible heat (warming the atmosphere) and latent heat (humidifying the atmosphere). This partitioning of energy depends on the surface resistances r_a and r_s as shown in Fig. 1. The aerodynamic resistance r_a is a function of wind speed and it depends on structural features of the vegetation; the stomatal resistance r_s depends on the number of

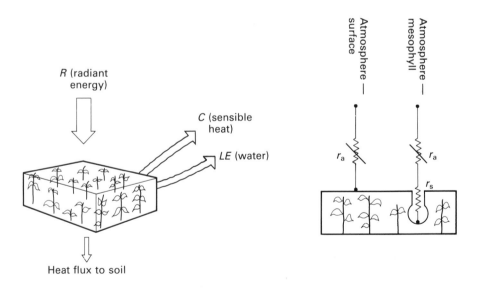

FIG. 1. The partitioning of absorbed energy R between sensible heat C and latent heat LE. Partitioning depends on the magnitude of the surface resistances r_a and r_s. Both the rate at which energy is absorbed and the magnitude of the resistances are likely to change when the vegetation changes.

stomata per area of ground, and on their degree of opening. Magnitudes of these resistances have been determined, so that, in principle at least, predictions may be made of the rates of transpiration and evaporation (Monteith 1973; Grace 1983; Miranda *et al.* 1984; McNaughton & Jarvis 1984). The general conclusion is that grassland, heathland and forest differ not so much in their transpiration rates, but considerably in the rates of evaporation from wet foliage, forests displaying much higher rates. It is much more difficult to make useful comparisons of water use in the field: experiments are in hand to measure the water yield of forested and non-forested catchments in Wales and Scotland, but so far the results at the Welsh and Scottish sites are contradictory (Blackie 1987).

The aerodynamic resistance alone governs exchanges between surface and atmosphere, such as the dissipation of heat and the capture of small particles and drops (aerosols). Recently, measurements of r_a for various vegetation types in the Cairngorms were made simultaneously (Wilson *et al.* 1987). They illustrate an important difference between tall and short vegetation, which has been noted previously (Grace 1981). Tall vegetation, coniferous forest in particular, displays very low values of r_a (Fig. 2). This implies that heat, reactive gases and aerosols

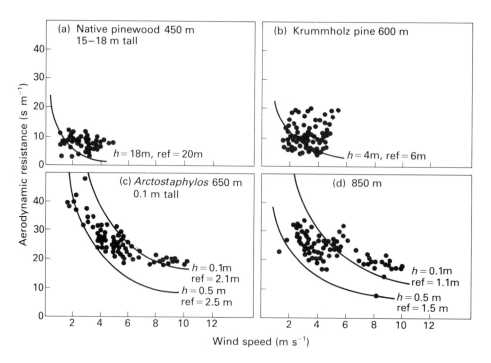

FIG. 2. Aerodynamic resistances to heat transfer from shoots at the top of the canopy to the atmosphere well above the canopy, for four vegetation types in the Cairngorm Mountains, at altitudes of 450, 600, 650 and 850 m. Curves are calculated from vegetation height (h) and height above the ground of the reference level (*ref*). See Wilson *et al.* (1987) for details (reproduced with permission from Blackwell Scientific Publications, Oxford).

are very effectively exchanged with the atmosphere; and vegetation with this structure may be said to be well coupled with the atmosphere (Monteith 1981). For the short vegetation, the values of r_a are much larger than those for tall vegetation, but still rather small in relation to the short vegetation of the lowlands, where wind speeds are often low. Also, topography may influence the aerodynamic resistance, so that even short vegetation at its edges, or on rocks that stand up from the landscape, may display low values of r_a.

This strong coupling between upland vegetation and atmosphere helps to explain several phenomena which are normally ascribed in the ecological and forestry literature to 'exposure'. These include the characteristics of the upland microclimate and the capture of particulate materials by upland vegetation (see Lee, Tallis & Woodin 1988; Gee & Stoner 1988).

DEPOSITION OF POLLUTION IN THE UPLANDS

The combination of high wind speeds and aerodynamically rough surfaces in the uplands create environments where the deposition of materials from the atmosphere to the surfaces of vegetation can be much larger than in the lowlands. It is well known from forest hydrology that trees capture mist and cloud from the atmosphere, and thus a forest receives more water than is recorded on a rain gauge at an open site near by, the so-called 'occult precipitation' (Rutter 1975). The occult deposition of pollutants in cloud water is now recognized as an important input in upland areas where hill cloud is frequent, but the following discussion points out that many other mechanisms of cloud water deposition may also be important, and that the dry deposition of gases and particles may be larger than at low elevations.

Fig. 3 illustrates some of the pathways for pollutant deposition in the uplands. Air reaching the hills may contain pollutant gases and particles transported considerable distances. As the air is forced upwards it cools until condensation level is reached; in moist maritime air this may be at only 200 m or even lower. Water then condenses around natural and man-made particles, and the cloud drops grow by further condensation and collisions while the air moves over the hill. Within hill cloud, drops may grow by one or two orders of magnitude, to 1–10 μm, but these sizes are insufficiently large to rain out significantly, and so rain gauges on hills do not record periods of cloud. When the air descends to the other side of the hill, the drops evaporate, leaving their particulate nuclei.

During their journey through the cap cloud, drops may be transferred to the surface by the following mechanisms. The direct impaction of cloud drops on foliage is efficient in strong winds when canopies are rough (small r_a) or when there are isolated shrubs or other plants through which the air flows. Impaction efficiency increases rapidly with drop or particle size and wind speed, and so drops above the condensation level impact much better than dry particles below this.

The importance of occult precipitation as a water input to cloud-forests of

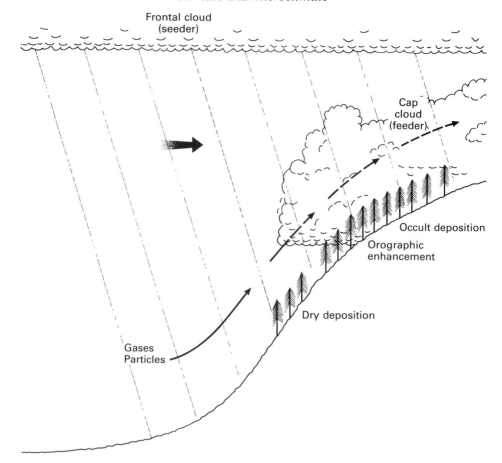

Fig. 3. Pathways for pollution deposition in the uplands (see text).

otherwise arid regions is well established (Rutter 1975), but in the wet climate of the British Isles such an input is unlikely to be hydrologically very significant. However, the occult deposition of insoluble and soluble material in cloud water may be an important, currently unquantified, input of acidity, heavy metals and other pollutants and nutrients (Unsworth & Crossley 1987). Lovett, Reiners & Olsen (1982) estimated that occult deposition accounted for between 60 and 80 per cent of the wet deposition of acidity, sulphate and nitrate to a balsam fir forest in New Hampshire. Good estimates cannot be made for occult deposition in the British Isles because the frequency of occurrence of polluted cloud, the climatology of cloud base height and cloud frequency, and the rates at which cloud water deposits on various types of vegetation are all unknown.

Figure 3 also illustrates another mechanism by which cloud drops are transferred to the surface: orographic enhancement, when rain falling from higher cloud falls through a cap cloud and scavenges cloud drops from the lower system. In strong winds this scavenging can increase the rate of frontal rainfall from 2 mm h^{-1} to 10 mm h^{-1} in the hills (Browning & Hill 1981). Cap cloud which forms in air of the surface boundary layer is likely to be polluted when this air layer has come over industrial areas upwind. In contrast, higher, frontal cloud is formed in air which may have cleaner origins and is less likely to contain local pollutants. Consequently, rain falling in the uplands which has scavenged cap cloud often contains larger concentrations of pollutants than rain at low levels where there is no orographic enhancement. If there is little or no orographic scavenging, then the amount of acid deposition will increase approximately linearly with rainfall. However, in areas where cap cloud is common, rain may contain two to three times the valley concentration of pollutants (Choularton & Hill, in press; Fowler et al., in press), in which case acid deposition may be substantially larger than has been estimated from rainfall amounts and chemistry measured in valleys. It is not yet established whether, and to what extent, enhanced wet deposition in the uplands occurs.

Hill cloud is an obvious indicator that the atmospheric environment in the uplands is different from that at low levels, but there may also be invisible alterations in the atmospheric composition that enhance the pollution exposure. For example, from measurements in central Europe and USA, it is becoming clear that photochemical ozone remains rather constant in concentration with time at mountain sites, whereas it decreases at night in the lowlands; taken in conjunction with the influence of larger wind speeds, and increased turbulent transport, this implies that there may be greater dry deposition of ozone in the uplands. Implications of this increased exposure of upland vegetation have not been explored. Ozone concentrations over upland Britain are currently unknown. Deposition of other gases such as nitric acid vapour or ammonia may also be enhanced in the uplands for similar reasons. Dry deposition of deliquescent particles increases as wind speed and relative humidity increase; thus the uplands may be preferential sites for this deposition, which is currently very difficult to measure.

SOME MICROCLIMATOLOGICAL OBSERVATIONS

Surface and tissue temperatures

As a consequence of its low aerodynamic resistance, the temperature and humidity at the surface of a conifer canopy is not very different from that of the atmosphere as a whole. Dwarf shrubby vegetation has a much higher value of r_a and therefore heat is dissipated at a much lower rate by convection. Differences between tall and short vegetation are illustrated by comparison of vegetation types in the Cairngorm mountains (Wilson et al. 1987). The temperatures of terminal meristems of dwarf vegetation (Arctostaphylos uva-ursi) were compared with those of trees in the native pinewood at a somewhat lower altitude. By day, the dwarf vegetation displayed much higher surface-to-air temperature excesses, and higher absolute temperatures than the tall vegetation. At night, when net radiation was negative, the short vegetation was significantly colder than the air, and colder than the tall vegetation.

Several authors have previously noted the large temperature excesses that occur in dwarf versus tall vegetation (Salisbury & Spomer 1964; Körner & Cochrane 1983; Green, Harding & Oliver 1984; Körner & Diemer 1987). In arctic and alpine conditions very high surface temperatures have been recorded, sometimes with tissue temperatures close to being lethal. Temperatures reached in these circumstances are highly dependent on wind speed, as one might expect from the energy balance equations (Grace 1983). As the wind speed increases, so the temperature excess diminishes. For this reason, the temperatures of plants in the lee of hills, boulders and rocks may be considerably higher than those elsewhere.

Surface temperatures can be conveniently sensed remotely, using infra-red imagery or as spot readings with an infra-red thermometer. Such measurements reveal that the change in surface temperature with altitude may greatly exceed the change in air temperature, at least when comparison is made of surfaces all of the same structural type (Fig. 4a). This occurs mainly because of the higher

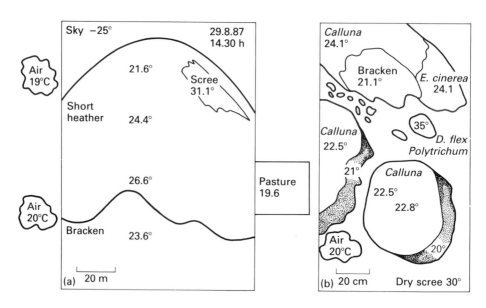

FIG. 4. Surface temperatures measured remotely with an infra-red thermometer on 29 August 1987 in the Pentland Hills, Scotland. The air temperature was 20 °C and the sky was clear of cloud and haze. In both cases the aspect was SW and the slope steep so that the solar beam was within 20 degrees of normal incidence (except for the pasture, shown in the appended window). (a) Shows variation over the entire hillside. (b) Shows a small patch of heathery vegetation on a dry scree slope.

wind speeds (and thus low r_a) prevailing at the high altitudes. The infra-red thermometer also reveals substantial differences between the surface temperatures of different vegetation types, which can be related to variations in albedo, slope and/or wetness. Thus, in Fig. 4a we see that the scree slope (dry, no latent heat loss) was much warmer than the adjacent *Calluna* (which dissipates some energy as latent heat through transpiration), and that nearby damp pasture (on the right of Fig. 4a) was a lot cooler, probably because of the lower angle of incidence between the solar beam and the ground, but perhaps because rates of transpiration were rather high. On a finer scale (Fig. 4b) there are also large differences in surface temperatures, which must be very important in the ecology of small life forms, such as annual plants and soil invertebrates.

Microclimate and shelter effect

Vertical profiles of wind speed, radiation, humidity and temperature immediately above and within the vegetation are characteristically very steep in the case of dwarf vegetation on sunny days (Delany 1953; Barclay-Estrup 1971; Grace & Marks 1978; Cernusca & Seeber 1981). There is an unfortunate tendency for authors to publish only data collected on sunny days, as in these conditions the relationship between vegetation structure and the development of microclimate is more easily detected; but it should be remembered that sunny days are the exception and for much of the time the solar energy flux is rather low in the British uplands. By contrast, the profiles in and above forests do not display strong gradients even when the sun is shining, a consequence of the higher rates of turbulent mixing between air within the canopy and air in the atmosphere as a whole, as discussed in an earlier section.

The importance of the temporal and spatial variation in the climatological variables, on the scales of hours and centimetres, to the lives of the animal components of the ecosystem is presumably immense, but has been little studied. Animal ecologists have sometimes combined climatological variables with a knowledge of thermal physiology to define a *climate space* in which a particular species is capable of surviving.

Grace & Easterbee (1979) found that the vertical profile of wind speed immediately above the vegetation could be estimated from the height of the canopy, provided that the terrain was fairly flat and horizontal. In this way it was possible to assess the shelter effect afforded to larger animals such as deer or sheep in the standing or sitting postures. But much of the spatial variation in wind speed over upland vegetation is caused by uneven topography and by the patchiness of the vegetation itself.

A common sight in the uplands is a small block of conifer plantation, established to provide shelter for sheep. These blocks provide useful shelter when they are mature, not only against convective heat loss, but also against radiative heat loss at night from animals sheltering inside the plantation (see Grace &

Easterbee (1979) for estimates based on deer). There is considerable shelter in the lee of a plantation, as illustrated by the study of Roberts (1972), although it seems likely that other elements in the landscape, particularly gullies and rocks, may themselves be adequate natural shelter.

CLIMATE AND PLANT GROWTH

The performance of most vascular plants is reduced as altitude is increased, and many native species display marked changes in morphology and anatomy when growing at high altitude. Commonly, upland and mountain forms have thicker leaves, thicker cuticles and, perhaps surprisingly, a higher maximal stomatal conductance (Woodward 1986; Körner & Diemer 1987). In some cases it has been possible to distinguish upland races which are short in stature, as in *Calluna vulgaris*. In other cases, lowland species have a closely related mountain counterpart, as in crowberry (*Empetrum nigrum* is found up to 800 m, but is often replaced by *E. hermaphroditum* above this altitude). Many species flower less at high altitude, and a higher proportion of the flora at high altitudes displays the curious phenomenon of vivipary, producing plantlets instead of seeds, as in *Festuca vivipara*.

The wetness of the uplands is beneficial for bryophytes and lichens, and, although depauperate in vascular plants, the British uplands have a rich bryophyte and lichen flora by European standards.

In assessing the upland environment for plant growth, the role of *exposure* is often emphasized. By this is usually meant exposure to the wind but, although wind does have important effects on plants, especially on their anatomy (Grace 1977), it seems likely that the main influence of this exposure is on surface temperature. In the most windswept sites where the vegetation is made up of dwarf shrubs and cushion plants, the surface temperatures are likely to be highly dependent on wind speed; by day, a sheltered microsite is likely to be significantly warmer than an exposed one. This point explains why small scale shelter is so effective, and why the distribution of plants on a mountain plateau is influenced so much by rocks and crevices.

Temperature influences rates of cell division and expansion, as well as the rate of photosynthesis, and in upland Britain the temperature is for much of the year below the threshold required for growth (Grace, in press). Thus an important difference between upland and lowland climates may be not just the mean summer temperatures, but the duration of the period when growth is possible. This period cannot easily be derived from meteorological data, as Manley (1952) and others have tried to do, because different species have different thresholds for growth and because the temperatures of the meristematic regions of mountain plants are dependent on radiation and wind speed as well as on the temperature of the air.

The strong influence of temperature on growth of upland species in their vegetative phase has been demonstrated for the case of native herbaceous plants (Arnold 1974; Graves & Taylor 1986), sown pasture grasses (Hunter & Grant

1971; Munro & Davies 1973), native trees (Millar 1965; White 1974; Hughes *et al.* 1984) and exotic conifers in plantations (Ford, Milne & Deans 1987). Attempts have sometimes been made to derive 'lapse rates' to express the rate of change of growth with respect to altitude: Hunter & Grant (1971) found a decline of 0·1 per cent m^{-1} in the yield of the grass crop in southern Scotland.

Water stress is not usually an important factor in the British uplands during summer, but in the late winter and early spring desiccation of foliage can occur if the ground is frozen or even just cold, especially in periods of sunny and dry weather (Watson, Miller & Green 1966). This phenomenon was recognized long ago in the Alps, where it is believed to be a major factor limiting tree growth at high altitudes (Tranquillini 1982).

As for the climatic limitation of species, it should be borne in mind that factors which limit the vegetative growth of the plant are not necessarily those which limit the reproductive success. Southern or lowland species are frequently prevented from spreading north or to high altitude by a failure to set fertile seed, as shown in the cases of stemless thistle and small-leaved lime (Pigott 1974; Pigott & Huntley 1981). Other cases where seed production becomes diminished and probably erratic from year to year as the distribution limit is approached include *Pinus sylvestris* (Miller & Cummings 1982) and cereals (Prince 1976). Once seed has been shed, the thermal conditions for germination and establishment may differ appreciably according to aspect and altitude (Rorison 1981).

It is not always poor summers that restrict growth and reproductive success. Many lowland species show a westerly distribution in the British Isles, being unable to stand extreme cold in winter (Salisbury 1926; Iversen 1944; Grace 1987). In contrast, many mountain plants can withstand very low temperatures indeed in the winter, provided they have undergone hardening by progressive exposure to increasingly low temperatures (Sakai & Eiga 1985). After budbreak in the spring, their tissues may no longer be hardy, and late frosts may be very damaging.

CONCLUDING REMARKS

Changes in climate with altitude in the British uplands are large, and the upland climate imposes major restrictions on plant growth and reproduction. Plant temperature is not always closely coupled with air temperature, nor are plant responses to temperature linear. Consequently, records of mean temperature at weather stations are not very useful for understanding plant distribution and success. More needs to be known about variation between and within species in responses to temperature.

Geographers have devoted considerable effort to radiation topoclimatology as a means of assessing the influence of slope and aspect on solar energy receipt. In view of the cloudiness of the British Isles and the consequent lack of direct sunshine, such approaches are likely to be of little value for explaining plant distribution and success in the uplands.

In addition to climatic stresses in the uplands, plants and soils are also likely to receive more wet- and dry-deposited pollutants than lowland sites with comparable air quality. Little is known, however, of the sensitivities of many of our native plant species to such exposure.

REFERENCES

Arnold, S.M. (1974). The relationship between temperature and seedling growth of two species which occur in Upper Teesdale. *New Phytologist*, **73**, 333–344.

Ballantyne, C.K. (1983). Precipitation gradients in Wester Ross, North-west Scotland. *Weather*, **38**, 379–387.

Barclay-Estrup, P. (1971). Cyclical processes in a heath community III. Microclimate in relation to the *Calluna* cycle. *Journal of Ecology*, **59**, 143–166.

Blackie, J.R. (1987). Balquhidder catchments, Scotland: the first four years. *Transactions of the Royal Society of Edinburgh*, **78**, 227–246.

Browning, K.A. & Hill, F.F. (1981). Orographic rain. *Weather*, **36**, 326–329.

Cernusca, A. & Seeber, M.C. (1981). Canopy structure, microclimate and the energy budget in different alpine plant communities. *Plants and Their Atmospheric Environment* (Ed. by J. Grace, E.D. Ford & P.G. Jarvis), pp. 75–81. Blackwell Scientific Publications, Oxford.

Chandler, T.J. & Gregory, S. (1976). *The Climate of the British Isles.* Longman, London.

Choularton, T.W. & Hill, T.A. (in press). Cloud microphysical processes relevant to cloud chemistry. *Acid Deposition Processes at High Elevations* (Ed. by M.H. Unsworth & D. Fowler). Reidel, Dortrecht.

Delany, M.J. (1953). Studies on the microclimate of *Calluna* heathland. *Journal of Animal Ecology*, **22**, 227–239.

Ford, E.D., Milne, R. & Deans, J.D. (1987). Shoot extension in *Picea sitchensis* II. Analysis of weather influences on daily growth rate. *Annals of Botany*, **60**, 543–552.

Fowler, D., Leith, I.D., Cape, J.N., Jones, A., Choularton, T.W. & Gay, M.J. (in press). Wet deposition and altitude: the role of orographic cloud. *Acid Deposition Processes at High Elevations* (Ed. by M.H. Unsworth & D. Fowler). Reidel, Dortrecht.

Gee, A.S. & Stoner, J.H. (1988). The effects of afforestation and acid deposition on the water quality and ecology of upland Wales. *This volume.*

Grace, J. (1977). *Plant Response to Wind.* Academic Press, London.

Grace, J. (1981). Some effect of wind on plants. *Plants and Their Atmospheric Environment* (Ed. by J. Grace, E.D. Ford & P.G. Jarvis), pp. 31–56. Blackwell Scientific Publications, Oxford.

Grace, J. (1983). *Plant–Atmosphere Relationships.* Chapman & Hall, London.

Grace, J. (1987). Climatic tolerance and the distribution of plants. *New Phytologist*, **106 (Suppl.)**, 113–130.

Grace, J. (in press). Temperature as a determinant of plant productivity. *Plants and Temperature; 42nd Symposium of the Society for Experimental Biology* (Ed. by S. Long & F.I. Woodward). Cambridge University Press, Cambridge.

Grace, J. & Easterbee, N. (1979). The natural shelter for red deer (*Cervus elaphus*) in a Scottish glen. *Journal of Applied Ecology*, **16**, 37–48.

Grace, J. & Marks, T.C. (1978). Physiological aspects of bog production at Moor House. *Production Ecology of British Moors and Montane Grasslands* (Ed. by O.W. Heal & D.F. Perkins), pp. 38–51. Springer-Verlag, Berlin.

Graves, J.D. & Taylor, K. (1986). A comparative study of *Geum rivale* L. and *G. urbanum* L. to determine those factors controlling their altitudinal distribution 1. Growth in controlled and natural environments. *New Phytologist*, **104**, 681–691.

Green, F.H.W. & Harding, R.J. (1980). Altitudinal gradients of soil temperature in Europe. *Transactions of the Institute of British Geographers*, **5**, 243–254.

Green, F.H.W., Harding, R.J. & Oliver, H.R. (1984). The relationship of soil temperature to vegetation height. *Journal of Climatology*, **4**, 229–240.

Harding, R.J. (1979a). Radiation in the British Isles. *Journal of Applied Ecology*, **16**, 161–170.

Harding, R.J. (1979b). Altitudinal gradients of temperature in the northern Pennines. *Weather*, **34**, 190–201.

Hughes, M.K., Schweingruber F.H., Cartwright, D. & Kelly, P.M. (1984). July–August temperature at Edinburgh between 1721 and 1975 from tree-ring density and width data. *Nature*, **308**, 341–344.

Hunter, R.F. & Grant, S.A. (1971). The effect of altitude on grass growth in east Scotland. *Journal of Applied Ecology*, **8**, 1–20.

Iversen, J. (1944). *Viscum, Hedera* and *Ilex* as climate indicators. *Geologisiska Föreningens i Stockholm Förhandlinger*, **66**, 463–483.

Jarvis, P.G. (1981). Stomatal conductance, gaseous exchange and transpiration. *Plants and Their Atmospheric Environment* (Ed. by J. Grace, E.D. Ford & P.G. Jarvis), pp. 175–204. Blackwell Scientific Publications, Oxford.

Jarvis P.G., James, G.B. & Landsberg, J.J. (1976). Coniferous forest. *Vegetation and the Atmosphere, Vol. 2* (Ed. by J.L. Monteith), pp. 171–240. Academic Press, London.

Körner, Ch. & Cochrane, P. (1983). Influence of plant physiognomy on leaf temperature on clear mid-summer days in the snowy mountains of south-eastern Australia. *Acta Oecological/Oecologia Plantarum*, **4**, 117–124.

Körner, C. & Diemer, M. (1987). *In situ* photosynthetic responses to light, temperature and carbon dioxide in herbaceous plants from low and high altitude. *Functional Ecology*, **1**, 179–194.

Lee, J.A., Tallis, J.H. & Woodin, S.J. (1988). Acidic deposition and British upland vegetation. *This volume.*

Lovett, G.M., Reiners, W.A. & Olsen, R.K. (1982). Cloud droplet deposition in subalpine balsam fir forests. *Science*, **218**, 1303–1304.

McNaughton, K.G. & Jarvis, P.G. (1984). Using the Penman–Monteith equation predictively. *Agricultural Water Management*, **8**, 263–278.

Manley, G. (1952). *Climate and the British Scene.* Collins, London.

Manley, G. (1969). Snowfall in Britain over the past 300 years. *Weather*, **24**, 428–437.

Millar, A. (1965). The effect of temperature and daylength on the height growth of birch *Betula pubescens* at 1900 feet in the northern Pennines. *Journal of Applied Ecology*, **2**, 17–29.

Miller, G.R. & Cummings, R.T. (1982). Regeneration of Scots pine *Pinus sylvestris* at a natural treeline in the Cairngorm mountains, Scotland. *Holarctic Ecology*, **5**, 27–34.

Miranda, A.C., Jarvis, P.G. & Grace, J. (1984). Transpiration and evaporation from heather moorland. *Boundary-layer Meteorology*, **28**, 227–243.

Monteith, J.L. (1973). *Principles of Environmental Physics.* Arnold, London.

Monteith, J.L. (1981). Coupling of plants to the atmosphere. *Plants and Their Atmospheric Environment* (Ed. by J. Grace, E.D. Ford & P.G. Jarvis), pp. 1–29. Blackwell Scientific Publications, Oxford.

Munro, J.M.M. & Davies, D.A. (1973). Potential pasture production in the uplands of Wales 2. Climatic limitations on productions. *Journal of the British Grassland Society*, **28**, 161–169.

Pigott, C.D. (1974). The responses of plants to climate and climatic change. *The Flora of a Changing Britain* (Ed. by F. Perring), pp. 32–44. Classey, London.

Pigott, C.D. & Huntley, J.P. (1981). Factors controlling the distribution of *Tilia cordata* at the northern limit of its geographical range. III. Nature and causes of seed sterility. *New Phytologist*, **87**, 817–839.

Prince, S.D. (1976). The effect of climate on grain development in Barley at an upland site. *New Phytologist*, **76**, 377–389.

Roberts, D.G. (1972). The modification of geomorphic shelter by shelterbelts. *Research Papers in Forest Meteorology* (Ed. by J.A. Taylor), pp. 134–146. Cambrian News, Aberystwyth.

Rorison, I.H. (1981). Plant growth in response to variations in temperature: field and laboratory studies. *Plants and Their Atmospheric Environment* (Ed. by J. Grace, E.D. Ford & P.G. Jarvis), pp. 313–332. Blackwell Scientific Publications, Oxford.

Rorison, I.H., Sutton, F. & Hunt R. (1986). Local climate, topography and plant growth on Lathkill Dale NNR 1. A twelve-year summary of solar radiation and temperature. *Plant, Cell and Environment*, **9**, 49–56.

Rutter, A.J. (1975). The hydrological cycle in vegetation. *Vegetation and the Atmosphere, Vol. 1,*

Principles (Ed. by J.L. Monteith), pp. 111–154. Academic Press, London.

Sakai, A. & Eiga, S. (1985). Physiological and ecological aspects of cold adaptation of boreal conifers. *Plant Production in the North* (Ed. by A. Kaurin, O. Juntilla & J. Nilson), pp. 157–170. Norwegian University Press, Oslo.

Salibury, E.J. (1926). The geographic distribution of plants in relation to climatic factors. *Geographical Journal*, **57**, 312–335.

Salisbury, F.B. & Spomer, G.G. (1964). Leaf temperatures of alpine plants in the field. *Planta*, **60**, 497–505.

Stewart, J.B. (1971). The albedo of a pine forest. *Quarterly Journal of the Royal Meteorological Society*, **97**, 561–564.

Tranquillini, W. (1982). Frost-drought and its ecological significance. *Physiological Plant Ecology II, Encyclopedia of Plant Physiology, Vol. 2b* (Ed. by O.L. Lange, P.S. Nobel, C.B. Osmand & H. Zeigler), pp. 379–400. Springer-Verlag, Berlin.

Unsworth, M.H. & Crossley, A. (1987). The capture of wind-driven cloud by vegetation. *Pollutant Transport and Fate in Ecosystems* (Ed. by P. Coughtrey, M. Martin, M.H. Unsworth), pp. 125–137. Blackwell Scientific Publications, Oxford.

Watson, A., Miller, G.R. & Green, F.H.W. (1966). Winter browning of heather (*Calluna vulgaris*) and other moorland plants. *Transactions of the Botanical Society of Edinburgh*, **40**, 195–203.

White, E.J. (1974). Multivariate analysis of tree height increment on meteorological variables, near the altitudinal tree limit in northern England. *International Journal of Biometeorology*, **8**, 199–210.

Wilson, C., Grace, J., Allen, S. & Slack, F. (1987). Temperature and stature: a study of temperatures in montane vegetation. *Functional Ecology*, **1**, 405–413.

Woodward, F.I. (1986). Ecophysiological studies on the shrub *Vaccinium myrtillus* L. taken from a wide altitudinal range. *Oecologia*, **70**, 580–586.

Acidic deposition and British upland vegetation

J. A. LEE, J. H. TALLIS AND S. J. WOODIN*

Department of Environmental Biology, The University, Manchester M13 9PL

SUMMARY

1 Soil acidification is a natural phenomenon in upland Britain which may be accelerated by processes resulting from anthropogenic activity including atmospheric pollution.

2 Acidification of some lakes in Scotland and Wales probably results from atmospheric pollution. There is little evidence from studies of lake acidification in regions remote from industrial and urban areas of effects of pollutants on terrestrial plant communities.

3 Atmospheric pollutants (most probably sulphur dioxide and its solution products) have drastically affected the vegetation of the southern Pennines over wide areas, notably destroying the *Sphagnum* spp. and *Racomitrium lanuginosum* cover of the blanket bogs in the nineteenth and twentieth centuries.

4 Sulphur dioxide emissions associated with lead smelting may also have contributed to local vegetation damage in other upland areas before and following the Industrial Revolution.

5 Although sulphur dioxide concentrations have decreased during the last few decades, nitrate deposition has increased approximately fourfold during the past 120 years in the southern Pennines.

6 The supply of combined nitrogen from the atmosphere is now supra-optimal for the growth of ombrotrophic *Sphagnum* species in the southern Pennies, and may be influencing more remote upland regions.

7 High concentrations of nitrate, ammonium and other pollutants in cloud and mist droplets may make mountain summit vegetation particularly vulnerable to relatively long-range atmospheric pollution.

INTRODUCTION

Since the last (Devensian) glaciation when the land surface of Britain was either covered by ice or affected by periglacial activity, approximately 14,000 years of weathering have created soils over most of the uplands which are strongly leached and acidic. To an extent this mimics the pattern of events observed in other interglacial periods (Sparks *et al.* 1969), suggesting that weathering processes over many thousands of years in a wet climate can produce naturally acidic soils on the

*Present address: Chief Scientist Directorate, Nature Conservancy Council, Northminster House, Peterborough PE1 1UA.

hard, resistant rocks which make up most of the parent materials in upland Britain. Evidence that natural acidification has occurred in the Flandrian (Postglacial) period comes from hydroseral succession in closed lake basins. In these little disturbed ecosystems Walker (1970) demonstrated that bog is the usual end-point of succession. However, over most of upland Britain the presence of human activity since the last glaciation will have affected acidification processes in many ways, and thus an interpretation of events as being entirely natural is not tenable. Over many centuries (since Mesolithic times in some areas — Jacobi, Tallis & Mellars 1976) forest clearance will have accelerated leaching processes by increasing the volume of precipitation reaching the soil surface and, more importantly, by the removal of nutrient recycling of the deeply rooted broad-leaved trees. Such destruction may also have enhanced the spread of wet acidic soils through paludification over many gently sloping hills. Similarly, burning and grazing in the cool oceanic climate of much of upland Britain will have accelerated the impoverishment of soils.

In the last two millennia, a further influence on acidification in the uplands has come to the fore, the rise of atmospheric pollution; and even more recently, within the present century, the widespread planting of alien conifer species has enhanced acidification processes on some soils (Gee & Stoner 1988). Atmospheric pollution has been widely implicated in the accelerated acidification of fresh waters and in damage to forests in parts of Europe and North America, but until recently relatively little attention has been paid to the effects of acidic deposition on British upland ecosystems.

HISTORICAL EVIDENCE

Amongst the earliest reports of atmospheric pollution are those associated with the burning of sea coal in some British towns in the thirteenth century (Brimblecombe 1976), but some evidence from parts of the uplands going back perhaps a millennium before this can be obtained by an examination of peat deposits. It is known that in parts of the British uplands lead was mined and smelted from at least Roman times. Lee & Tallis (1973) showed that in a southern Pennine peat deposit initial enhanced concentrations of lead were datable to the Roman period, although these concentrations were much lower than those observed in peat formed since the Industrial Revolution. The site was at least 10 km from sources of ore and likely smelting sites, suggesting that some regional atmospheric pollution must have been present in Roman times if it is assumed that lead deposited on these peatlands is virtually immobile (Livett, Lee & Tallis, 1979). The principal lead ore is galena, and its smelting releases sulphur dioxide into the atmosphere. Lead is also released by the burning of fossil fuels. In lead-mining districts some contamination of peat may occur through small particles of wind-blown ore, but elsewhere it can be assumed that wherever lead is detected in peat cores its deposition would have been accompanied by an enhanced atmospheric supply of sulphur pollutants. Lee & Tallis (1973) and Livett, Lee &

Tallis (1979) observed massive accumulations of lead in peat from upland regions of Britain close to urban or industrial centres, but much smaller accumulations in north-west Scotland. These data reveal a marked gradation in lead pollution in the uplands certainly as early as the eighteenth century, and presumably this was at least in part reflected in the gradient of sulphur deposition. There is evidence to suggest that emissions from lead smelting did affect the vegetation at some distance from the furnaces. Thus complaints by local farmers for loss of stock and pasture were frequent. Raistrick & Jennings (1965) record one such complaint for 1770 in the Pennines:

> the Reek of Smoak of a Lead Mill is of such a Nature as not very readily to mix with the air, and frequently is driven in an unbroken Current upon the Earth's Surface to a Miles Distance, nay we have ourselves seen the Smoak of Lead Mills at the Distance of 6 or 7 miles; it seems therefore that the Mill Reek is capable of affecting or having the appearance of affecting so large a surface that Mr Falls complaint was ever so well founded.

However, it is difficult to evaluate to what extent the damage to vegetation was the result of lead or sulphur dioxide emissions, or of an interaction between them. But the progressive introduction of flues to smelt mills during the late eighteenth and early nineteenth centuries, to reclaim lead which would otherwise have been lost to the atmosphere, must have proportionately decreased the lead to sulphur emissions quite dramatically. It is difficult also to evaluate how wide spread, if at all, was the ecological damage in the uplands attributable to this source of sulphur dioxide. But in the Pennines, at least, the numbers of smelt mills and their proximity to otherwise little-disturbed ecosystems makes it most unlikely that sulphur emissions had no effect on the vegetation before the major initial impact of the Industrial Revolution at the beginning of the nineteenth century. And, indeed, this source of sulphur dioxide was so great that, belatedly, smelting companies introduced scrubbing of emissions to limit corrosion in and around the smelt mills.

Evidence for urban and industrial emissions having an effect on the acidification of remote upland areas of Britain from the time of the Industrial Revolution comes from the investigations of Battarbee and his co-workers (Battarbee et al. 1985). These workers have demonstrated, through studies of diatom assemblages in lake sediments, the acidification of some Galloway lochs dating from the mid-nineteenth century which cannot readily be ascribed to other forms of human interference such as changes in land-use practice. These workers have also found similar evidence for acidification in some Welsh lakes (R. W. Battarbee, personal communication). However, there is no evidence arising from these studies to indicate that acidic deposition has markedly affected the growth of terrestrial plants within these lake catchments. Nevertheless it is possible that, elsewhere in the uplands, acidic deposition may have enhanced natural acidification processes, thus accelerating the spread of calcifuge species into some soils.

Evidence for the effects of atmospheric pollution on upland plant communities is best documented for the southern Pennines. This region, extending to at least 50,000 hectares, contains flat-topped hills of 500–600 m in altitude, on which extensive peat deposits have been formed during the last 7000 years (Tallis 1985). These hills are surrounded by towns which were early centres of the Industrial Revolution and which showed spectacular growth in population during the first half of the nineteenth century. Thus Manchester, the largest and most important of them, grew from 90,000 inhabitants in 1800 to 400,000 in 1840. The proximity of these towns to the southern Pennine upland area must have resulted in it being heavily contaminated by urban air pollution, and this can be demonstrated today by the widespread abundance of soot particles in the surface 10 cm of peat deposits in the region. However, until relatively recently, the concentrations of atmospheric pollutants in the region could only be guessed at since the few available measurements were almost always confined to the towns and, particularly during the nineteenth century, they were few and far between anyway. Data from Smith (1872) suggest that high concentrations of sulphur dioxide were widespread in the region during the mid-nineteenth century. His estimates for Buxton give a value of 1197 µg m^{-3} sulphur dioxide, approximately half those he observed in Manchester. Ferguson and Lee (1983) showed that the present-day annual mean concentrations of sulphur dioxide on a mire surface at Holme Moss, southern Pennines, are approximately half those in the surrounding towns. If this relationship pertained from the mid-nineteenth to the mid-twentieth centuries, then it would seem likely that a large part of the southern Pennine uplands was exposed to mean annual concentrations of sulphur dioxide of several hundred µg m^{-3} for at least a century. Most published fumigation experiments which have employed >200 µg m^{-3} sulphur dioxide have shown marked detrimental effects of the gas on plant growth (Roberts 1984), and it is likely that this component of atmospheric pollution alone must have adversely affected plant growth and distribution in the region.

The demise of plants during the nineteenth century in the region was linked by some naturalists to atmospheric pollution. Grindon (1859) associated the disappearance of lichens with atmospheric pollution, and the loss of a number of bryophyte species was ascribed to the same cause (Nowell 1866). Moss (1913) recorded the disappearance of several bog plants since the late eighteenth century based on the records of earlier naturalists, and the present-day bog surface is extremely species-poor, being dominated largely by *Eriophorum vaginatum* L., with *E. angustifolium* Honck. in the wet hollows. *Empetrum nigrum* L. is conspicuous on the edge of erosion gullies and *Vaccinium myrtillus* L and *Deschampsia flexuosa* (L.) Trin. may be locally abundant on the drier peats. But the most remarkable feature of these bogs is the almost complete absence of dominant bryophyte species. Thus *Sphagnum* species and *Racomitrium lanuginosum* (Hedw.) Brid. are absent or almost so from the bog surface. This can be shown to be a recent phenomenon, since over wide areas much of the underlying peat contains abundant *Sphagnum* and *Racomitrium* remains. In any one peat

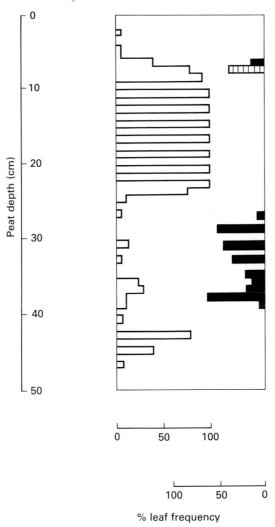

FIG. 1. The distribution of bryophyte remains in a peat sample from Holme Moss, southern Pennines. (□) *Sphagnum* Spp., (■) *Racomitrium lanuginosum*, (▥) Hypnoid mosses.

core the abundance of *Sphagnum* species can be shown to have fluctuated in the past but, when the *Sphagnum* species have diminished, the deposits usually contain abundant *Racomitrium* (or other bryophyte) remains (Fig. 1). This fluctuation in species abundance has been interpreted in terms of responses to climatic change from wetter to drier periods, and to a lesser extent to drainage effects resulting from erosion (Tallis 1987), but no such explanation can account for the absence of all these species, even from uneroded peat surfaces, at the present day. Tallis (1964) first demonstrated that the disappearance of *Sphagnum* species in the southern Pennines was correlated with the appearance of soot in the peat profile, and the widespread occurrence of soot in the peat pro-file, combined with the suspected concentrations of sulphur dioxide in the region during the nineteenth and early twentieth centuries, makes atmospheric pol-lution the most likely cause of this large vegetation change.

EXPERIMENTAL EVIDENCE

The widespread vegetation change in the southern Pennine blanket mires occurred before experimental ecology was established as a discipline; thus evidence for an involvement of atmospheric pollution in this phenomenon can only be sought retrospectively. Although widespread soot contamination affected the region, giving rise to what Wilson (1900) described as 'the great smoke cloud of the north of England', it would seem that the prevalence of sulphur dioxide was the more likely phytotoxic agent. In a controlled fumigation experiment, Ferguson, Lee & Bell (1978) showed that air containing 131 μg m^{-3} sulphur dioxide caused a marked reduction in the growth of most *Sphagnum* species tested (Fig. 2); wet deposition of sulphur pollutants at high concentrations also inhibited growth. The most resistant species to wet and dry deposited sulphur pollutants was *Sphagnum recurvum* P. Beauv., the only species at all widespread in the region today and whose habitats are minerotrophic rather than ombrotro-phic. In similar experiments Crittenden (1975) showed that the dominant *Eriophorum vaginatum* was not adversely affected by *c.* 300 μg m^{-3} sulphur dioxide.

The observed sensitivity of the plants to sulphur pollutants thus correlates with their present-day distribution, adding weight to the involvement of atmos-pheric pollution as a determinant of the present vegetation. Further evidence comes from experiments involving plants of the aquatic species *Sphagnum cuspidatum* Hoffm. from small, possibly relict, southern Pennine populations (Studholme 1988). He showed that plants from these populations were more tolerant of sulphite in solution than plants from sites in North Wales and the northern Pennines, which are more remote from urban and industrial influences, suggesting the importance of sulphur dioxide as an ecological factor in the southern Pennines. There is as yet no experimental evidence on the sensitivity of *Racomitrium lanuginosum* to atmospheric pollutants, but the abundance of potentially suitable sites in the southern Pennine mires for its establishment strongly suggests that pollution is the most likely cause of its exclusion. This is currently being examined.

Ombrotrophic mires are likely to be amongst the most sensitive ecosystems to atmospheric pollution because the plants are naturally adapted to utilize the atmospheric supply of elements, which is typically small except where this is perturbed by anthropogenic activity. However, there is some evidence that grasslands in the southern Pennines have also been affected. Thus Bell & Mudd (1976) showed that *Lolium perenne* L. from pastures in the Rossendale uplands to the north of Manchester was more tolerant to sulphur dioxide than the S23

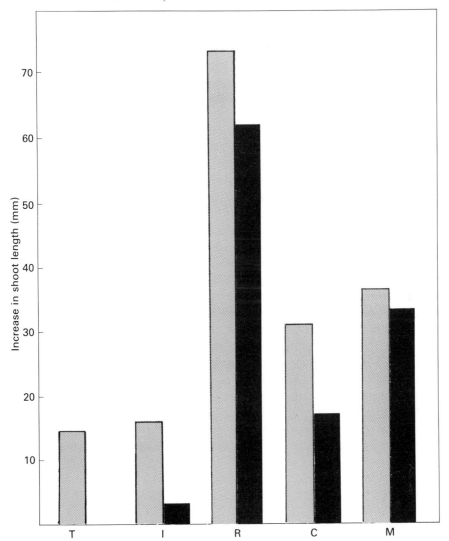

FIG. 2. The growth of *Sphagnum* species fumigated in clean air (9 µg m³ SO₂) (▒) and in polluted air (131 µg m⁻³ SO₂) (■) in perspex cabinets out of doors. T, *Sphagnum tenellum*; I, *S. imbricatum*; R, *S. recurvum*; C, *S. capillifolium*; M, *S. magellanicum*. Data from Ferguson, Lee & Bell (1978).

variety, and the species poverty, particularly of cryptogams, of acidic grasslands and woodlands in the region is strongly suggestive of the elimination of some species by atmospheric pollutants.

Another form of evidence to establish the involvement of atmospheric pollution is reintroduction experiments. In the early 1960s J.H. Tallis (unpublished) attempted a number of introduction experiments of *Sphagnum* species into the southern Pennine blanket mires without success. These experiments were repeated in the late 1970s by Ferguson & Lee (1983), by which time the sulphur dioxide concentration in the region had fallen appreciably (Ferguson & Lee 1983) to *c.* 40μg m^{-3}. Of the six species introduced to a high water-table site in non-eroded blanket peat, only *Sphagnum recurvum* thrived, although in a later introduction *S. papillosum* Lindb. also survived. An examination of element accumulation showed a massive increase in tissue nitrogen (from 10 to 25 mg g^{-1} over 12 months), with smaller accumulations of sulphur (from 1·7 to 3·0 mg g^{-1}) and some other elements (Ferguson *et al.* 1984). Taken together, these data and other evidence suggest that sulphur pollution is of declining importance in the southern Pennine blanket mires, but that nitrogen supply may be critical.

The decline in sulphur pollution (Ferguson & Lee 1983) has, to an extent, been counteracted by an increase in nitrogen deposition. Lee *et al.* (1987) estimated that the bulk deposition of nitrate in the region had increased approximately fourfold since the 1860s, and it can be shown that the atmospheric supply of combined nitrogen alone is capable of affecting the concentration of this element in *Sphagnum*. Press, Woodin & Lee (1986) demonstrated that, when *Sphagnum cuspidatum* from a relatively unpolluted site in North Wales was floated on unpolluted bog water in polypropylene containers placed on the bog surface in the southern Pennines, the plants showed a large and rapid increase in total tissue nitrogen (Fig. 3). In another experiment, the total nitrogen concentration of the moss increased by more than 10 per cent in ten days. The evidence from these and other studies (Lee *et al.* 1987) in the southern Pennines is that the atmospheric supply of combined nitrogen in the region is now supra-optimal for the growth and metabolism of ombrotrophic *Sphagnum* species. This raises the possibility that nitrate deposition may be becoming an important ecological factor in other upland regions more remote from urban and industrial centres.

UPLAND COMMUNITIES AT RISK FROM ACIDIC DEPOSITION

The southern Pennines is not the only region of Britain that can be shown to have been affected by acidic deposition. Chambers, Dresser & Smith (1979) showed that the surface deposits of blanket peat in South Wales were heavily contaminated with soot, and there is some evidence of a decline in abundance of *Sphagnum* species in this region, which may be related to atmospheric pollution (R. Woods, pers. comm.). Similar peat deposits in northern England and southern Scotland can be shown also to have been subjected to appreciable atmospheric pollution, and even more remote sites in northern and western Scotland have received some pollution loading (Livett, Lee & Tallis 1979). The unique feature of the southern Pennines is that past atmospheric pollution there, notably sulphur deposition, can be linked directly by experimental evidence to vegetation change; and present-day nitrogen deposition can be shown to be

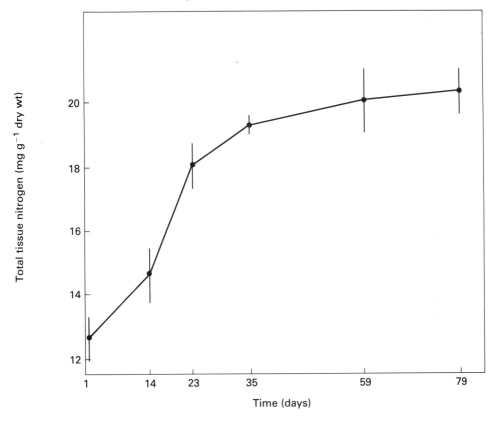

FIG. 3. The tissue nitrogen concentration (mg g^{-1} dry weight) of *Sphagnum cuspidatum* plants transplanted from an 'unpolluted' site in North Wales into a southern Pennines blanket mire. Plants were floated on bog water from the 'unpolluted' mire and received only wet and dry deposition from the atmosphere. Vertical bars are ± 1 S.E. ($n = 4$). There was no detectable change in the nitrogen content of plants in North Wales over the same period. Data from Press, Woodin & Lee (1986).

affecting plant growth in sensitive ombrotrophic mire communities. But what is the evidence that these and other communites in less grossly polluted regions are being affected by acidic deposition?

Although there are many pollen diagrams for remote regions of Scotland and Wales (Birks 1988), these generally contain insufficient detail in deposits dated to the last few centuries to allow any interpretation of vegetation change in relation to atmospheric pollution. There is no unequivocal evidence from other British upland regions in which annual mean concentrations of both sulphur dioxide and nitrogen oxides are less than 20 µg m^{-3} that these gases are directly affecting plant growth. Fumigation experiments with vascular plants at these concentrations have shown no generally adverse effect on plant growth, although it is very difficult to reproduce upland climatic conditions in controlled fumigations. Possibly even at these concentrations the more sensitive cryptogams are being affected in the field. It is also unlikely that ozone is an important pollutant in the British uplands with their high incidence of cloud cover.

For many years it has been realized that occult deposition (wind-driven cloud and fog particles) contains high concentrations of ions (Eriksson 1952; Grace & Unsworth 1988), normally approximately 3–5 times higher than those found in rain, but on occasions >10 times. However, only relatively recently have the ecological effects of this been considered. The high incidence of cloud cover makes the uplands particularly exposed to pollutants concentrated in this form of atmospheric deposition. Dollard, Unsworth & Harvey (1983), working in the northern Pennines, estimated that at Great Dun Fell occult deposition was 18 per cent of the sulphate and 10 per cent of the nitrate deposited by rainfall; and these workers observed pH of cloud water as low as 3·5. Other workers have also shown high acidity in cloud water, with concentrations of nitrate of up to 12,000 µmol dm^{-3} (Waldman *et al.* 1982); and there is growing concern about the harm which these very high concentrations may cause to vegetation (Weathers *et al.* 1986). In laboratory experiments plants frequently show visual lesions when sprayed with solutions of pH <3·5.

In Britain, most work on occult deposition has been devoted to its quantification, and little attempt has been made to examine its direct injury to vegetation in the field. However, habitats which are likely to be particularly at risk are mountain summits in which *Racomitrium* heath is an important vegetation type (Thompson, Galbraith & Horsfield 1987). Circumstantial evidence from the southern Pennines suggests that the dominant of this vegetation *R. lanuginosum*, is sensitive to atmospheric pollution. This moss, which has leaves with long hair points, is well adapted to intercept fog droplets, and is probably very largely dependent on the atmospheric supply of solutes for growth. The perturbation of this solute supply by pollutants may be particularly detrimental to its growth. Other plants characteristic of summit heaths, e.g. *Carex bigelowii* Torr, ex Schwein., may also be adapted to a low solute supply despite being much more dependent on the soil solution for element supply. The skeletal soils which support this plant have poor element cycling, and this is at least in part conditioned by low temperatures and short growing seasons. Under these conditions the supply of nitrogen for plant growth may be particularly limiting, and adaptation to low available nitrogen supply may be an important part of growth in this environment. The increasing deposition of combined nitrogen as a result of atmospheric pollution may mean a supra-optimal supply of this element to vascular plants on the impoverished soils of British mountains. This could lead to an increased sensitivity of plants to climatic stresses such as frost and drought (Davison & Barnes 1987) with potentially catastrophic effects. Much more attention should be placed on the effects of acidic deposition on these exposed high-altitude sites. As yet much must be conjecture, but the combination of prolonged exposure to high solute concentrations superimposed on a naturally nutrient-poor soil and atmospheric supply would seem to indicate a marked perturbation of these ecosystems. There is circumstantial evidence for a dramatic decline in *Racomitrium* heath in North and West Britain during the last 50 years (Thompson *et al.* 1987). To what extent, if at all, this can be ascribed to atmospheric pollution is uncertain, but at least it deserves thorough investigation.

REFERENCES

Battarbee, R.W., Flower, R.J., Stevenson, A.C. & Rippey, B. (1985). Lake acidification in Galloway: a palaeoecological test of competing hypotheses. *Nature*, **314**, 350–352.

Bell, J.N.B. & Mudd, C.H. (1976). Sulphur dioxide resistance in plants: a case study of *Lolium perenne*. *Effects of Air Pollutants on Plants* (Ed. by T.A. Mansfield), pp. 87–103, Cambridge University Press, Cambridge.

Birks, H.J.B. (1988). Long-term ecological change in the British uplands. *This volume.*

Brimblecombe, P. (1976). Attitudes and responses towards air pollution in medieval England. *Journal of the Air Pollution Control Association*, **26**, 941–944.

Chambers, F.M., Dresser, P.Q. & Smith, A.G. (1979). Radiocarbon dating evidence on the impact of atmospheric pollution on upland peats. *Nature*, **282**, 829–831.

Crittenden, P. (1975). *The effects of air pollution on plant growth with special reference to sulphur dioxide.* Ph.D. thesis, University of Sheffield.

Davison, A.W. & Barnes, J. (1987). How are the effects of air pollutants on agricultural crops influenced by the interaction with other limiting factors? *Direct Effects of Dry and Wet Deposition and Forest Ecosystems.* Proceedings of the COST Workshop, Göteborg.

Dollard, G.J., Unsworth, M.H. & Harvey, M.J. (1983). Pollutant transfer in upland regions by occult precipitation. *Nature*, **302**, 241–243.

Eriksson, E. (1952). Composition of atmospheric precipitation. 1. Nitrogen compounds. *Tellus*, **4**, 215–232.

Ferguson, P. & Lee, J.A. (1983). Past and present sulphur pollution in the southern Pennines. *Atmospheric Environment*, **17**, 1131–1137.

Ferguson, P., Lee, J.A. & Bell, J.N.B. (1978). Effects of sulphur pollutants on the growth of *Sphagnum* species. *Environmental Pollution, Ser. A.*, **21**, 59–71.

Ferguson, P., Robinson, R.N., Press, M.C. & Lee, J.A. (1984). Element concentration in five *Sphagnum* species in relation to atmospheric pollution. *Journal of Bryology*, **13**, 107–114.

Gee, A.S. & Stoner, J.H. (1988). The effects of afforestation and acid deposition on the water quality and ecology of upland Wales. *This volume.*

Grace, J. and Unsworth, M.H. (1988). Climate and microclimate of the uplands. *This volume.*

Grindon, L.H. (1859). *The Manchester Flora.* W. White, London.

Jacobi, R.M., Tallis, J.H. & Mellars, P.A. (1976). The southern Pennine Mesolithic and the ecological record. *Journal of Archaeological Science*, **3**, 307–320.

Lee, J.A & Tallis, J.H. (1973). Regional and historical aspects of lead pollution in Britain. *Nature*, **245**, 216–218.

Lee, J.A. Press, M.C., Woodin, S.J. & Ferguson, P. (1987). Responses to acidic deposition in ombrotrophic mires in the UK. *Effects of Atmospheric Pollutants on Forests, Wetlands and Agricultural Ecosystems* (Ed. by T.C. Hutchinson & K.M. Meema), pp. 549–560, Springer-Verlag, Berlin.

Livett, E.A., Lee, J.A. & Tallis J.H. (1979). Lead, zinc and copper analyses of British blanket peats. *Journal of Ecology*, **67**, 865–892.

Moss, C.E. (1913). *Vegetation of the Peak District.* Cambridge University Press, Cambridge.

Nowell, J. (1866). Notes on some rare mosses at Todmorden. *Naturalist*, **3**, 1–3.

Press, M.C., Woodin, S.J. & Lee, J.A. (1986). The potential importance of an increased atmospheric nitrogen supply to the growth of ombrotrophic *Sphagnum* species. *New Phytologist*, **103**, 45–55.

Raistrick, A. & Jennings, B. (1965). *A History of Lead Mining in the Pennines.* Longman, Green, London.

Roberts, T.M. (1984). Long-term effects of sulphur dioxide on crops: an analysis of dose–response relations. *Philosophical Transactions of the Royal Society of London, B*, **305**, 299–316.

Smith, R.A. (1872). *Air and Rain, the Beginnings of a Chemical Climatology.* Longman, Green, London.

Sparks, B.W., West, R.G., Williams, R.B.G. & Ransom, M. (1969). Hoxnian interglacial deposits near Hatfield, Herts. *Proceedings of the Geological Association,* **80,** 243–267.

Studholme, C. (1988). *The growth and physiology of* Sphagnum cuspidatum *in relation to atmospheric pollution.* Ph.D. thesis, University of Manchester.

Tallis, J.H. (1964). Studies on the southern Pennine peats III. The behaviour of *Sphagnum. Journal of Ecology,* **52,** 345–353.

Tallis, J.H. (1985). Erosion of blanket peat in the southern Pennines: new light on an old problem. *The Geomorphology of North-west England* (Ed. by R.H. Johnson), pp. 313–336. Manchester University Press, Manchester.

Tallis, J.H. (1987). Fire and flood at Holme Moss: erosion processes in an upland blanket mire. *Journal of Ecology,* **75,** 1099–1129.

Thompson, D.B.A., Galbraith, H. & Horsfield, D.H. (1987). Ecology and resources of Britain's mountain plateaux: land-use issues and conflicts. *Agriculture and Conservation in the Hills and Uplands* (Ed. by M. Bell & R.G.H. Bunce), pp. 22–31. Institute of Terrestrial Ecology, Grange-over-Sands.

Waldman, J.M., Munger, J.M., Jacob, J.J. & Hoffman, M.R. (1982). Chemical composition of acid fog. *Science,* **218,** 677–679.

Walker, D. (1970). Direction and rate of some British post-glacial hydroseres. *Studies in the Vegetational History of the British Isles* (Ed. by D. Walker & R.G. West), pp. 117–141. Cambridge University Press, Cambridge.

Weathers, K.C., Likens, G.E., Bormann, F.H., Eaton, J.S., Bowden, W.B., Andersen, J.L., Cass, D.A., Galloway, J.N., Keene, W.C., Kimball, K.D., Huth, D. and Smiley, D. (1986). A regional acidic cloud/fog water event in the eastern United States. *Nature,* **319,** 657–658.

Wilson, A. (1900). On the great smoke cloud of the north of England and its influence on plants. *Report of the 70th Meeting of the British Association for the Advancement of Science,* p. 930. John Murray, London.

LAND-USE CHANGES
AND THEIR IMPACTS

Rural policy factors in land-use change

A. MOWLE AND M. BELL*

Nature Conservancy Council, 12 Hope Terrace, Edinburgh EH9 2AS and
*National Farmers Union, Agriculture House, Salters Lane South, Houghton-le-Skerne,
Darlington, County Durham DL1 2AA

SUMMARY

1 Economic enterprises in the British uplands often operate on the margins of financial viability.

2 The major land uses, hill farming and forestry, are the subject of public policy and attract high levels of subsidy. The nature of these enterprises is heavily influenced by the precise terms of the subsidy, so it is not surprising that changes in these terms can lead to significant changes in land use.

3 The paper explores recent policy developments, identifying for ecologists the factors likely to condition significant land-use change in Britain's uplands and examining the balance between the interests of the environment and the social and economic interests of rural communities.

THEMES OF CONTINUITY AND CHANGE IN UPLAND POLICY

To the ecologist, Britain's uplands hold special interest because they are relatively undisturbed and therefore more natural than the lowlands. It should not be surprising that rural communities suspect that such interest is inherently against their own interests as consequent conservation measures may constrain development and economic activity. This polarization of interests is the cause of continuing public controversy, perhaps obscuring an underlying common interest. The burgeoning cost and surplus difficulties of the European Communities' (EC) Common Agricultural Policy (CAP) are likely to produce new circumstances in which the continuity of traditional rural activity and environmental needs might be more closely allied. Rural policy decisions will be of great consequence in the future use of the uplands.

Our focus on 'rural policy factors' must begin with recognition of the sectoral and uncoordinated nature of rural policy in Britain. Agriculture, forestry, nature conservation, amenity, rural services and regional development policies have traditionally been drawn up and implemented in isolation one from the other. This has led directly to conflict where the concerns of one interest cannot be accommodated within the policy instruments of another. The recent controversy over conservation of peatland ecosystems in Sutherland and Caithness is a classic example. The only instruments available to nature conservation interests are for

165

site designation, either as Sites of Special Scientific Interest (SSSIs) or National
Nature Reserves (NNRs). Neither instrument is suited to the protection of the
vast areas of ground involved. The instruments of forestry policy, grant aid and
tax relief, are unable to provide a strategic framework for coordinated development. The local rural communities and economic development interests,
represented by the Highland Regional Council and Highlands and Islands
Development Board, have no established locus in either conservation or forestry
development instruments. It was recognition of such shortcomings of the sectoral
approach which was responsible, in part, for the introduction of 'reasonable
balance' duties into forestry and agriculture policies in 1985 and 1986 respectively. The way in which such a balance may be achieved in practice is still far
from clear.

This paper concentrates on those aspects of rural policy which have had, and
will have, most direct effects on land-use change in the uplands. The complexity
and interconnection of the many strands of rural policy is such that six key
themes can be identified for our primary concern with changing patterns of land
use.

The first theme is the continuity of commitment to support farming in the
uplands, a concern predating EC accession. We must remember that the EC
Directive (75/268), providing special assistance for Less Favoured Areas (LFAs),
is a link in a chain of arrangements originating in the 'hill cow line' of 1943 and in
the subsequent Hill Farming Act 1946. The LFA, although strictly an agricultural
definition, provides a convenient definition of 'upland' as a widely recognized
boundary encompassing the areas in which conditions of economic disadvantage
deriving from land quality and distance from markets may be found. For present
purposes, therefore, in the west and north of Britain the uplands go down to sea
level.

The second theme concerns the multiple stated aims for agriculture and
forestry policies in the LFAs. Economic efficiency of agricultural production has
been linked to social considerations, while forestry policy, under attack for its in-
ability to meet financial return targets, calls on additional aims including social
support for rural communities and import-saving arguments. More recently,
conservation considerations have also been used to justify policy instruments.
Indeed, it can be argued that the very existence of LFAs and special support
measures runs counter to the concept of a 'common market' (Tracy 1987; Bowler
1985). Until recently, it was generally thought that supporting agriculture fed
through into the rural economy generally and maintained the landscape and
habitat of upland areas. The Scott Committee, describing farmers and foresters as
the natural landscape gardeners of Britain, represented such a view (Scott 1942;
Addison 1939).

The third theme, now widely challenged, has been the support of agriculture
by encouraging increased output. This normally took the form of headage
payments for livestock and high percentage rates of grant for capital investment
in land reclamation, buildings and handling facilities. During the 1960s and

1970s the economic production theme dominated policy. This was a response to the strategic economic problems facing upland agriculture as living standards and accompanying expectations increased with burgeoning economic growth in the European economy. The relative contribution of agricultural output to gross domestic product (GDP) tended to decline as population numbers stabilized and the demand for food was satisfied. Farmers had to increase either their output or the return per unit of output to maintain their relative incomes and living standards. The easiest option was to increase output. European agriculture fell well short of self-sufficiency, so LFA measures to increase livestock numbers and stocking rates, often linked to land improvement, could be justified.

The fourth theme concerns similarities and differences between Britain and her EC partners. The pattern of urban versus rural economic development was a general problem across Europe (Arkleton Trust (1982) compared the many measures adopted throughout the EC to assist special areas). In its original conception, the Mansholt plan envisaged a massive transfer of employment out of agriculture. It led to the three Directives of 1972 (72/159, 72/160 and 72/161), which established a framework of structural policy within the CAP which accepted in part the Mansholt thesis, aiming to modernize those who stayed while supporting those leaving farming. European nations were largely unwilling to accept a large-scale rural exodus (Tracy 1982; Bergmann 1985), which had already taken place in Britain, where average farm size is much larger than in the other EC member states. In Britain, where agricultural employment was already below 5 per cent, the 1975 LFA Directive confirmed special support to maintain the agricultural foundation of rural communities in the uplands. However, the tools were to be Hill Livestock Compensatory Allowances (HLCAs), with no limit per farm (see Smith (1985) for the contrast with French arrangements), and the Farm and Horticulture Development Scheme (FHDS). The success of this policy framework is open to debate. Certainly the long-term decline in the number of farmers continued, with consolidation into larger units (Clark 1979) and transfers of land into forestry. It can be argued that without these measures the rate of decline would have been even more rapid, although models have indicated that investment drives out labour (Traill 1982a; Bowers & Cheshire 1983). The policy may have failed to provide adequate incomes for smaller farms on the one hand and have facilitated considerable environmental damage through overgrazing and uneconomic land reclamation on the other.

The increasingly significant fifth theme is the idiosyncratic nature of the British forestry industry, which has developed as a completely separate activity from agriculture. This is not the case in other European countries, where the farmer-forester is a familiar occupation. Manson (1987), for example, compared and contrasted Grampian and the Lozère. The characteristics of British forestry have been detailed elsewhere (Clout 1984; Stewart 1985), but note the concentration of British forestry on large plantations of exotic conifers on bare ground in the uplands, with the acquisition of land made available as hill farm enterprises have been abandoned. The price differential between land cleared for planting and that sold as rough grazing is distorted by agricultural support on the one hand, but far more by the separate support measures in forestry policy, which are largely founded upon tax relief and therefore not available to low-income upland farmers (Tompkins 1986; Stewart 1986). We are not dealing with market economics in any sensible use of the phrase (Bowers 1983).

Finally, the sixth theme is the surplus output of the CAP and the consequent financial crisis. The CAP has been in financial difficulty before, but not when facing a future of permanent ability to produce food beyond the requirements of the market. The various instruments of agricultural policy are therefore being reviewed with three main objectives. In descending order these are: (i) reducing surpluses, and thus ameliorating budgetary difficulties and international friction caused by export of EC produce; (ii) maintaining the Treaty of Rome commitment to support rural communities; and (iii) achieving environmental ends and overcoming some of the externalities of intensive agriculture. In the UK these concern landscape or habitat loss, but in other countries water pollution from nitrate and intensive livestock effluents are bigger issues. These objectives have obvious inherent contradictions, and proposed instruments emerging must be accepted by a Council of Ministers dominated by national infighting. Nevertheless, there are signs of a move away from the traditional single-objective measures.

POLICY INSTRUMENTS IN THE UPLANDS AND THEIR ECOLOGICAL IMPACT

The conservation value of the uplands has been diminished as farmers and others, responding to the financial signals of changing support measures, have modified their enterprises. Unfortunately, little serious research into these interactions has taken place. This review draws upon only a scanty published literature.

Farm development schemes

The farm capital grant schemes of the 1960s and early 1970s were largely replaced on accession to the EC with the FHDS, later the Agriculture and Horticulture Development Scheme (AHDS). This required the farmer to put together a development plan which would raise him to a 'reference income' based on national averages for industrial workers. In the LFAs, 70 per cent grants were offered for associated land reclamation, drainage and fencing. Grants at lower rates were offered for buildings and livestock handling facilities. Farmers therefore increased stock numbers, using grants to provide housing, more grazing and winter fodder, notably in the form of silage. In theory, this increased level of activity would be self-sustaining. In practice, several factors prevented this sustainability from being achieved. Farmers often had to increase loans to finance their contribution, and purchase of stock and machinery, interest on capital and higher rents ate into farm incomes (Harvey 1985). Subsidies on capital investment may have accelerated the loss of employed labour on farms (Traill 1982b; Slee 1981). Most important, the lack of support for maintenance of reclaimed land often led to a decline in the productive capacity of the land. By the early 1980s, many farmers who had embarked on the FHDS were no better off in net income terms.

In areas with a high FHDS uptake there was a marked loss of conservation interest, especially of semi-natural vegetation previously used for rough grazing. This loss resulted from the drainage, reseeding and fertilizer application enabling the pastures to produce silage or carry higher stocking. Many of these areas had survived earlier reclamation efforts and thus were marginal prospects for profitable development. These changes affected botanical and landscape features and animal populations, notably *Crex crex* (corncrake), certain raptors and wading birds (NCC 1984).

The fourth and sixth themes listed above have been dominant in the recent history of investment support. The 1972 Directives, which had a 10-year life, have been replaced by the Agricultural Structures Regulation 797/85. Capital assistance under a farm plan continues, but the Agricultural Improvement Scheme (AIS) provides lower rates and emphasizes certain environmental investments. No grant aid is available for reclamation of rough ground, the rates of grant for land improvement are much lower and the highest rates of grant are for conservation measures (field boundaries, shelterbelts and pollution control). The Regulation switched support mainly from assisting those larger and more businesslike farms, which were capable of achieving an income comparable with urban areas, to farms which had lower returns and might never generate a high income, but which needed help even to continue. The tailoring of aid to smaller farmers is an important refinement of policy now encountered in a number of instruments.

Livestock support

The HLCA system provides an annual payment per ewe and per breeding cow. There is no limit on the number of eligible animals per farm and, while there are limits per hectare, these do not constrain stocking in most parts of the LFAs; some larger units are collecting at least £20,000 per year (MacEwan & Sinclair 1983; Crabtree et al. 1987). The HLCA is an easy target for criticism, being indirect income support unresponsive to need or to managerial performance. But if HLCAs were not available much more hill land would be afforested.

HLCAs in the beef sector are supplemented by the Suckler Cow Premium, which is paid on traditional beef herds and indirectly by intervention buying. This is outweighed, however, by the combination of an Annual Ewe Premium and a Sheepmeat Variable Premium which has guaranteed comparatively high prices for the sheep flock. The effect has been a rapid increase in sheep in many areas, exacerbating the long decline in hill cattle numbers. The cost of the sheepmeat regime is rapidly approaching beef support expenditure even though the sector only produces one-tenth of the tonnage of meat (Pooley, pers. comm.). In practice, these subsidies represent the annual disposable income of many LFA farms. The gross annual level of these payments in Britain is in excess of £100 million.

The effects of these payments on conservation interests derive from the increased stocking of rough grazing, with consequent alteration of semi-natural vegetation, notably loss of *Calluna vulgaris* (heather) and, more subtly, the imbalance in the sheep:cattle ratio, which probably contributes to the spread of *Pteridium aquilinum* (bracken). Cattle certainly require access to winter fodder, either produced on the farm or bought in, and are as a consequence more difficult to manage. As eight ewes are roughly equivalent in grazing terms to one beef cow (MAFF 1983), the direct cash subsidy is higher for sheep. The market for lambs has been relatively stable for a number of years, while beef market returns have been poor due to the sale of animals from dairy herds consequent on milk quotas. Many ecologists feel this imbalance urgently requires to be rectified, but the new extensification proposals will seek reductions of 20 per cent or more in beef numbers, while no similar constraints will be applied to sheep.

Proponents of the sheepmeat regime argue that, as self-sufficiency has not yet been reached, the present expansion can continue. Such statements ignore the continued heavy imports of New Zealand lamb which are locked into the whole complex of international trade agreements. It is unrealistic to ignore this, so in practice lamb production has already encountered the same limits as other stock. None the less, MAFF see the UK as having a comparative advantage in sheepmeat production and think a modest increase is reasonable to expect. Although a review of the sheepmeat regime is under way, there seems little chance that the outcome will help the sheep:cattle ratio imbalance.

Development programmes

Continuing economic and other problems in Europe in the late 1970s led to experimentation and research. Three 5-year Integrated Development Programmes (IDPs) were started in 1981/82, one in the Western Isles of Scotland, one in Lozère, France, and one in the Luxemburg province of Belgium. Ten research projects into integrated rural development were commissioned, three in Britain (Ulbricht 1986). One of these was in the English Peak District, which, although small in scale,demonstrated the willingness of some farmers to consider alternatives to the traditional production-oriented forms of support (Parker 1984).

The £20 million Western Isles IDP aimed to improve working and living conditions, largely through agricultural developments, fish farming and fish processing. The IDP was required to 'guarantee that any measures proposed are compatible with the protection of the environment', but no money was available for positive conservation measures, or to ensure that the guarantee was honoured.

The land improvement and livestock marketing elements of the IDP sought to improve the quality and quantity of livestock leaving the islands. To achieve this, production of winter keep from the land, and the feeding quality of the grazing, would have to be increased. This might involve the introduction of silage production on hay meadows and other land used for traditional rotational cropping, with additional fertilizer, drainage and reseeding. It is not surprising that conservationists were concerned that the delicate balance of crofting agriculture with a rich wildlife resource was threatened. The lack of positive conservation measures presented crofters with few options; they either accepted the strategy of the IDP or opted out. The Nature Conservancy Council undertook an assessment of individual project applications, but was unable to influence the result in the absence of alternatives, except on those particularly rich sites designated as SSSIs. In practice the integrated programme was unable to meet its own requirement that measures should be compatible with protection of the environment (NCC 1986a).

The IDP did not, in practice, upset the balance of crofting and wildlife in any measurable way. Almost all the money allocated for land improvement was spent on replacing fences which, funded at 85 per cent rates of grant at standard cost and erected by crofters themselves, added up to an effective form of employment subsidy. Thus the IDP's success as a development measure is the subject of debate. Houston (1987) was unable to establish firm evidence that the IDP had 'improved the working and living conditions in the islands'.

Under article 18 of the Structures Regulation (797/85), approval was given for an Agricultural Development Programme (ADP) for the Scottish Islands. This reflects current fashion and thinking in Brussels, avoiding the notion of integration, recognizing the futility of increasing output in an era of surpluses and making provision for positive conservation measures. The ADP, announced in autumn 1987, will be centred on a farm development plan. Support may help intensify production on existing inbye, but there will be no aid for reclamation. The plan will also provide a framework for conservation measures with annual payments for positive management of 'areas of particular importance for the natural environment'. However, these positive developments must be set against the exclusion of support for local processing and marketing which is logically required if island farmers and crofters are to secure economic development. Like the Structures Regulation from which it is derived, the ADP displays confusion about progress in the LFAs in an era of surplus. The ADP will be a test case for the feasibility of reconciling the three objectives of structural policy.

Part-time farming: the crofting example

The IDP and ADP focused attention on crofting agriculture, traditionally part-time in areas where pluriactivity is an adaptation to economic disadvantage. For years, the main instruments of agricultural policy pointed crofters toward full-time farming; part-timers were excluded from certain measures, such as FHDS, and some were forced to become spare-time crofters or give up altogether. This has changed with a resurgence of interest in part-time farming as a wider solution to the economic problems of agriculture. The formation, in 1985, of the Scottish Crofters Union marked the emergence of an organized political lobby for the crofting way of life. Their director, James Hunter (1987), has advocated crofting as a model for agriculture even in the lowlands.

There are, however, other factors. Crofters are eligible for capital grants under the nationally funded Crofting Counties Agricultural Grants Scheme (CCAGS). This provides high rates of grant, beyond the EC legal limits set in the Structures Regulation (a special derogation has been obtained by the British Government). These grants are still paid for reclamation and reseeding, an example of the present confusion. In addition, crofting communities are not generally enthusiastic about conservation, not recognizing the value of the conservation resource on their land and resenting outside interference and attempts to influence land management. It remains to be seen whether the ADP can help to change such attitudes.

Support for commercial afforestation

The largest land-use change taking place in Britain's uplands is the use of abandoned agricultural land for afforestation. In 1986/87, 43,000 ha of agricultural land were cleared for afforestation in Scotland (Hansard 1987). Although grant aid from the Forestry Commission (FC) under the Forestry Grant Scheme is available, the main incentive for the investor is the opportunity to set establishment costs against an income tax liability from other sources. In addition, when the growing plantation is sold after 10 to 15 years, the proceeds are treated as capital for tax purposes and are largely free of any tax liability. The land is normally bought by absentee owners and the turnover of ownership is notably higher than for more traditional uses (Tompkins 1986; RSPB 1987).

Large-scale forest development, using close-planted exotic conifers, is damaging to the environment (NCC 1986b). Such plantations affect river systems and peatlands, eliminate characteristic open-ground moorland and upland bird communities and affect the landscape.

Nature conservation and amenity policies

The main instrument of public policy for nature conservation is site safeguard, via NNRs and SSSIs administered by the NCC. While such measures, and their associated management agreements, can aid protection of extant wildlife interest in the most important areas, less than 10 per cent of the area of Britain as a whole, site safeguard tends to divert attention from the needs for wildlife protection in the countryside at large (Mowle 1986). These measures will tend to prevent land-use change; a similar conclusion may be drawn about amenity measures, which are also largely site-oriented (National Parks and Areas of Outstanding Natural

Beauty (AONBs) in England and Wales, National Scenic Areas (NSAs) in Scotland). The shortcomings of the National Park system have been explored by MacEwan & MacEwan (1982, 1987).

One major problem deriving from the application of a site safeguard approach is the basis of management agreements which assume basically stable economic circumstances for the individual occupier. The rationale of such agreements for conservation remains founded on compensation or payments based on profits forgone. This sustains a presumption that maximized output is the proper strategy and that individuals confine themselves to economically rational decisions. It does not provide an adequate basis for agreement in circumstances where price levels are such that no profit is notionally forgone, as can be found in upland land use. In such cases, the compensatory rationale breaks down. Unless payments for conservation can be for positive management, there may then be a question mark against the future of the farm as an entity, its associated habitats and the farmer's contribution to social structures. There must come a point at which some farmers receive more under conservation-oriented payments than they would from efficient cost-effective production. Critics argue that direct controls would be cheaper, more cost-effective, or both.

Environmentally Sensitive Areas

ESAs did not originate as measures intended to tackle the problems of upland agriculture. However, of the first round of ESA designations, several are within or partly within the LFAs (see Smith 1988). Neither their principle nor their contribution can be ignored. The essence of an ESA lies in the coincidence on the ground of high conservation value, for landscape, wildlife or archaeological reasons, with active agricultural management. The concept originated in the broads grazing marshes of East Anglia, threatened with conversion to cereal production via drainage, ploughing and intensive management. Flat-rate payments per hectare were made each year to farmers in return for an undertaking, freely entered into, to maintain the grazing regime.

This flat-rate approach was taken as the model for wider application in England and Wales. In Scotland, exploration of possible candidate areas revealed that this approach was not feasible. Many Scottish farms, especially in the LFAs, have land of variable quality, from small areas of low-lying arable ground which are disproportionately important agriculturally because they can yield winter fodder for livestock, to high or steep mountain land which is virtually unusable. It was necessary to devise a method of payment which was sensitive to these variations and also to the commitment of individual farmers. The ESA agreement is built around a farm-conservation plan; the farmer receives a modest annual flat-rate payment in return for a standard set of conditions and the preparation of a plan which identifies certain features of value for conservation on the farm. A second tier of payments is then made based on the extent of such features and for positive management in accordance with standard guidelines.

ESAs are therefore a major departure in agricultural support measures. They are the first substantial payments for purposes other than production. They also introduce annual payments for agreement to abide by an agreed management plan. There is a long way to go before their success can be judged and major problems remain, such as the lack of a method either to restrain overstocking or to prevent transfer to forestry. But will ESAs be seen as picturesque curiosities set apart from 'real agriculture'? The possibility of any farmer having the opportunity to enter an ESA-type agreement may soon have to be taken seriously by Government.

The EC 'extensification' scheme and future proposals

Even beyond an agreement broadly to maintain agriculture at present levels, as in the ESA, is the concept of reducing output. EC Regulation 1706/87 gives the UK Government 9 months to devise a scheme which, *inter alia*, offers assistance to farmers who reduce cereal acreage or beef numbers by at least 20 per cent. In cereals, this is not extensification (lowered production over all the land) but setaside. In beef, farmers in the scheme may build up their sheep enterprise, damaging the delicate economy of the hills. Agriculture departments and farmers' unions are understood to share this concern. More broadly, the Regulation alludes to the wide objectives set out in the sixth theme, and it is likely that MAFF will face opposition should they produce a scheme based purely on commodity factors and lacking environmental sensitivity (NCC 1987).

There are two further EC proposals with significance for the uplands. The first would have introduced ceilings on HLCA payments per holding. UK agricultural interests eventually transposed this into an innocuous statement that 'member states may vary the compensatory allowance on the basis of the economic situation of the holding', as the French already do (Smith 1985). Second, EC have also produced proposals for a 'prepension scheme' to encourage older farmers to leave the industry. The marginal nature of upland farming makes it potentially liable to considerable response to such schemes or their knock-on effects. Ecological change is driven by more than simply biological factors.

THE DEBATE ON POLICY AND LAND-USE CHANGE

A number of books have been published criticizing agricultural and forestry policy from a broadly environmental viewpoint (Shoard 1980; Norton-Taylor 1982; Bowers & Cheshire 1983; Pye-Smith & North 1984; Pye-Smith & Rose 1984; Drabble 1985). Two notable features are that the uplands tend to be seen as part of a broad countryside issue (there is comparatively little recognition of their special features and heterogeneity), and most space tends to be devoted to listing negative impacts and only a little to some single-issue solutions reflecting the authors' special concern. Other publications have focused on policy options for the uplands (Parry 1982; MacEwan & Sinclair 1983; Sinclair 1984a, b; RSPB

1984; Woods 1984; Parry & Sinclair 1985; Dixon 1987), although reports often deal either with England and Wales or with Scotland (TRRU 1981; Parker 1984; Sayce 1984; CPRE 1985a, b, c; CPRE & CNP 1985; NCC 1986a). Hodge (1985) provided a research review and Lowe *et al.* (1986) a useful series of case studies. By 1986, the debate had become fairly clear and well informed, but the more recent EC proposals have led to a re-examination of basic principles and the introduction of new issues (NCC 1987; Shoard 1987).

The concept of pushing up output to help support people on farms had attracted broad support since the 1940s. Lord Boyd-Orr's keen interest in scientific research on hill farming was because

> We might get all the mutton we need imported from Australia, New Zealand and the Argentine; we might do without our sheep on the hills, but our national character could ill afford to lose the shepherd. (Fraser 1954)

Conservationists have argued that the price was too high. In cases such as Exmoor they had some success, but until the EC ran into its cost and surplus crisis their criticisms had little effect. Now the shortcomings of both agriculture and forestry policies have combined to create pressures to cut support at three levels, from a national government keen to reduce exchequer cost, from the EC unbalanced by agricultural overspend, and from world trade with summit undertakings to reduce farm protection multilaterally. Traditional upland farming, as one of the weakest sectors with fewest options, might fare badly whilst the CAP readjusts. There could be negative impacts for communities, landscapes and wildlife habitats alike. Sinclair (1984b) considered the withdrawal of agricultural support, saying

> The farmed landscape would be ranched; its vernacular artefacts would decay; and a whole rural society, complete with its special skills and cultures, would wither away within a generation. If financial incentives to forestry were then to be continued, most of the freehold land now used as rough grazing by hill farmers would become a conifer plantation.

If, as a nation, we treasure the uplands for their naturalness, we must recognize the fact that their present condition is greatly affected by public policy, which is having to come to grips with agricultural surpluses. Because land is the factor of production which carries a range of public goods and externalities, the popular debate has focused on the concept of 'surplus acres'. Table 1 summarizes five studies; they are analysed in more detail by Tinker (1988). There are two immediately striking aspects, namely (i) the fact that the uplands receive special support because of their LFA status, if retained, suggests that there will be new 'marginal' areas on poor ground outside the LFAs; and (ii) if public expenditure in the uplands is sustained, the cost would increase as lowland producers of surplus products affected by price cut or quota move into traditional upland enterprises such as sheep and store cattle.

Support for LFAs runs counter to the aim of a 'common market' that output

TABLE 1. Comparison of studies of farmland requirements. Sources: Edwards (1986), Gould (1986), CAS (1986), North (1986a, b) and Agriculture EDC (1987)

Source of study	Area studied	Definition employed	At what date	Area range (million ha)	Main 'surplus' estimate (million ha)
Wye College	UK	Area available for other uses	2000 with sensitivity tests	1–6	3–4
Gould's	Great Britain	Surplus to needs	1990 & 2000		1·1 & 2·6
Reading	England and Wales	Low gross margin/area equivalent of the reduction in intensity required	5 years of various scenarios	0·2–2·2	1·3 (price pressure) 1·9 (quotas)
North	Great Britain	Farmland requirement and surplus	2015	5·4–5·72	5·5
Agriculture EDC	UK	Land displaced from tillage	Mid-1990s	nil–1	0·7

should come from the areas best suited to each product. Certain types of farming are deliberately maintained in areas where they cannot be justified alongside the modern structure of the industry. As long as there was scope for increased output from all regions this worked fairly effectively. Higher-value products, such as milk and cereals, utilized better land. Hill farms gained from the underlying price support as well as from HLCAs (Whitby 1977; Evans & Felton 1987). Already, former dairy land is moving into beef and sheep following quota and, ironically, this may produce a temporary boom as lowland farmers buy hill stock to build herds and flocks. Although the studies in Table 1 assume that uplands will be protected in some way because the small, poor hill farmer is a politically popular figure, none have spelt out the mechanism for how this will be achieved when lowland sheep rearing can produce perhaps five lambs per unit area against one in the hills.

Land in the hills is also moving into forestry, but the release of lowland ground potentially gives the forester a greater choice of sites. We envisage specific measures to aid farmers in the hills, slowing the pace and amending the pattern of afforestation, but the few options which seem politically feasible are unlikely to alter the fundamental pattern. Taking a broad view, the interests of conservationists and rural communities seem closely allied when set against exchequer disciplines, European infighting and world trade pressures. There is a strong argument for adoption of multi-objective policies which seek to accommodate both conservation and development aims.

POLICY FACTORS FOR THE FUTURE

The introduction of the 'reasonable balance' duty for agriculture departments in Section 17 of the Agriculture Act 1986 may have been the most significant recent change in the climate of policy-making. This is not particularly onerous, merely requiring Ministers to have regard to and endeavour to achieve a reasonable balance between the following four considerations: (i) the promotion and maintenance of a stable and efficient agricultural industry; (ii) the economic and social interests of rural areas; (iii) conservation and enhancement of the natural beauty and amenity of the countryside, etc.; and (iv) the promotion of the enjoyment of the countryside by the public. The significance of this duty lies in its implications for the sectoral divisions of British land use. Agricultural policy has had multiple objectives in the past, but certainly from *Food from Our Own Resources* (MAFF 1975) all other considerations were secondary to increasing agricultural output. Now the main thrust of policy is to seek an accommodation between the three main factors, economic efficiency, environmental conservation and social equity.

Economic efficiency has traditionally been judged in terms of labour productivity, which has accelerated the decline in farm employment. In Scotland, most of which lies in the LFAs, two-thirds of the farm labour force have gone since 1950 (Bryden 1985). On many farms, there is no longer any employed labour and further increases in labour productivity are unlikely. There are different definitions of efficiency (CAS 1980), but given the prominence of farm income per labour unit as a measure of progress, relative to an economy in which average incomes increase well ahead of inflation, the problem will remain. How can farmers, especially in the uplands, maintain their living standards over the next 10 to 20 years? The Islands ADP is attempting to turn farmers' attention to increasing value added to each unit of output. The Scottish Crofters Union advocates an expansion of part-time farming, hence deriving a higher proportion of income from sources other than farming. Certainly the solution does not lie in increasing stock numbers per man unless the number of farmers is going to continue a rapid decline.

Environmental objectives are, at least superficially, more clear-cut. The wildlife and landscape of the uplands has suffered, and continues to suffer, serious losses. This is not a side-effect of interplay of market forces, nor is it a deliberate act of policy. It is the accidental result of policies aiming to maintain the income position of people living and working in rural areas relative to the majority who live and work in the towns. Conservationists broadly agree on what the ecological requirements of policy are, if not on how they may be achieved. Losses of semi-natural vegetation in the uplands must be halted and, preferably, reversed. Other management which adversely affects animal populations must be controlled. The consensus is less clear where, as in the Western Isles, the high conservation interest is intimately linked to an established form of agriculture. ESA-type payments, to maintain an existing regime of management, can help in the short term, but what opportunities do they offer for long-term development of the rural community? On the Uist machairs, both development and abandonment of agriculture would lead to a loss of botanical and ornithological value. The crofter could be paid direct income support — a salary in all but name — to maintain this artificial regime. Similarly in Islay, the dung from outwintered cattle is thought to provide a vital invertebrate food supply for *Pyrrhocorax pyrrhocorax* (chough). How far can agricultural policy go to maintain this regime when economic production trends are to inwintered stock or no cattle at all?

The social strand of policy is perhaps the most interesting of all. Why subsidize farmers and crofters in this way when no similar support has been available to coal-miners, steelworkers, shipbuilding yards and other declining industries which are at least as important to their respective local communities? Certainly the £100 million per year paid out in HLCAs cannot be justified solely in agricultural terms. The 'reasonable balance' duty also refers to 'enjoyment of the countryside by the public'. The urban taxpayer ultimately pays for agricultural support to upland farmers, so demands for access and amenity cannot be ignored. But the leisure time and money of the urban population to visit the uplands generally runs ahead of that of rural communities. Such questions of equity and distributional justice raise issues well beyond the traditional confines of sectoral rural policies.

The main obstacle to integration is the sectoral approach to rural policy. There are good administrative reasons for the approach. The Treasury wish to control the allocation of resources and needs to know, for example, that money voted to agriculture departments is going to agriculture and not, say, to rural schools or transport subsidies, conservation, etc. The 'reasonable balance' duty, the introduction of ESAs and the conservation measures in the Islands ADP place all of this in question. It is not surprising that rural sector policies in 1987 are characterized by confusion, with a breakdown of these sectoral divisions in prospect.

Policy-makers and practical farmers are having to come to terms with policies and programmes of implementation which are genuinely multi-objective, requiring a degree of integration. This will require a framework if the result is not to be chaos. Some urge the formation of a Ministry of Rural Affairs, but could this provide the sensitivity to local conditions which is required? A more appropriate solution might be to introduce a regional element into policies such as agriculture, forestry and conservation, which at present are the province of national agencies. The missing element is the 'honest broker' to mediate between these interests; local authorities are the only extant institutions with the potential for this role. Perhaps all that is required is an extension of the structure plan process to include rural sector issues, resulting in agreed strategies which the national agencies would be obliged to abide by. This would provide a framework in which the complex terms of the 'reasonable balance' duty can be worked out in practice, the work of the different agencies coordinated and the multiple conflicts of interest resolved.

CONCLUSIONS

Conservation and development are rarely seen to be coincident in rural communities. Important sociological factors, including linguistic and racial differences, life styles and values, and feelings of locality or rurality (Bradley & Lowe 1984; Redcliffe-Maud 1969), may overlay economic interests. Sometimes, as in the rejection by local people of a Welsh Rural Development Board with pre-emptive powers in the land market, local sensibility and economic interest may be opposed. A leader of the Havasupai people of Arizona pleaded with the American Congress:

> The environmentalists are city people thinking only about taking these lands away from our people, so they can come up once or twice a year and have some recreation on them. Recreation! We are talking about survival while they talk about recreation. Where does the greed of these people stop! (Highwater 1976)

The question now is whether a period when the EC has reached effective food self-sufficiency offers greater or less hope for conservationists and rural communities to share a common future. In theory the answer could be 'yes', but perhaps rural policy is not yet attuned to grasp the opportunities.

Upland policy, at least since the 1940s, has barely been distinguished among the functional divisions of rural policy. The dominant theme in agriculture was increased output. By 1979, the goals were adjusted towards efficiency and cost of production, recognizing that the environment might be a constraint in certain limited circumstances. In 1980, when the established timber production basis of forestry policy was restated, passing reference was made to the environment. Policy has come a long way since then, through the 1981 Wildlife and Countryside Act and its amendments, with new duties on Ministers to secure a reasonable balance (Lowe et al. 1986; Mowle 1986).

There has been no fundamental shift of attitudes towards some new consensus. The aim of additional, more efficient and cost-effective output provided an underlying rationale which has not been replaced. In the medium to long term, the ESA type of agreement should be developed and broadened to form the basis of a range of management payments to assist ecological objectives. These should be more explicitly related to the maintenance of the rural population and conservation of the natural environment. McLaughlin (1987) pointed out that rural deprivation affects individuals, not areas. ESAs represent a welcome step in the right direction; without measures of this kind in the uplands, an extensive ranching system of farming might survive, but large areas of land would be transferred to coniferous afforestation.

Recent EC pronouncements, for example the European Parliament on the Single European Act (1987), continue to stress the importance of rural life and the environmental dimension. Our EC partners tend to see it as an assumed part of farm policy; 'retaining the rural population is almost part of the social contract'

(Tracy 1987). Does this not mean that we can rely on the EC? Unfortunately not; the principal decision-making body, the Council of Ministers, is dominated by macro-economic considerations of national advantage (Bowler 1985; Fennel 1987). It still seems unreasonable to most people to expect transnational negotiations on economic policy instruments to have more than a minimal regard for environmental factors. Even less can they take cognizance of the sub-regional, local or even field specificity of much conservation interest. If upland ecologists are interested in recording and studying habitat change then the next few years seem set to give them plenty of scope.

ACKNOWLEDGMENT

MB would like to thank ESRC and NERC for the award of a joint fellowship, based at ITE Merlewood.

REFERENCES

Addison, Lord (1939). *A Policy for British Agriculture.* Victor Gollancz, London.

Agriculture EDC (1987). *Directions for Change.* National Economic Development Office, London.

Arkleton Trust (1982). *Schemes of Assistance to Farmers in the Less Favoured Areas of the EEC.* Arkleton Trust, Langholm, Dumfries.

Bergmann, D. (1985). *5th Asher Winegarten Memorial Lecture. A Future for European Agriculture.* National Farmers Union, London.

Bowers, J. (1983). *Do we need more forests?* Discussion Paper No. 137, School of Economics, University of Leeds.

Bowers, J.K. & Cheshire P. (1983). *Agriculture, the Countryside and Land Use.* Methuen, London.

Bowler, I.R. (1985). *Agriculture under the Common Agriculture Policy.* Manchester University Press, Manchester.

Bradley, T. & Lowe, P. (1984). *Locality and Rurality: Economy and Society in Rural Regions.* Geo Books, Norwich.

Bryden, J.M. (1985). Scottish Agriculture 1950–1980. *The Economic Development of Modern Scotland 1950–1980* (Ed. by R. Saville), pp. 141–162. John Donald, Edinburgh.

CAS (1980). *The Efficiency of British Agriculture.* Report No.7, Centre for Agricultural Strategy, Reading.

CAS (1986). *Countryside Implications for England and Wales of Possible Changes in the Common Agricultural Policy.* Report to Department of the Environment, Centre for Agricultural Strategy, Reading.

Clark, G. (1979). Farm amalgamations in Scotland. *Scottish Geographical Magazine,* **95**, 93–107.

Clout, H. (1984). *A Rural Policy for the EEC?* Methuen, London.

CPRE (1985a). *Environmental Implications of Future CAP Pricing Policies.* Council for the Protection of Rural England, London.

CPRE (1985b). *How to Help Farmers and Keep England Beautiful.* Council for the Protection of Rural England, London.

CPRE (1985c). *Environmental Protection and Agriculture.* Council for the Protection of Rural England, London.

CPRE & CNP (1985). *Farming for the Countryside.* Council for the Protection of Rural England & Council for National Parks, London.

Crabtree, J.R., Evans, S., Revell, B.J. & Leat, P.M.K. (1987). *Agricultural Structures Policy and Nature Conservation in Upland Grampian: Pilot Study.* Research and Survey in Nature Conservation No.5, Nature Conservancy Council, Peterborough.

Dixon, J. (1987). Ecology and management of improved, unimproved and reverted hill grasslands in mid Wales. *Agriculture and Conservation in the Hills and Uplands* (Ed. by M. Bell & R.G.H. Bunce), pp. 32–37. Institute of Terrestrial Ecology, Grange-over-Sands.

Drabble, P. (1985). *What Price the Countryside?* Joseph, London.

Edwards, A.M. (1986). *An Agricultural Land Budget for the UK.* Wye College, Ashford.

Evans, S. & Felton, M. (1987). Hill Livestock Compensatory Allowances and upland management. *Agriculture and Conversation in the Hills and Uplands* (Ed. by M. Bell & R.G.H. Bunce), pp. 66–72. Institute of Terrestrial Ecology, Grange-over-Sands.

Fennel, R. (1987). Reform of the CAP: shadow or substance? *Journal of Common Market Studies,* **26,** 61–77.

Fraser, A. (1954). *Sheep Farming.* Crosby Lockwood, London.

Gould, L. (1986). *Changes in Land Use in England, Scotland and Wales 1985 to 1990 and 2000 report.* Report to Nature Conservancy Council, Laurence Gould Consultants, Warwick.

Hansard (1987). *Written Reply by John Mackay to Parliamentary Question from Archie Kirkwood.* Hansard, 29 April 1987, Column 146.

Harvey, D.R. (1985). *Milk Quotas: Freedom or Serfdom?* Centre for Agricultural Strategy, Reading.

Highwater, J. (1976). *Fodor's Indian America.* Hodder & Stoughton, London.

Hodge, I. (1985). *The Changing Countryside: a Review of Research.* Economic and Social Research Council, London.

Houston, G. (1987). *An Interim Assessment of the IDP for Agriculture and Fish Farming in the Western Isles.* Report commissioned by the Highlands and Islands Development Board, Inverness.

Hunter, J. (1987). Against the grain. *Scotland 2000* (Ed. by K. Cargill), pp. 61–86. BBC Publications, London.

Lowe, P., Cox, G., MacEwan, M., O'Riordan, T. & Winter, M. (1986). *Countryside Conflicts: the Politics of Farming, Forestry and Conservation.* Gower, Aldershot.

MacEwan, A. & MacEwan, M. (1982). *National Parks: Conservation or Cosmetics.* Allen & Unwin, London.

MacEwan, A. & MacEwan, M. (1987). *Greenprints for the Countryside: the Story of Britain's National Parks.* Unwin, London.

MacEwan, M. & Sinclair, G. (1983). *New Life for the Hills.* Council for National Parks, London.

McLaughlin, B.P. (1987). The rhetoric and the reality of rural deprivation. *Journal of Rural Studies,* **2,** 291–307.

MAFF (1975). *Food from Our Own Resources.* Command 6020, HMSO, London.

MAFF (1983). *The Livestock Units Handbook.* Booklet 2267, Ministry of Agriculture, Fisheries & Food Publications, Alnwick.

Manson, B. (1987). *The Institutional Basis of Rural Development.* Arkleton Trust, Langholm, Dumfries.

Mowle, A. (1986). *Nature Conservation in Rural Development.* Focus on Nature Conservation No.18, Nature Conservancy Council, Peterborough.

NCC (1984). *Nature Conservation in Great Britain.* Nature Conservancy Council, Peterborough.

NCC (1986a). *Agriculture and Environment in the Outer Hebrides.* Report to the European Commission, Nature Conservancy Council, Edinburgh.

NCC (1986b). *Nature Conservation and Afforestation in Britain.* Nature Conservancy Council, Peterborough.

NCC (1987). *Conversion and Extensification of Production: Implications and Opportunities for Nature Conservation.* Nature Conservancy Council, Peterborough.

North, J.J. (1986a). The new realities — threat or opportunity? *RURAL: 4th Annual Conference* (Ed. by F. Raymond), pp. 6–10. RURAL, Stonesfield, Oxfordshire.

North, J.J. (1986b). 13th Bawden Lecture. Use and management of the land: current and future trends. *Proceedings of the British Crop Protection Conference,* **1,** 3–14.

Norton-Taylor, R. (1982). *Whose Land is it Anyway?* Turnstone Press, London.

Parker, K. (1984). *The Tale of Two Villages.* Peak Park Planning Board, Bakewell.

Parry, M. (1982). *Surveys of Moorland and Roughland Change.* Department of Geography, Birmingham University.

Parry, M. & Sinclair, G. (1985). *Mid-Wales Uplands Study.* Countryside Commission, Cheltenham.

Pye-Smith, C. & North, R. (1984). *Working the Land.* Temple Smith, London.

Pye-Smith, C. & Rose, C. (1984). *Crisis and Conservation.* Penguin, London.

Redcliffe-Maud, Lord (1969). *Local Government in England.* Report of Royal Commission, Command Paper 4040, HMSO, London.

RSPB (1984). *Hill Farming and Birds.* Royal Society for the Protection of Birds, Sandy.

RSPB (1987). *Forestry in the Flow Country of Sutherland and Caithness.* Royal Society for the Protection of Birds, Sandy.

Sayce, R.B. (1984). *Integrated Rural Development: the Radnor/Eden Study.* Rural Planning Research Trust, Didcot.

Scott, J. (1942). *Report of the Committee on Land Utilisation in Rural Areas.* Command 6378, HMSO, London.

Shoard, M. (1980). *The Theft of the Countryside.* Temple Smith, London.

Shoard, M. (1987). *This Land is Our Land: the Struggle for Britain's Countryside.* Paladin, London.

Sinclair, G. (1984a). *The Upland Landscapes Study.* Report to Countryside Commission, Environment Information Services, Dyfed.

Sinclair, G. (1984b). *Agricultural Changes in the Less Favoured Areas of the UK.* Environment Information Services, Dyfed.

Slee, R.W. (1981). Agricultural policy and remote rural areas. *Journal of Agricultural Economics,* **32,** 113–121.

Smith, M. (1985). *Agriculture and Nature Conservation in Conflict — the Less Favoured Areas of France and the UK.* Arkleton Trust, Langholm, Dumfries.

Smith, R.S. (1988). Farming and the conservation of traditional meadowland in the Pennine Dales Environmentally Sensitive Area. *This volume.*

Stewart, P. (1985). British forestry policy — time for a change? *Land Use Policy,* **2,** 16–29.

Stewart, P. (1986). *Against the Grain.* Report commissioned by Council for the Protection of Rural England, London.

Tinker, P.B. (1988). Efficiency of agriculture industry in relation to the environment. *Environmental Management in Agriculture* (Ed. by J.R. Park), pp. 7–20. Belhaven Press, London.

Tompkins, S.C. (1986). *The Theft of the Hills: Afforestation in Scotland.* Ramblers Association, London.

Tracy, M. (1982). *Agriculture in Western Europe,* 2nd edn. Granada, London.

Tracy, M. (1987). *Structural Policy under the CAP.* Arkleton Trust, Langholm, Dumfries.

Traill, B. (1982a). The effect of price support policies on agricultural investments, employment, farm incomes and land values in the UK. *Journal of Agricultural Economics,* **33,** 369–385.

Traill, B. (1982b). Taxes, investment incentives and the cost of agricultural inputs. *Journal of Agricultural Economics,* **33,** 1–12.

TRRU (1981). *The Economy of Rural Communities in the National Parks of England and Wales.* Research Report No. 47, Tourism & Recreation Research Unit, University of Edinburgh.

Ulbricht, T. (Ed.) (1986). *Integrated Rural Development — Proceedings of a European Symposium.* Nationale Raad voor Landbouwkundig, Onderzoek.

Whitby, M.C. (1977). Subsidy shifts across the store sheep market. *Journal of Agricultural Economics,* **28,** 1–9.

Woods, A. (1984). *Upland Landscape Change: a Review of Statistics.* Countryside Commission, Cheltenham.

Farming and the conservation of traditional meadowland in the Pennine Dales Environmentally Sensitive Area

ROGER S. SMITH

Department of Agricultural and Environmental Science, University of Newcastle upon Tyne, Newcastle upon Tyne NE1 7RU

SUMMARY

1 The Pennine Dales Environmentally Sensitive Area (ESA) comprises eight separate blocks of pasture and meadowland in the North Pennines and the Yorkshire Dales. The main nature conservation objective for the meadowland is the maintenance of its floristic diversity.

2 The management prescription proposed for this ESA is discussed in the context of the grassland types and the evaluation for their conservation interest, the effects of fertilizers and grazing on the botanical composition of the sward and the relevance of proposed cutting dates.

3 The prescription is seen as an inevitable compromise between the urgent need for conservation measures and the need for adequate scientific information.

INTRODUCTION

The Agriculture Act 1986 enables the Agriculture Ministers in the UK to designate Environmentally Sensitive Areas (ESAs) where traditional farming methods are necessary to maintain the wildlife, landscape and archaeological interest. Each area is set up separately, boundaries are identified and a management prescription is defined. By June 1988 there were 18 ESAs in the UK; five in Scotland (Breadalbane, Stewartry, Whitlow/Eildon, the machair of the Uists and Benbecula and Loch Lomond), one in Northern Ireland (Mourne), two in Wales (Cambrian Mountains and the Lleyn Peninsula) and ten in England (Somerset Levels, West Penwith, South Downs, Broads, North Peak, Shropshire Borders, Test Valley, Breckland, Suffolk River Valleys and Pennine Dales).

This paper describes the wildlife interest in the meadows of the Pennine Dales ESA and the ecological basis for the management prescription developed to maintain this interest. Emphasis is placed upon the use of existing ecological knowledge where there is considerable urgency over the conservation issues. Improvement of the management prescription is then possible through the implementation of medium- to long-term monitoring and research programmes.

The particular interest of the Pennine Dales ESA centres on the traditionally managed hay meadows, although the conservation of stone walls, field barns and woodland is also important. The wildlife interest of the meadows is due to the very high diversity of plant species in vegetation which also shows variation related to drainage and soil pH. The ESA consists of eight separate areas: Weardale (2780 ha), Rookhope (460 ha) and Teesdale (3270 ha), in the North Pennines Area of Outstanding Natural Beauty, and Swaledale with Arkengarthdale (5570 ha), Dentdale with Deepdale (1364 ha), Upper Wharfedale (1400 ha), Langstrothdale (460 ha) and Waldendale (850 ha), in the Yorkshire Dales National Park. Within these areas farmers may voluntarily enter into a five-year agreement with the Ministry of Agriculture to manage their land in accordance with the ESA management prescription. For this they receive an annual payment of £100 ha^{-1}.

Gradients of altitude, soil drainage and soil pH occur throughout the ESA. The management superimposed upon these natural environmental gradients varies from traditional low input/low output farming to that which uses high inputs of mineral fertilizer on reseeded swards to give high yields of hay or silage. The latter gives uniform grass-dominated swards with few herbs, and tends to be practised on the better, more productive land in the lower reaches of the dales. Traditional management is concentrated in the Dale heads. Elsewhere it is often associated with elderly farmers who farm along more traditional lines.

The range of vegetation types has been described by J Rodwell (National Vegetation Classification). As well as a general 'old grassland' type, characterized by *Cynosurus cristatus* and *Centaurea nigra*, he recognizes a very distinctive type of 'northern' meadow, characterized particularly by the presence of *Geranium sylvaticum* and *Anthoxanthum odoratum*. Local variants of these types have been recognized (Alcock 1982; Smith 1985) to give a range of meadows, including (i) calcareous meadows on thin limestone soils; (ii) wet, slightly acid meadows on boulder clay or peat at the dale heads; (iii) neutral grassland with three variants, on damp alluvial soils in the dale bottoms — the 'northern' meadow, on mineral soils on the dale sides — the 'buttercup' meadow, and on dry sandy soils in valley bottoms; (iv) heavily fertilized 'northern' meadows; and (v) reseeded grassland dominated by *Lolium perenne* and *Phleum pratense*. Species nomenclature follows (Clapham, Tutin & Warburg 1981).

This range of grassland types is defined from an analysis of records of the relative abundance of vascular plant and pteridophyte species in 1 m^2 quadrats, located at random from visually uniform areas of vegetation (communities). The extent of each type is not known because such surveys have not been extensively carried out in any of the dales, although stratified sample surveys by Bunce *et al.* (1985) suggested that there were 1074 fields with diverse grass swards in the Yorkshire Dales National Park. Attempts by the Nature Conservancy Council (NCC) to improve on this estimate in their 'phase 1' surveys have used species thought to be indicative of the main differences between traditional and modern swards. A list of 34 species characteristic of old meadow communities, plus 13 more rare species, is used to grade meadows directly for their conservation value (Table 1). The species are not weighted, the assessment being based on a count of the number of species in this list. If a meadow contains less than 5 of these indicator species it is considered to be of little value for wildlife conservation. The

TABLE 1. Indicator species used by the Nature Conservancy Council to identify herb-rich meadows in the Pennine Dales (after Alcock 1982). Species characteristically associated with different vegetation types (after Smith 1985): 1, meadows on thin limestone soils; 2, slightly acidic, wet meadows on peat or boulder clay; 3, 'buttercup' meadows; 4, 'northern' meadows

Species characteristic of old meadow communities:

Ajuga reptans[1,2,3]	*Geum rivale*
Alchemilla spp.[4]	*Hypochoris radicata*[1,2]
Alopecurus geniculatus	*Knautia arvensis*
Anemone nemorosa[2]	*Lathyrus pratensis*[4]
Avenula pratense	*Leontodon hispidus*[1]
A. pubescens[1]	*Lotus corniculatus*[1,2]
Briza media[1,2]	*Luzula campestris*[1]
Caltha palustris[2]	*Lychnis flos-cuculi*
Carex caryophyllea	*Plantago media*
C. flacca[1,2]	*Primula veris*
C. panicea[2]	*Prunella vulgaris*[1,2,3]
Conopodium majus[4]	*Sanguisorba minor*
Euphrasia spp.[1,2,3]	*S. officinalis*
Filipendula ulmaria[2,3]	*Saxifraga granulata*
Galium verum	*Stachys officinalis*[1]
Geranium pratense	*Stellaria graminea*
G. sylvaticum[4]	*Succisa pratensis*[1,2]

Rare species of old meadows:

Botrychium lunaria	*Meum athamanticum*
Cirsium helenioides	*Ophioglossum vulgatum*
Dactylorhiza fuchsii	*Platanthera bifolia*
D. maculata	*Polygonum bistorta*
D. majalis subsp. *purpurella*	*Primula farinosa*
Gymnadenia conopsea	*Trollius europaeus*
Listera ovata	

presence of between 5 and 10 species suggests that there may well be some wildlife interest. The greatest value is attached to meadows which have more than 10 indicator species. Alcock (1982) found only 185 meadows, with a total area of 291·7 ha, with more than 10 indicator species in the Yorkshire Dales National Park.

The method is very dependent upon the number of vegetation types within a meadow, as indicated by the presence in Table 1 of species associated with a range of environmental conditions. There are wetland species, such as *Alopecurus geniculatus, Caltha palustris, Lychnis flos-cuculi* and *Filipendula ulmaria,* damp alluvial grassland species such as *G. sylvaticum* and *Sanguisorba officinalis,* dry grassland species such as *Galium verum,* and calcareous indicator species such as *Plantago media, Sanguisorba minor* and *Briza media.* In a survey of 30 meadows in Teesdale the NCC have shown that three or four vegetation types are found in a single field. These included uncut banks and wet flushes as well as the area cut for hay.

TABLE 2. The extent of meadowland of varying conservation interest in the Pennine Dales ESA (from maps provided by the Nature Conservancy Council)

Area	Improved meadow (less than 5 indicator spp.)		Intermediate meadow (5–10 indicator spp.)		SSSI quality meadow (more than 10 indicator spp.)	
	Area (ha)	% ESA	Area (ha)	% ESA	Area (ha)	% ESA
Weardale and Rookhope	460	14·2	159	4·9	47	1·4
Swaledale with Arkengarthdale	1170	21·0	350	6·3	170	3·1
Wharfedale	no data		54	3·9	60	4·3
Langstrothdale	no data		26	5·5	46	10·0
Waldendale	Limited data owing to access difficulties. 14 ha (1·7% of the dale) of SSSI quality					
Dentdale with Deepdale	Limited data with only 13·4 ha (1·0%) of SSSI quality meadow known					
Teesdale	No data from systematic surveys but some sites outside the NNR are known					

Swaledale with Arkengarthdale has been the most completely surveyed by the NCC for meadowland indicator species, with about 90% of the meadows visited and categorized. It is evident that traditional meadows of value for their wildlife interest are relatively infrequent (Table 2), although more may be found as surveys progress (Clayden & Slater 1986).

The conservation objectives for the management of meadowland are primarily directed to the conservation of the range of traditional plant communities with their high plant species richness. This is greatest in the more uncommon vegetation types on limestone, peat and boulder clay with a mean of 31 species m^{-2} and up to 42 species m^{-2} (Smith 1985). The variants of Rodwell's *C. cristatus–C. nigra* 'old grassland' range in species richness from 17 to 41 species m^{-2}, with a mean number of about 27 to 28 species m^{-2}. The *G. sylvaticum* 'northern' meadow has from 14 to 34 species m^{-2}, with a mean species richness of 22 or 24 species m^{-2} from different surveys (Alcock 1982; Smith 1985). Improved variants of the different neutral grasslands have about 20 species m^{-2} with a preponderance of competitive grasses such as *L. perenne*, *Alopecurus pratensis* and *Poa trivialis*.

The NCC place particular emphasis on the maintenance of 'natural' vegetation in meadows. Their more detailed 'phase 2', quadrat-based surveys evaluate meadows using a triple weighting system based on 263 species that are thought to be typical of more 'natural' grassland (Clayden & Slater 1986). Those of particular note are *Botrychium lunaria*, *Cirsium helenioides*, *Coleoglossum viride*, *G. sylvaticum*, *Gymnadenia conopsea*, *Ophioglossum vulgatum*, *Polygonum bistorta* and *Trollius europaeus*. The weightings for such species are summed and the highest values are used to define the most 'natural' meadows. These

TABLE 3. The correlation coefficients between NCC 'naturalness' scores and the species richness of various grassland types (calculated from data presented by Alcock (1982)). Significance is indicated by ***$P \leqslant 0.001$, **$0.01 \geqslant P > 0.001$, and *$0.05 \geqslant P > 0.01$.

Meadow type	Number of species in community	Sample size	Number of species m^{-2}	Sample size
Improved 'old grassland'	0·714***	50	0·193 ns	50
'Old grassland'	0·859***	27	0.522*	23
'Northern' meadow	0·817***	99	0·449***	90
Improved 'northern' meadow	0·767***	36	0·490**	35
Ryegrass meadow	0·747***	16	0·563*	16
Wet, slightly acid meadow	0·911***	27	0·496**	26

meadows are also the most species-rich as naturalness and species richness are closely correlated (Table 3), enabling either to be used to identify meadows of particular conservation significance. The direct use of species richness, however, avoids arguments as to exactly what constitutes a 'natural' species (Usher & Gardner 1988), a dubious concept as all meadows are anthropogenic and the most desirable, in conservation terms, are better defined as traditional.

The maintenance of traditional swards is usually sought through the continuation of the traditional management regime, where the predominant use of the meadows is for the grazing of sheep and cattle during the autumn, winter and spring. Cattle are often put in immediately after the hay is cut, to eat the uncut edge vegetation. The aftermath growth will be grazed intensively by cattle in the autumn and by sheep in November, when the ewes and rams are put in for mating. Grazing will also be heavy in the late spring during the lambing period. Livestock are removed from the meadows during mid-May to allow the growth of a hay crop. This growth is enhanced by the addition of dung from the barns in which the cattle have been overwintered. In the past the barn in each meadow was used to store the hay, this being placed in the loft above the byres. With the building of larger, modern farm buildings around the farmsteads these field barns are now often redundant. The nutrient cycle is, therefore, usually a closed system as far as the farming regime is concerned. The date of hay-cutting each year is dictated by the weather. This can be any time between late June and early September, with farms in the dale heads tending to cut later than those lower down the dale.

ESA MANAGEMENT

The management prescription

The successful conservation management of the meadows depends upon the correct identification of the ecological factors responsible for the coexistence of many species and the incorporation of these factors into a list of approved and proscribed actions for the farmer or land manager to follow (the prescription).

Current ecological models explain the coexistence of species through the spatial or temporal partitioning of resources or regeneration opportunities (equilibrium models) or through dynamic gradients caused by disturbance (non-equilibrium models) (Grime *et al.* 1987). These involve a consideration of the phenology of the constituent species, the significance of the relative absence of the more vigorous, competitive species, the heterogeneity of the soil environment, the level of soil fertility, soil disturbance and soil pH. Those parts of the current management prescription that are directed towards maintaining the diversity of the meadow swards oblige the farmer

(i) to maintain grassland and not to plough, level or reseed the land, and to cultivate only with a chain harrow or a roller;

(ii) to exclude stock from the meadows at least seven weeks before the first cut for hay or silage;

(iii) not to cut grass for hay or silage in any year before
    (a) 1 July in Dentdale with Deepdale,
    (b) 8 July in Wharfedale, Langstrothdale or Waldendale, and
    (c) 15 July in Swaledale with Arkengarthdale, Teesdale, Weardale and
        Rookhope;

(iv) to wilt and turn grass before removal if it is cut for silage;

(v) to retain existing fertilizer regimes as long as they are below 20 units of nitrogen, 10 units of phosphate and 10 units of potash per acre per year (this is equivalent to 24·9 kg ha^{-1} of nitrogen and 12·4 kg ha^{-1} of phosphate and potash);

(vi) not to apply slurry or poultry manure to the meadows;

(vii) to apply only light dressings of farmyard manure, with a total of no more than 10 tons per acre (25·19 t ha^{-1});

(viii) not to use pesticides;

(ix) to use herbicides only to control bracken, nettles, spear thistle, creeping or field thistle, curled dock, broadleaved dock or ragwort (they should only be applied with a weedwipe or by spot treatment, and only Asulam should be used to control bracken);

(x) not to use lime, slag or any other substance to reduce soil acidity; and

(xi) neither to install new drainage systems nor substantially to modify existing drainage systems.

Cutting dates for hay or silage

The cutting dates specified in section (iii) of the management prescription are a matter of particular concern to those farmers who might enter into an ESA agreement. There are few detailed data on the phenology of the swards. The relevance of the stipulated dates can, therefore, only be assessed against general ecological considerations. The variation in the timing of growth and flowering of the different species is evident from general observation. Species such as *Anemone nemorosa* flower early (late May), *L. perenne* and *Dactylis glomerata* flower late (mid-July), and *G. sylvaticum* flowers in late June. The balance of species in the sward at hay time is very different from the balance earlier on. All species should persist in the sward as long as they are able to set seed, as long as their seed bank persists in the soil and as long as germination sites continue to be created. The frequency with which some species set seed is likely to be dictated in part by the timing of the hay cut. If this is before seed set over an appreciable length of time then those species which do not have a persistent seed bank are likely to disappear from the sward. Annual species will probably be the first to disappear if there is no bare ground for seed germination. The management prescription attempts to ensure that seed is set by excluding stock from the meadows for at least seven weeks prior to cutting and delaying the date of the cut to early July. This stipulation of a cutting date is, however, probably unnecessary. Most of the early-flowering species will set seed in each year. It is only the late-flowering species that are endangered by an early cut. If these are long-lived perennials there will probably be sufficient opportunity for one individual to set seed in its lifetime when natural variation in the climate occasionally delays the hay cut.

More data on the ecology of the individual species is needed if the impact of different cutting dates is to be properly assessed. Data are particularly needed on the longevity of the species, the size of their seed banks and their seed output. However, one can argue that cutting dates in the past have been dictated by the climate, in line with the old adage 'make hay while the sun shines'. This has pro-duced the characteristic species composition that is valued by conservationists. It has meant that species such as *G. sylvaticum* are characteristic members of the vegetation, whereas *Geranium pratense*, which flowers later, is absent from the meadows and restricted to the uncut edges of fields and roadside verges. Traditional management did not use mineral fertilizers and a late cutting date was probably inevitable because of the slow growth of the vegetation. However, the prescription allows mineral fertilizer to be applied, when an early cutting date to remove the more vigorous, competitive species might be advantageous for the conservation interest.

The effects of grazing livestock

Whilst the management prescription requires the farmer to cut the sward, no mention is made of grazing, despite the fact that it is a major use of meadowland during autumn, winter and spring. The little direct evidence that exists shows that vegetation change is rapid when both cutting and grazing cease; the most vigorous species take over and diversity is lost. Shelterbelts recently planted in meadowland provide some evidence for this. It has to be assumed that the vegetation in these shelterbelts was originally the same as that in adjacent meadows and that the notch planting of the trees does not in itself influence the vegetation between the trees in the years immediately after planting (Table 4). *D. glomerata* and *Holcus lanatus* become dominant, contributing about 75 per cent to the vegetation cover in the uncut, ungrazed exclosures.

TABLE 4. A comparison between a traditionally managed meadow and grassland in a two-year-old shelterbelt

	Shelterbelt		Meadow		Significance of difference between means
	Mean	SE	Mean	SE	
Number of 25 × 25 cm quadrats	30		36		—
Species richness	5·9	0·45	19·9	0·62	$P < 0.001$
Diversity (Shannon index)	0·52	0·03	1·01	0·02	$P < 0.001$
Equitability	0·70	0·03	0·78	0·01	$0.05 > P > 0.01$

Grazing reduces competitive grasses such as *D. glomerata* and enables the less competitive herbs to survive. The grasses make a relatively large contribution to the sward in the autumn and winter and in the winter many of the herbs are dormant (Duffey *et al.* 1974). Heavy grazing results in the different species being eaten in the proportion in which they occur in the sward (Arnold 1964). The grasses, therefore, bear the bulk of the autumn and winter grazing pressure. This also occurs when grazing is more selective, as some of the more competitive species in the sward are also very palatable, e.g. *D. glomerata.* The very act of cutting the sward for the hay crop also selectively debilitates those species that comprise the hay crop. The competitive grasses are often present in the greatest quantity at this time.

Spatial variation in soil conditions is also probably created by dung and urine patches deposited by grazing livestock (MacLusky 1960). Its significance for the coexistence of different plant species is suggested by the fact that variation in soil nutrient status in 5 cm diameter soil cores from traditional meadowland has been found to be considerable (Table 5). Ordination of plant species abundance data from 20 cm² circular quadrats in old, well-drained, meadowland at Ravenstonedale, Cumbria, shows that the main gradient of species change (the first ordination axis) is significantly correlated with the available nitrogen level in the soil ($r = 0.78$, $P < 0.001$). Similar detailed vegetation and soil correlations have been shown by Grime (1979) for grassland on thin limestone soils, the species distributions being related to the depth of soil above the fissures and depressions in

TABLE 5. The soil nutrient status of a traditionally managed Dales buttercup meadow at Bowberhead, Ravenstonedale, Cumbria (data provided by P. Hirst)

	Available nitrogen (ppm)	Extractable phosphorus (ppm)	Extractable potassium (ppm)
Range	134–424	7–36	46–223
Mean	260·4	16·1	111·8
SD	76·1	6·4	41·6

the limestone bedrock. This results in soil patches of varying droughtiness and is likely to be important in explaining the very high diversity of species in the limestone hay meadows of upper Wharfedale.

Considerable disturbance occurs to the soil surface during both haymaking and grazing. Bare soil is created by the wheels of agricultural equipment and the scarifying effect of the machinery used to turn the hay. The subsequent trampling and soil disturbance that occurs with grazing livestock opens up more sites later in the year. Some species take advantage of these niches at different seasons. Those which have a chilling requirement, e.g. *Rhinanthus minor,* will delay germination until the spring, taking advantage of bare soil created over winter by grazing livestock and earthworm activity.

The effects of soil pH

The management prescription's restriction on the use of lime and basic slag, to prevent soil pH being raised, is not supported by evidence from the long-term Park Grass Experiment at Rothamsted. Modern alterations to the old fertilizer regime were made by Thurston, Williams & Johnston (1976) to investigate the effects of lime on the swards. The considerable reduction in species richness on those plots originally fertilized with ammonium sulphate was quickly reversed as the soil pH increased. The general ecological model developed by Grime (1979) associates high soil pH with high species richness of the vegetation. The basis of his 'humpbacked' model is that high numbers of species per unit area are found at moderate levels of production ($3\cdot5$–$7\cdot5$ t ha^{-1}), the numbers at these levels increasing with soil pH. Values greater than a pH of $6\cdot5$ are associated with a species richness in excess of 40 species m^{-2}. The range of soil pH in meadowland in the Yorkshire Dales is generally between $5\cdot3$ and $6\cdot3$. This needs to be maintained by the addition of lime, particularly when the high rainfall in the upper dales inevitably leads to the long-term leaching of bases, with the eventual production of an organic mor horizon in the soil and the build-up of peat, a process which can occur directly on carboniferous limestone.

The use of fertilizers

Nutrient availability has been shown to be important for the maintenance of a number of grassland types. Relatively low levels of nitrogen and phosphorus are associated with chalk downland (Smith, Elston & Bunting 1971), *Kobresia*-rich swards in Teesdale (Jeffrey & Pigott 1973), dune grassland (Willis 1963), upland *Festuca–Agrostis* grassland and *Molinia* moorland (Jones 1967). Long-term agricultural experiments (Brenchley & Warrington 1958; Pawson 1960; Jones 1967) have been used to show how the species composition of grasslands can be manipulated by fertilizer applications (Harper 1971; Rorison 1971; Smith 1987).

The control of fertilizer use is, therefore, an important element of the ESA management prescription and this can be seen in the sections which deal with mineral fertilizers, farmyard manure, slurry and poultry manure. It is often assumed that the best regime for maximizing the species diversity is the complete absence of fertilizers, with the exception of light dressings of organic manures. Such manures break down slowly and do not provide the large flush of nitrogen and phosphorus in the early spring that is available with mineral fertilizers and which promotes the growth of the more vigorous competitive species. The management prescription recommends the use of organic manures but excludes poultry manures and slurry because of their large amounts of available nutrients. Appropriate organic manures should prevent the nutrient status becoming too low, when the 'humpbacked' model (Grime 1979) would predict that species will be lost.

Meadow soils are moderately fertile, with hay yields varying, according to the management regime, from 3·5 to 7·0 t ha^{-1}. Without additional fertilizer the growth after the hay cut varies between 1·9 and 2·9 t ha^{-1} (Shiel & Batten 1988). Soils from meadows in the Yorkshire Dales contain about 6 per cent organic matter, with carbon:nitrogen ratios between 16 and 19 and available nitrogen levels of between 118 and 167 mg l^{-1} (Smith 1985). The maintenance of this moderate fertility is important to the farmer for reasonable yields of hay as well as for the maximization of the species richness of the sward. The problem is to decide how much additional mineral fertilizer can be used without increasing the abundance of competitive species, with the associated loss of other species. This may differ according to the needs of landscape and wildlife conservation. The former may be satisfied by colourful meadows which may not be as diverse as those required by the latter. There is little information about the detailed effects of different fertilizer regimes on the detailed species composition of upland meadows. Direct experimentation on this type of grassland has been mainly agricultural and directed towards the measurement of yield rather than towards plant species composition and diversity (Davies 1969).

A fertilizer trial on meadow turfs

Methods

A fertilizer trial was set up in 1984, well before ESAs were designated, to test the effects of fertilizers on turfs taken from a *G. sylvaticum/D. glomerata* 'northern' meadow near Sedbergh, Cumbria. The fertilizer levels were chosen in 1984 to represent the standard fertilizer recommendations then given by the MAFF to Dales farmers (3 cwt acre^{-1}, approximately 377 kg ha^{-1}, 20:10:10 NPK fertilizer). The intention was to detail the changes in species composition, species richness and diversity that occurred with this and lower inputs. The ESA management prescription has subsequently superseded this with its maximum recommended fertilizer level set at one-third of the previously recommended level.

The turfs were 25 cm square in size and were taken with 25 cm depth of soil. Four levels of nitrogen (0, 25, 50 and 75 kg ha^{-1}, applied as ammonium nitrate) and three of phosphorus (0, 17 and 34 kg ha^{-1}, applied as calcium hydrogen orthophosphate) were used on three replicates of each treatment. The hay crop was cut in late June or early July and hand-sorted into its constituent species. These were air-dried at 70 °C and weighed. Changes in the turfs were assessed from the species richness (number of species per turf), species diversity (Shannon index) and species composition, the latter by means of an ordination (Hill 1979).

Results

There were no significant differences between the treatments in any one year, except for 1985 when there were fewer species at the highest nitrogen level (0·05 > P > 0·01). The species richness dropped progressively throughout the four years from about 18 to 13 species per turf (Fig. 1). It was not significantly

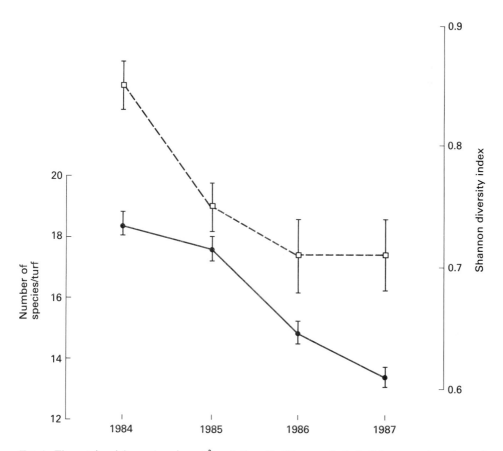

FIG. 1. The species richness (species m^{-2}) and diversity (Shannon index) of the vegetation of a turf trial from a *Geranium sylvaticum–Dactylis glomerata* 'northern' meadow. The lines represent the Shannon diversity index (– – – – □ – – –) and the species richness (——— • ———). The total length of the error bars is twice the standard error

different between 1984 and 1985, but was between 1985 and 1986 ($P < 0.001$) and 1986 and 1987 ($0.01 > P > 0.001$). The Shannon diversity index also dropped, levelling off in 1987, 27 per cent lower than it had been at the start of the experiment. The only significant drop in the Shannon index in one year occurred between 1984 and 1985 ($0.01 > P > 0.001$).

Species found only in the first year were *Prunella vulgaris, Euphrasia officinalis, R. minor, Cerastium fontanum, Cardamine pratensis, Bellis perennis, Avenula pubescens* and *Bromus hordeaceus*. These were lost from the turfs and there was a rise in the relative abundance of species such as *Rumex acetosa* and *D. glomerata*. The mean ordination position of each of the nitrogen and phosphorus levels in each year shows a two-way trend (Fig. 2). There is an ordering of the nitrogen and phosphorus treatments along the second axis, although this breaks down slightly in the latter part of the experiment, and a time trend along the first axis. This is greatest between the first and second years and has ceased after the third year. The corresponding species ordination is given in Fig. 3.

DISCUSSION

In theory the ideal conservation management plan should be based upon a complete understanding of the ecological factors that influence the conservation interest of a site. This is a counsel of perfection that is probably never attainable. Conservationists have to work with current ecological knowledge because of the urgency of many conservation issues, where rapid changes in land-use practice are causing major losses of species and habitats (Nature Conservancy Council 1984). Grasslands are a good example where the causes of floristic diversity have been attributed to a number of factors and where the relative importance of each one is still debated (Grime *et al.* 1987). The conservation of traditional meadowland cannot wait for the resolution of this ecological debate. Management techniques have to be developed from current ecological theory, examples from similar sites and ecosystems, anecdotal information and experimentation where possible. Management experiments need to be geared towards the rapid resolution of questions which ask 'What happens if we change x?' For example, what happens to the vegetation diversity if we change the grazing or the fertilizer regime? The variables measured in such experiments or in monitoring schemes should be capable of being related to the features that conservationists are concerned to maintain, e.g. sward character and diversity. Such management trials are not directed towards the elucidation of basic ecological principles, and extrapolation outside the context of the experiment can lead to mistakes when different conditions prevail. Ecologists are not, therefore, always attracted to such experimentation, although conservationists need the data to help with management plans on specific sites.

The small-scale fertilizer experiment described is an example of a management trial. It had limited objectives directed to a specific problem. The use of turfs limited its relevance to the field situation but it was cheap to run and did

Axis 1

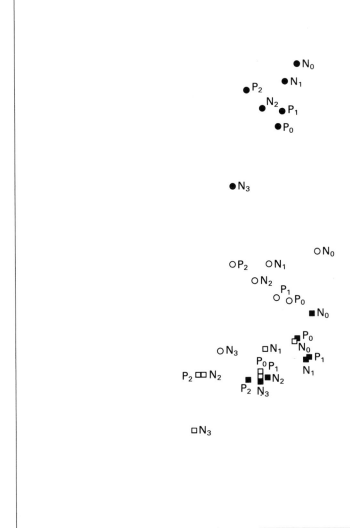

Axis 2

FIG. 2. Mean positions of the sample ordination scores for each nitrogen and phosphorus level in each year. Symbols are as follows:
N_0, control plots; N_1, 25 kg ha^{-1} nitrogen; N_2, 50 kg ha^{-1} nitrogen; N_3, 75 kg ha^{-1} nitrogen; P_0, control plots; P_1, 17 kg ha^{-1} phosphorus; P_2, 34 kg ha^{-1} phosphorus; •, 1984; ○, 1985; □, 1986; and ■, 1987

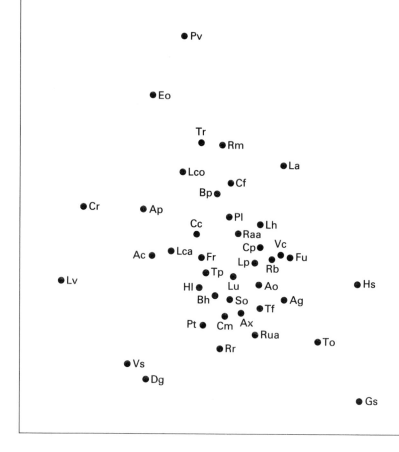

FIG. 3. Ordination of the species in the turf trial over the four years. Species abbreviations are as follows:

Ac	*Agrostis capillaris*	Lh	*Leontodon hispidus*
Ag	*Alchemilla glabra*	Lv	*Leucanthemum vulgare*
Ax	*Alchemilla xanthochlora*	Lco	*Lotus corniculatus*
Ao	*Anthoxanthum odoratum*	Lu	*Lotus uliginotus*
Ap	*Avenula pubescens*	Lca	*Luzula campestris*
Bp	*Bellis perennis*	Pl	*Plantago lanceolata*
Bh	*Bromus hordeaceus*	Pt	*Poa trivialis*
Cr	*Campanula rotundifolia*	Pv	*Prunella vulgaris*
Cp	*Cardamine pratensis*	Raa	*Ranunculus acris*
Cf	*Cerastium fontanum*	Rb	*Ranunculus bulbosus*
Cm	*Conopodium majus*	Rr	*Ranunculus repens*
Cc	*Cynosurus cristatus*	Rm	*Rhinanthus minor*
Dg	*Dactylis glomerata*	Rua	*Rumex acetosa*
Eo	*Euphrasia officinalis*	So	*Sanguisorba officinalis*
Fr	*Festuca rubra*	To	*Taraxacum officinale*
Fu	*Filipendula ulmaria*	Tp	*Trifolium pratense*
Gs	*Geranium sylvaticum*	Tr	*Trifolium repens*
Hs	*Heracleum sphondylium*	Tf	*Trisetum flavescens*
Hl	*Holcus lanatus*	Vc	*Veronica chamaedrys*
Lp	*Lathyrus pratensis*	Vs	*Vicia sepium*
La	*Leontodon autumnalis*		

provide some insight into management factors. The lack of any significant effect of the fertilizer was unexpected but may be attributed to the spatial heterogeneity of the vegetation and the size of the turfs; this points to soil heterogeneity as one possible mechanism for the maintenance of a high overall species richness in this vegetation and to the need for larger-scale experimentation. The time trend, unrelated to the fertilizer treatments, was also unexpected. Some of the species in the turfs in 1984–1985 are annuals that could not survive without the provision of bare soil. Others are relatively low-growing and are unlikely to have been able to withstand the competition for light in the dense vegetation which existed early in each year in the absence of grazing, e.g. *Trifolium repens, E. officinalis* and *B. perennis.* Generally the species that thrived throughout the experiment were those vigorous perennials that can compete successfully for light and nutrients and which have no short-term need for bare ground for seed germination. This time trend in all treatments suggests that other environmental factors are acting on all the turfs regardless of their fertilizer treatment. The most likely factors are the absence of soil disturbance by farm machinery and grazing livestock, and the cessation of predation by such grazing stock. This suggests that highly diverse swards might be maintained even with moderate fertilizer inputs as long as the competitive species are prevented from dominating the sward by appropriate grazing regimes and as long as the soil is sufficiently disturbed to create germination niches. There is no direct evidence for this.

Improvements to the ESA management prescription will come with further knowledge, both of the basic ecological mechanisms at work and of the impacts of different management techniques. The latter should enable management decisions to be made in the short term whereas the former can only come up with solutions in the medium to long term. Particular attention needs to be placed on the effects of soil pH, the effects of grazing under different fertilizer regimes and the effects of different cutting dates on sward composition. An understanding of the effects of these factors may enable the adoption of a tiered system of management incentives such as exists already for some of the other ESA areas, e.g. the South Downs. Enhanced payments could be made for more positive conservation gains through the creation of herb-rich, attractive swards from existing improved swards (Wells, Bell & Frost 1981) or for the optimal conservation management of existing swards through, for instance, the cessation of the use of any fertilizer other than organic manures.

ACKNOWLEDGMENTS

R. S. Shiel and W. Stelling helped to set up the turf trial, which was funded by a small grant from the Northumberland National Park Committee. The Ministry of Agriculture supplied maps of the ESA. Much assistance was given by various officers of the Nature Conservancy Council through the provision of access to in-house reports and maps of their 'phase 1' survey of the Yorkshire Dales

National Park, Weardale and Rookhope. P. Hirst provided the data on soil heterogeneity.

REFERENCES

Alcock, M.R. (1982). *Yorkshire grasslands: a botanical survey of hay meadows within the Yorkshire Dales National Park.* Project Report Number 10, England Field Unit, Nature Conservancy Council, Banbury.

Arnold, G.W. (1964). Factors within plant associations affecting the behaviour and performance of grazing animals. *Grazing in Terrestrial and Marine Environments* (Ed. by D.J. Crisp), pp. 133–154. Blackwell Scientific Publications, Oxford.

Brenchley, W.E. & Warrington, K. (1958). *The Park Grass Plots at Rothamsted, 1856–1949.* Rothamsted Experimental Station, Harpenden.

Bunce, R.G.H., Crawley, R.V., Gibson, R.A. & Pilling, R. (1985). *Composition of enclosed grasslands.* Yorkshire Dales National Park Committee, Bainbridge.

Clapham, A.R., Tutin, T.G. & Warburg, E.F. (1981). *Excursion Flora of the British Isles,* 3rd edn. Cambridge University Press, Cambridge.

Clayden, D. & Slater, I. (1986). *Survey and Evaluation of Mesotrophic Grassland in South Cumbria.* Nature Conservancy Council, Blackwell.

Davies, H.T. (1969). Manuring of meadows in the Pennines. *Experimental Husbandry,* **18,** 8–24.

Duffey, E., Morris, M.G., Sheail, J., Ward, L.K., Wells, D.A. & Wells, T.C.E. (1974). *Grassland Ecology and Wildlife Management.* Chapman & Hall, London.

Grime, J.P. (1979). *Plant Strategies and Vegetation Processes.* Wiley, Chichester.

Grime, J.P., Mackey, J.M.L., Hillier, S.H. & Read, D.J. (1987). Floristic diversity in a model system using experimental microcosms. *Nature,* **328,** 420–422.

Harper, J.L. (1971). Grazing, fertilizers and pesticides in the management of grasslands. *The Scientific Management of Animal and Plant Communities for Conservation* (Ed. by E. Duffey & A.S. Watt), pp. 15–23. Blackwell Scientific Publications, Oxford.

Hill, M.O. (1979). DECORANA — *a* FORTRAN *Program for Detrended Correspondence Analysis and Reciprocal Averaging.* Report of the Section of Ecology and Systematics, Cornell University.

Jeffrey, D.W. & Pigott, C.D. (1973). The response of grasslands on sugar-limestone in Teesdale to application of phosphorus and nitrogen. *Journal of Ecology,* **61,** 85–92.

Jones, L.I. (1967). *Studies on Hill Land in Wales.* Technical Bulletin No. 2, Welsh Plant Breeding Station, Aberystwyth.

MacLusky, D.S. (1960). Some estimates of the areas of pasture fouled by the excreta of dairy cows. *Journal of the British Grassland Society,* **15,** 181–188.

Nature Conservancy Council (1984). *Nature Conservation in Great Britain.* Nature Conservancy Council, Peterborough.

Pawson, H.C. (1960). *Cockle Park Farm.* Oxford University Press, London.

Rorison, I.H. (1971). The use of nutrients in the control of the floristic composition of grassland. *The Scientific Management of Animal and Plant Communities for Conservation* (Ed. by E. Duffey & A.S. Watt), pp. 15–23. Blackwell Scientific Publications, Oxford.

Shiel, R.S. & Batten, J.C. (1988). Redistribution of nitrogen and phosphorus on Palace Leas meadow hay plots as a result of aftermath grazing. *Grass and Forage Science,* **42,** 353–358.

Smith, C.J., Elston, J. & Bunting, A.H. (1971). The effects of cutting and fertilizer treatments on the yield and botanical composition of chalk turf. *Journal of the British Grassland Society,* **26,** 213–219.

Smith, R.S. (1985). *Conservation of Northern Meadows.* Yorkshire Dales National Park Committee, Bainbridge.

Smith, R.S. (1987). The effect of fertilizers on the conservation interest of traditionally managed upland meadows. *Agriculture and Conservation in the Hills and Uplands* (Ed. by M. Bell & R.G.H. Bunce), pp. 38–43. Institute of Terrestrial Ecology, Grange-over-Sands.

Thurston, J.M., Williams, E.D. & Johnston, A.E. (1976). Modern developments in an experiment on permanent grassland started in 1856: effects of fertilizers and lime on botanical composition and crop and soil analyses. *Annales Agronomiques*, **27**, 1043–1082.

Usher, M.B. & Gardner, S.M. (1988). Animal communities in the uplands: how is naturalness affected by management? *This volume.*

Wells, T.C.E., Bell, S. & Frost, A. (1981). Creating attractive grasslands using native plant species. Nature Conservancy Council, London.

Willis, A.J. (1963). Braunton Burrows: the effects on the vegetation of the addition of mineral nutrients to the dune soils. *Journal of Ecology*, **51**, 353–374.

Hill vegetation and grazing by domesticated herbivores: the biology and definition of management options

S. A. GRANT AND T. J. MAXWELL
Macaulay Land Use Research Institute, Bush Estate, Penicuik, Midlothian EH26 OPY

SUMMARY

1 Grazing by domesticated herbivores has a major impact on the dynamics and succession of upland vegetation. This paper examines the information needed to provide objective criteria for the management of grazing livestock with a view to managing the vegetation.

2 The grazing system is examined in terms of flow and partitioning of material. Factors which are altered as a result of grazing management decisions and which influence various process rates are enumerated. The importance of sward conditions as driving variables is highlighted.

3 It is suggested that a research approach based upon the control and manipulation of sward conditions would both advance understanding and provide a flexible basis for defining grazing management options.

4 Examples are given for studies of foraging behaviour, of sward responses to grazing and of intake and feeding value.

5 Ways in which the information gained can be collated to arrive at management decisions are outlined.

INTRODUCTION

The trends in agriculture and land use in the hills and uplands and their implications for ecological research have been reviewed by Eadie (1984). Three priority areas for research can be identified. First, there is a clear need to monitor vegetational change in the semi-natural vegetation of the hills and uplands. Second, there is a need to evaluate the costs and benefits of changing the emphasis of management between objectives for production and conservation. And, third, there is a need for knowledge and understanding of the ecology of soil–plant–animal relationships on which to base predictions about the consequences of management manipulation for vegetational change, most particularly grazing animal manipulation. It is with the third area that this paper is concerned.

In discussing the factors affecting vegetation, Ball *et al.* (1982) comment that the physical environment sets limits to the range of vegetation types which can occur, but that within these limits management modifies the type and occurrence of the vegetation. The broad trends in response to soil pH, drainage and grazing

201

pressure have received much attention over the years, and there have been attempts to quantify the stocking rates at which particular vegetation types might be maintained (King & Nicholson 1964; Institute of Terrestrial Ecology 1978; Ball *et al.* 1982; Miles 1985; Mowforth & Sydes, in press; Sydes & Miller 1988). Although such schemes are useful as pointers to likely trends in succession on a regional basis, they cannot be used to provide management guidelines at the level of the individual farm. In practice, stocking rates are quoted for whole farm enterprises, involving a range of inputs and management systems, so that the figures rarely reflect actual stocking levels on the semi-natural rough grazings. Furthermore, hill pastures are vegetationally heterogenous; the distribution of grazing activity on different vegetation types changes through the year (Hunter 1962) and is strongly influenced by the combination of types on an individual hill (Armstrong, Grant & Hodgson 1987). As yet we have an inadequate knowledge and understanding of the biology of grazing systems to predict the effects of an alteration in overall stocking rate on the balance of grazing pressure exerted on each of the sensitive component vegetation types. The aim of this paper is to discuss the research approach and the nature of the information needed to define the scope for manipulating the floristic composition of upland vegetation by grazing management.

RESEARCH APPROACH

Limitations of stocking rate studies

A major limitation of the stocking rate concept is that it equates animals to a unit area of land. As Hodgson (1985) has pointed out, this ignores the biological variation in space and time. The problems are exacerbated on semi-natural vegetation where the unfenced hill grazings are vegetationally heterogeneous and unique in character (Miles 1985). This is not to argue that stocking rate studies have no place on hill vegetation; much can be learned using this traditional approach in well-designed and properly monitored studies. However, there is a need to move forward from the basically descriptive and site-specific studies of the past (Malechek, Balph & Provenza 1986). An understanding of the biology of plant–animal relationships is seen as a vital stage in defining the scope for control and manipulation of grazing systems and is central to the philosophy which has been adopted over the years in our own institute (Hodgson & Grant 1981; Hodgson & Maxwell 1981; Hodgson 1985; Grant & Hodgson 1986).

The biology of grazing systems

The plant and animal interrelationships can be highlighted by viewing the grazing system in terms of the flow and partitioning of material along alternative pathways (Fig. 1). In this way attention is focused on the numerous attributes of the sward, which collectively describe the sward state and which have a direct impact on the various process rates. For example, herbage growth is influenced by

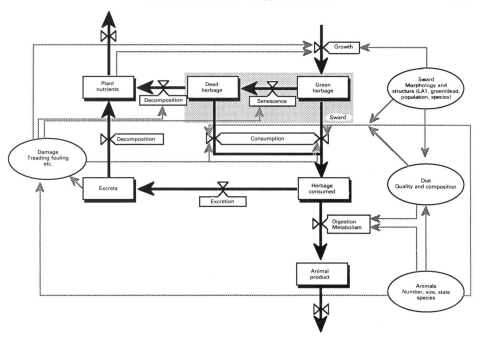

Fig. 1. Plant and animal interrelationships in grazing systems (adapted from Hodgson 1985). The bold lines and arrows indicate the flow and partitioning of material through the system; the factors which are altered by grazing management decisions and which influence the various rate processes (growth, consumption etc.) are circled; faint lines are used to indicate impacts and interrelationships.

sward attributes which affect light interception, ability to expand leaves, and the photosynthetic efficiency of those leaves; and herbage consumption is influenced by a variety of sward attributes, including the nature of the plant material (green or dead, species, digestibility) and canopy amount (height, biomass) and structure. Demand for food is determined by several animal factors (e.g. stocking rate, species and physiological state of the animal) which interact with the sward. This interaction influences diet selection and quality, which together determine the rate of consumption and herbage intake. The effect of stocking rate on herbage production and animal performance is largely mediated through changes in sward state. Sward state is influenced also by local variation in soils, climate and topography. In semi-natural vegetation these local factors, together with management history, contribute to the unique character of the vegetation and add to the problems of interpretation and extrapolation from stocking rate studies.

An alternative approach involves a detailed examination of the components of the system, taking particular account of sward conditions. Grazing experiments in which treatments are based on the direct manipulation of sward characteristics, previously identified to be of importance, offer better opportunities for progress.

The areas of focus for grazing research differ between intensive grazing systems on sown pastures and more extensive systems on semi-natural vegetation. In the former case food quality is high and pasture quantity limits output. Consequently, interest has focused on the ways in which manipulation of sward conditions affects the rates of herbage growth and consumption, losses to senescence and decomposition, and animal responses. Information of this sort has already resulted in improved grazing systems using sward surface height as a guide for grazing control (Hodgson, Mackie & Parker 1986).

On semi-natural vegetation the overall quantity of herbage is rarely limiting; the factors which require consideration are life form, seasonality of growth, and quality and quantity of herbage produced by the different plant species. Variation in these attributes between species results in uneven distributions of grazing activity in time and space, which have significant consequences for vegetation dynamics (Miles 1979). An ability to predict the consequences of altering grazing management for vegetation succession is therefore of primary and fundamental importance. Although differences of opinion often exist between agriculturalists, conservationists and those with an amenity or sporting interest as to what constitutes a satisfactory species composition, there is a common interest in acquiring sufficient understanding of plant–animal relationships for objective criteria to be drawn up to formulate management prescriptions which achieve specific management objectives (see also Sydes & Miller 1988).

In order to define the scope for manipulating species composition of semi-natural vegetation we need information from three specific research areas. First, we need an adequate understanding of foraging behaviour (i.e. which factors influence grazing choice in different animal species, both between vegetation types and between plant species within types). Second, we need to know how the plant species of major interest respond to different seasonal patterns and levels of use. And, third, we need to know how the feeding value of different vegetation types varies with season and management. The following are also necessary for the longer-term evaluation of management options: studies of the distribution of excreta, of plant and animal pathways of nutrient cycling, and of the effects of treading on both plants and soils.

RESEARCH ISSUES

In the following sections there is a brief discussion of the three main research issues in turn, with examples given to illustrate their usefulness. The integration and application of the knowledge gained is then considered.

Foraging behaviour

Foraging behaviour has been the subject of much speculation over the last 20 years. Attention has focused particularly on strategies and on evolutionary aspects, out of which the theory of optimal foraging has developed (Stephens & Krebs 1987). The theory proposes that animals make feeding choices in order to maximize their intake of energy or some limiting nutrient while minimizing energetic costs of foraging and exposure to predators, and that successful strategies should look to genetic fixation.

The assumptions underpinning the optimal foraging theory have been reviewed by Pyke (1984) and Stephens & Krebs (1987). Both highlighted the large number of assumptions involved for which validation is required. Pyke (1984) concluded that the theory could be useful only when its assumptions, mathematical development and testing are appropriate for the studies to which it is applied. Before we could hope to evaluate the theory in respect of free-ranging domestic livestock, we have much to learn about the precise nature of the various factors involved in grazing choice between and within vegetation types. An example of the level of complexity and approach appropriate to the examination of the optimal foraging theory is given by Illius (1986).

The aim of our studies of foraging behaviour is to gain sufficient understanding of the mechanics of diet selection and nutrient intake to predict how animals will respond in a heterogeneous environment. A first approach, whereby grazing choices are examined in the light of changing sward conditions, probably offers the best opportunity at present to collect sufficiently precise data to manage vegetation as well as to test optimal foraging theories. Two phases of study are needed, namely (i) the detailed monitoring of grazing choice and sward conditions at a series of points in time, which allows hypotheses to be established, and (ii) follow-up studies with experimentally manipulated swards, in order to test the hypotheses. Studies with single and with mixed animal species, grazing on vegetation ranging from a single community type to simple combinations of two types to more complex mixes, should follow on in time.

An example of this sequence is illustrated by our work on heather (*Calluna vulgaris*) which began with the grazing behaviour studies of Hunter (1954, 1962), who appreciated the need to relate to herbage characteristics the sheep's preferences for particular vegetation communities. Hunter examined the distribution of grazing sheep on selected parts of his study area in relation to the productive capacity of the vegetation (yield under cages) and to its botanical and chemical composition. Grazing intensity was found to be most closely related to pasture quality. Hunter differentiated between the preferred vegetation types, which occurred on the better or 'mull' soils, and the unpreferred types on the more acid, peaty or 'mor' soils. Heather communities belonged to the latter group, having a characteristically low comparative grazing intensity with the main period of use being in winter, but with a secondary peak round about July. It was suggested that availability of grazing on the preferred vegetation types was a major determinant of the pattern and level of use of the unpreferred types (Hunter 1954, 1962).

With respect to different patches of heather, age of stand affected both the seasonal pattern and the overall level of use. In Grant & Hunter's (1968) study of interactions between grazing and burning, one-quarter of each of a series of grazed plots was burned at 2-yearly intervals to provide four ages of heather. Sheep (*Ovis aries*) preferentially grazed the younger heather stands at all times of year. This behaviour, however, was primarily a response to differences in sward conditions; preference for younger heather was strongly developed at the lower grazing pressures, where old heather was able to increase rapidly in height and cover. At the highest grazing pressure the heather was maintained short, the current shoots had high nitrogen content and there was little preference for age of stand (Fig. 2).

Heather has poor feeding value and the sheep's diet must contain a proportion of grass to achieve acceptable production levels (lactation, lamb growth, reproduction and, indeed, survival) (Maxwell *et al.* 1986). The factors influencing grazing choice between sown grass and heather have been studied in two experiments (Milne & Grant 1987). In the first of these, on plots containing different proportions by area of grass, changes over time in grass biomass, heather utilization and grass/heather proportions in the diet were monitored during 2–4-week grazing periods in May, July and October. In a follow-up experiment, heather utilization during May–August was monitored on plots with 30 per cent by area of grass, but with the grass biomass maintained at 400, 950 or 1500 kg dry wt ha^{-1} by adjusting stock numbers as necessary.

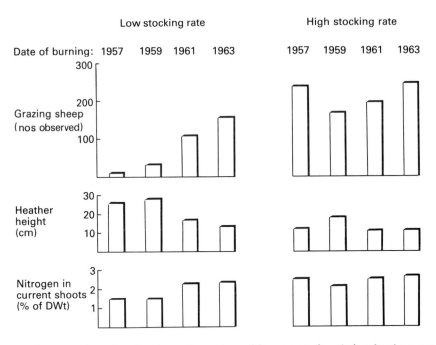

FIG. 2. Influence of time since burning and sward conditions on grazing choices by sheep grazing heather. Grazing sheep numbers are the sum of observations made at intervals during grazing periods from winter 1964/65 to summer 1967; heather height is as recorded in 1967; nitrogen contents (current season's shoots) are the means of estimates in September 1965 and May 1966.

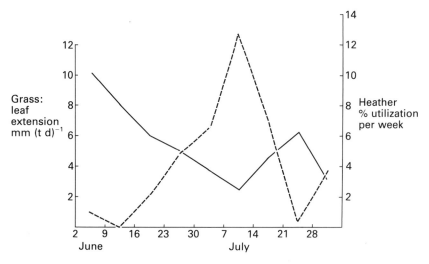

FIG. 3. Relationship between grass growth (mm tiller^{-1} day^{-1}, ——) and % of heather utilized per week (– – – – –). Grass biomass was held at around 950 kg dry wt ha^{-1}.

These studies indicated that grass biomass was a major determinant of heather utilization (thus confirming Hunter's suggestion). That relationship was modified by season, however, for sheep grazed less heather early in the season when grass, unlike heather, was growing rapidly. When grass biomass was controlled there was an inverse relationship between current growth rate of grass and heather utilization (Fig. 3). The proportion of grass influenced (i) the number of grazing days and (ii) the amount of grass remaining at the same levels of heather utilization.

A second example illustrating the usefulness of treatments based on the control of a particular sward characteristic is given by our comparisons of different animal species on *Nardus stricta* grassland. Such studies have the added advantage that grazing choices of different animal species can be compared against a common criterion. The work on *Nardus* grassland began with a comparative study of diet selection by sheep and cattle (*Bos taurus*) grazing together. The floristic composition of diet was related to concomitant and detailed records of sward conditions (Grant *et al.* 1985). This study showed that cattle consistently ingested more *Nardus* than did sheep and that the proportion of *Nardus* in the diets of both sheep and cattle was inversely and curvilinearly related to the height or biomass of the preferred between-tussock grasses.

This information has been used to design follow-up experiments in which plots grazed by different animal species are being managed so that the preferred between-tussock grasses are maintained at predetermined heights. Sheep and cattle are being compared at one site, and sheep and goats (*Capra hircus*) at another. Records of the proportion and closeness of grazing on *Nardus* leaves, of

TABLE 1. Utilization of *Nardus* by different animal species on separate plots (with grazing controlled in relation to maintaining the stated sward heights for the preferred between-tussock grasses). Means are for three years' records for July; SE of difference between means was calculated between years

Experiment 1		Animal species		SE of difference
	Cattle	Sheep	Sheep	(6 df)
Sward height	4·5 cm	4·5 cm	3·5 cm	
Proportion of leaves grazed	0·72	0·17	0·40	0·07
Grazed leaf lengths (cm)	3·6	9·0	5·7	0·57

Experiment 2		Animal species			SE of difference
	Sheep	Goats	Goats	Goats	(8 df)
Sward height	4·5 cm	4·5 cm	5·5 cm	6·5 cm	
Proportion of leaves grazed	0·38	0·64	0·47	0·17	0·13
Grazed leaf lengths (cm)	7·7	4·7	6·9	11·7	0·94

tissue turnover on *Nardus* tillers, of trends in sward species composition, and (for sheep and cattle site only) of diet composition, nutrient intake and animal live-weight changes are all being collected. Data for *Nardus* utilization in July show that cattle not only graze more *Nardus* than do sheep but that they also graze it more closely (Table 1). Goats are intermediate in behaviour, though nearer to cattle than sheep; there is a negative relationship between the height of the preferred between-tussock grasses and *Nardus* utilization by goats.

Sward responses

As with foraging behaviour there is a wealth of literature covering the theoretical aspects of plant strategy in relation to grazing tolerance/avoidance (Archer & Tieszen 1986; Malechek *et al.* 1986). Interesting though this literature is, the concentration on theory and search for evidence to illustrate concepts, rather than on the collection of more precise quantitative data of plant responses to given levels of utilization, limits its usefulness in providing guidelines for management decisions.

We have already argued that interest is centred on the effects of grazing on vegetation dynamics in the case of semi-natural hill vegetation. The difficulties of collecting and interpreting herbage production data from the sward as a whole have been discussed fully elsewhere (Grant & Hodgson 1986). In any event, the widely differing life forms, seasonality of growth and feeding value of the various hill species is such that the value attached to the dry-matter contribution of component species differs. Considerations such as the need to maintain a satisfactory species balance to ensure that feed is available for grazing animals all the year round, or constraints related to multiple land use, help to indicate the species on which our interest needs to focus and the desired direction of vegetation change.

The next step is to ensure that one has adequate knowledge of the species, growth form and tissue flow. This is essential and has been recognized by others as being so (Wells 1966) if experiments and measurements of plant responses to grazing or defoliation are to be appropriate. Our native species include a wide variety of growth forms, and even within types (e.g. graminoids) there is wide variation in patterns of carbon and nutrient allocation, and in leaf and tiller demography and phenology. Generalized approaches to the modelling of plant–animal relationships are unlikely to be helpful because of these variations.

In our work on heather we have assessed the impact of removing a range of equivalent proportions of the current season's shoots by weight (grazing) or length (cutting) at different times of year on the (i) productive capacity of the plant, (ii) morphology of the developing heather stand, and (iii) trends in floristic composition (Grant et al. 1978, 1982). This approach indicated that the level at which the current season's shoots are utilized is fundamental in controlling the long-term productivity of heather communities. It also highlighted why suscepti- bility of heather to damage by overgrazing varies with season and increases with age of stand.

The growth form of grasses (a sequence of leaves appearing from the base of the sward), with species and seasonal differences in the rates of growth, rates of senescence and tillering patterns suggested the choice of a different criterion to control sward productivity—that of sward height control. This approach is being used to investigate the scope for manipulating the floristic composition and nutritive value of *Nardus* grassland by controlled grazing. As *Nardus* is not a pre- ferred species, it is the height of the preferred between-tussock grasses which is managed. *Nardus* utilization is monitored (see above) and related to sward performance in terms of (i) tissue turnover rates on *Nardus* tillers, (ii) changes in sward morphology and canopy structure, and (iii) trends in species composition over time. In controlling *Nardus*, the early results (Grant et al. 1988; MLURI, unpublished) are encouraging. Leaf extension rates of *Nardus* tillers in the second year were certainly reduced on the cattle and goat treatments with the closest grazing of *Nardus*.

The potential of *Molinia caerulea* grassland to provide summer grazing for hill cattle is also being investigated. This work is investigating the proportion of current growth that can be utilized if *Molinia* dominance is to be maintained. *Molinia* is a preferred species during the summer months and the level of utilization can be controlled directly. Animal numbers are adjusted as necessary to maintain predetermined target lengths of *Molinia* laminae. These lengths were set following measurement of height and mass increments over the growing season in the absence of grazing, with the heights chosen approximating to the re- moval of one-third or of two-thirds of the leaf produced. Vegetation monitoring includes the collection of data on *Molinia* leaf extension rates and on amounts of energy and nutrient reserves per tiller as well as the annual monitoring of sward composition and structure (see also Torvell, Common & Grant 1988).

Feeding value

Any evaluation of management options for hill vegetation must take account of the potential nutritive value of the alternative vegetation types and the effect of sward conditions on levels of herbage intake. Space does not permit expansion on this topic, but data for the annual ranges in herbage intake on sample communities of the five main vegetation categories (Fig. 4) illustrate the substantial differences in feeding value between the grass and dwarf-shrub types. On the grass communities, both under-utilization (allowing a build-up of senescent tissue) and low height or biomass lead to reduced intake. In some circumstances nutritional penalties for the grazing stock may limit numbers of stock that can be carried to a figure below that set by sward considerations. A heather-clad hill with very little grass is one example of this.

DEFINING MANAGEMENT OPTIONS

Before attempting to outline some of the practical issues to be faced, it is worth listing some of the objectives that need to be considered. These are:

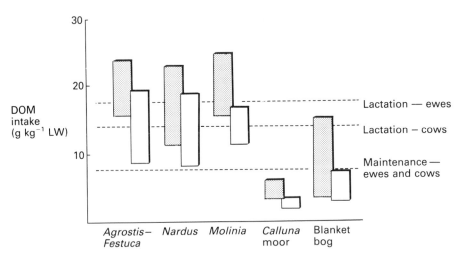

FIG. 4. Range of variation in the intake of digestible nutrients (digestible organic matter (DOM), g kg^{-1} live weight) by sheep (▨) and cattle (☐) grazing hill plant communities at different seasons of the year (from Hodgson & Eadie 1986).

(i) To provide farmed livestock with the maximum possible nutritional requirements from indigenous pasture throughout the year;
(ii) To maintain or create appropriate habitats for wildlife; and
(iii) To sustain or develop amenity areas and scenically attractive landscapes.

Inherent in these objectives is the need to manage the grazing of these resources at levels of herbage utilization which either sustain the existing balance between species or are designed to manipulate a shift in that balance to one which is more appropriate.

We have argued that in order to achieve one, or any combination, of these objectives under grazing management, it is necessary to find a simple criterion for each major vegetation type which, together with a knowledge of the physiology and seasonal patterns of growth of the main plant species, can be used to control the level of utilization and consequently the level of productivity. A number of examples have been given which indicate that some success can be achieved using this approach, and it is possible to begin to quantify for specific sites the levels of stocking needed to meet the criterion set for some plant communities. This information can be used as a guide to what might be achieved at other sites; more precise extrapolation depends on acquiring better quantitative information that relates plant production to its site, soil and climatic characteristics. But even without this more precise information progress in developing management strategies can be made. Whatever level of utilization is chosen, the consequences for some plant communities can be roughly estimated. Thus, if it is possible to define the conditions necessary to optimize the amenity or wildlife value of an area, then the effect of doing so on the agricultural production can be assessed reasonably accurately.

Good progress has been made in developing management strategies for heather moorland (Maxwell et al. 1986), using quantitative information from research (Milne, Bagley & Grant 1979; Grant et al. 1982; Milne & Grant 1987) which has led to the construction of a computer-based management model (for details see Sibbald et al. 1988). The output is in terms of sheep numbers, frequency of burning and amount of sown grassland, with assumptions made about sheep performance and winter feeding regimes. The model can be used to (i) provide the management framework to maintain existing balances in Calluna content and distribution; (ii) calculate the number of sheep likely to prevent overgrazing and to remove the risk of replacement of heather by native grass species; and (iii) avoid under-utilization and prevent a degenerative change towards scrub regeneration and the loss of the agricultural resource. It is important to extend this approach and to develop management decision rules for all our major indigenous vegetation types, which in principle need to be similar to those described for heather moorland.

Most hill grazings have at least four vegetation types present within a defined area. Earlier, reference was made to the need to examine the effects of changing sward conditions on the choice of grazing site of animals in single and mixed plant communities. This is an important research area but ultimately it will need to be complemented by research which examines grazing behaviour and foraging strategy in a wider context. Apart from the particular sward characteristics of plant communities, which affect the choice of grazing site, there are other factors that need to be taken into account. These have received little attention since the classic studies of Hunter (1962, 1964). The proportion of different vegetation types present, their position relative to one another and in relation to altitude, exposure and other physical and climatic factors, in interaction with social behaviour, are all likely to influence animals in terms of their occupancy and seasonal pattern and intensity of grazing of particular vegetation types and individual communities within types. Few quantitative data currently exist which describe this behaviour. While much progress can be made by understanding the effects of sward characteristics on choice of grazing site, the larger and more varied the area, the more important these other factors are likely to be.

However, as we acquire a more precise knowledge of the nutritional attributes of the major vegetation types and an increasing understanding of the factors which determine the choice of grazing site, there is the future prospect of being able to define more objectively the grazing area requirements for livestock. It should be possible to describe the proportion (by area) of each of the major vegetation types which would optimize levels of nutrition throughout the year from grazed herbage. This may require a significant adjustment of some boundaries of the existing grazings and the appropriate division of others. Conversely, given a defined grazing area, it should be possible to calculate much more precisely the sustainable numbers of livestock needed to obtain specific levels of animal performance and output, and at the same time to achieve defined amenity and wildlife objectives.

This paper has outlined a research approach towards developing the understanding needed for the management of semi-natural vegetation using domestic grazing animals. The approach is equally valid for studies of the consequences of grazing for vegetation dynamics in the case of the major wild herbivores, such as red deer (*Cervus elaphus*) or rabbits (*Oryctologus cuniculus*). In the wider context, given that conservationists, or those with a wildlife, sporting or amenity interest, can define the conditions they require (for example Sydes & Miller 1988; Hudson 1988), we believe that the above experimental approach can be used to provide the understanding whereby these conditions could be predictably achieved using grazing by domestic and other animals. It should be appreciated, however, that the objectives in managing vegetation for agricultural purposes may be quite different from those required for conservation, wildlife or other interests (see Sydes & Miller 1988). If the grazing management of agricultural animals has to be modified to meet conservation or wildlife objectives, it is likely that the economic performance of the agricultural enterprises would be adversely affected (see also Dixon 1988).

In determining future land-use policies it is reasonable to evaluate the consequences of alternative policies in cost–benefit terms. A more explicit and quantitative description of grazing management protocols for semi-natural vegetation provides a much more rational basis on which to carry out such an analysis. This does not imply, however, that a cost–benefit analysis would be the only criterion determining land use, but it does provide an important reference point, particularly since much of what may be positively achieved in the uplands is likely to be dependent upon government financial support.

REFERENCES

Archer, S.R. & Tieszen, L.L. (1986). Plant responses to defoliation: hierarchical considerations. *Grazing Research at Northern Latitudes* (Ed. by O. Gudmundsson), pp. 45–59. NATO ASI Series A 108, Plenum Press, New York.

Armstrong, R.H., Grant, S.A. & Hodgson, J. (1987). Grazing choices and hill management. *Efficient Sheep Production from Grass* (Ed. by G.E. Pollot), pp. 175–178. Occasional Symposium No. 21, British Grassland Society, Maidenhead, Berks.

Ball, D.F., Dale, J., Sheail, J. & Heal, O.W. (1982). *Vegetation Change in Upland Landscapes.* Institute of Terrestrial Ecology, Cambridge.

Dixon, J.B. (1988). Comparison of agriculture and conservation in the uplands of New Zealand and Wales. *This volume.*

Eadie, J. (1984). Trends in agricultural land use: the hills and uplands. *Agriculture and the Environment* (Ed. by D. Jenkins), pp. 13–20. Institute of Terrestrial Ecology, Monkswood.

Grant, S.A. & Hodgson, J. (1986). Grazing effects on species balance and herbage production in indigenous plant communities. *Grazing Research at Northern Latitudes* (Ed. by O. Gudmundsson), pp. 69–77. NATO ASI Series A 108, Plenum Press, New York.

Grant, S.A. & Hunter, R.F. (1968). Interactions of grazing and burning on heather moors and their implications in heather management. *Journal of the British Grassland Society,* 23, 285–293.

Grant, S.A., & Barthram, G.T., Lamb, W.I.C. & Milne, J. (1978). Effect of season and level of grazing on the utilization of heather by sheep. 1. Responses of the sward. *Journal of the British Grassland Society,* 33, 289–300.

Grant, S.A., Milne, J.A., Barthram, G.T. & Souter, W.G. (1982). Effects of season and level of grazing on the utilization of heather by sheep. 3. Longer-term responses and sward recovery. *Grass and Forage Science,* 37, 311–320.

Grant, S.A., Suckling, D.E., Smith, H.K., Torvell, L. Forbes, T.D.A. & Hodgson, J. (1985). Comparative studies of diet selection by sheep and cattle: the hill grasslands. *Journal of Ecology,* 73, 987–1004.

Grant, S.A., Torvell, L., Armstrong, R.H. & Beattie, M.M. (1988). The manipulation of *Nardus* pasture by grazing management. *Agriculture and Conservation in the Hills and Uplands* (Ed. by M. Bell & R.G.H. Bunce), pp. 62–64. Institute of Terrestrial Ecology, Grange-over-Sands.

Hodgson, J. (1985). The significance of sward characteristics in the management of temperate sown pastures. *Proceedings of the XV International Grassland Congress, Kyoto, Japan,* 15, 63–67.

Hodgson, J. & Eadie, J. (1986). Vegetation resources and animal nutrition in hill areas: agricultural and environmental implications. *Hill Land Symposium, Galway, 1984* (Ed. by M. O'Toole), pp. 118–133. An Foras Taluntais, Dublin.

Hodgson, J. & Grant, S.A. (1981). Grazing animals and forage resources in the hills and uplands. *The Effective Use of Forage and Animal Resources in the Hills and Uplands* (Ed. by J. Frame), pp. 41–57. Occasional Symposium No. 12, British Grassland Society, Maidenhead.

Hodgson, J. & Maxwell, T.J. (1981). Grazing research and grazing management. *The Hill Farming Research Organisation Biennial Report,* 1979–1981, 169–187.

Hodgson, J., Mackie, C.K. & Parker, J.W.G. (1986). Sward surface height for efficient grazing. *Grass Farmer,* 24, 5–10.

Hudson, P.J. (1988). Spatial variations, patterns and management options in upland bird communities. *This volume.*

Hunter, R.F. (1954). The grazing of hill pasture sward types. *Journal of the British Grassland Society*, **9**, 195–208.

Hunter, R.F. (1962). Hill sheep and their pasture: a study of sheep grazing in south-east Scotland. *Journal of Ecology*, **50**, 651–680.

Hunter, R.F. (1964). Home range behaviour in hill sheep. *Grazing in Terrestrial and Marine Environments* (Ed. by D.J. Crisp), pp. 155–171, British Ecological Society Symposium No. 4, Blackwell Scientific Publications, Oxford.

Illius, A.W. (1986). Foraging behaviour and diet selection. *Grazing Research at Northern Latitudes* (Ed. by O. Gudmundsson), pp. 227–236. Plenum Press, New York.

Institute of Terrestrial Ecology (1978). *Upland Land Use in England and Wales.* Countryside Commission Publication 111, Countryside Commission, Cheltenham.

King, J. & Nicholson, I.A. (1964). The grasslands of the forest zone. *The Vegetation of Scotland* (Ed. by C.H. Burnett), pp. 168–206. Oliver & Boyd, Edinburgh.

Malechek, J.C., Balph, D.F. & Provenza, F.D. (1986). Plant defense and herbivore learning: their consequences for livestock grazing systems. *Grazing Research in Northern Latitudes* (Ed. by O. Gudmundsson), pp. 193–208. NATO ASI Series A 108, Plenum Press, New York.

Maxwell, T.J., Grant, S.A., Milne, J.A. & Sibbald, A.R. (1986). Sytems of sheep production on heather moorland. *Hill Land Symposium, Galway, 1984* (Ed. by M. O'Toole), pp. 188–211, An Foras Taluntais, Dublin.

Miles, J. (1979). *Vegetation Dynamics.* Chapman & Hall, London.

Miles, J. (1985). The ecological background to vegetation management. *Vegetation Management in Northern Britain* (Ed. by R.B. Murray), pp. 3–20, Monograph No. 30, British Crop Protection Council, Croydon.

Milne, J.A. & Grant, S.A. (1987). Sheep management on heather moorland. *Efficient Sheep Production from Grass* (Ed. by B. E. Pollett), pp. 165–167. Occasional Symposium No. 21, British Grassland Society, Maidenhead, Berks.

Milne, J., Bagley, L. & Grant, S.A. (1979). Effect of season and level of grazing on the utilization of heather by sheep. 2. Diet selection and intake. *Grass and Forage Science*, **34**, 45–53.

Mowforth, M. & Sydes, C. (in press). *Moorland Management: a Literature Review.* Nature Conservancy Council, Peterborough.

Pyke, G.H. (1984). Optimal foraging theory: a critical review. *Annual Review of Ecological Systems*, **15**, 523–575.

Sibbald, A.R., Grant, S.A., Milne, J.A. & Maxwell, T.J. (1988). Heather moorland management — a model. *Agriculture and Conservation in the Hills and Uplands* (Ed. by M. Bell & R.G.H. Bunce), pp. 107–108. Institute of Terrestrial Ecology, Grange-over-Sands.

Stephens, D.W. & Krebs, J.R. (1987). *Foraging Theory.* Princeton University Press, Princeton, New Jersey.

Sydes, C. & Miller, G.R. (1988). Range management and nature conservation in the British uplands. *This volume.*

Torvell, L., Common, T.G. & Grant, S.A. (1988). Seasonal patterns of tissue flow and responses of *Molinia* to defoliation. *This volume.*

Wells, T.C.E. (1966). *Grazing Experiments and the Use of Grazing as a Conservation Tool.* Monks Wood Experimental Station Symposium No. 2, Nature Conservancy, Monks Wood.

Long-term changes in output and vegetation on a small Northern Ireland hill farm

J. H. McADAM AND S. M. CHANCE

Agricultural Botany Research Division and Greenmount College of Agriculture and Horticulture,
Department of Agriculture for Northern Ireland, Newforge Lane, Belfast BT9 5PX

INTRODUCTION

The Vogie hill farm (Co. Antrim) lies between 250 m and 390 m above sea level. It has a south-westerly aspect and a wide range of soil and vegetation types. The 136 ha farm was purchased in 1964 by the Department of Agriculture and was subsequently managed as an independent unit with minimal capital investment in reclamation and intensive land improvement. A vegetation map of the farm was prepared (Poole 1964) at the time of acquisition and detailed records of stock and economic performance have been maintained (Wilson 1986). In 1985 the vegetation of the farm was remapped using similar criteria to those used in 1964. The exercise represents the opportunity to study long-term vegetation changes and associated changes in output following minimal improvement on a small hill farm in Northern Ireland.

MATERIALS AND METHODS

Following a preliminary investigation into number of species and quadrat size, the vegetation was classified by selecting 141 2 m square quadrats from distinct communities along regular transect grids. For each quadrat, species were listed and given a cover/abundance value during July and August (Poole 1964). Using an ordination technique reported by Poore (1955), groups of common species were selected (Poole 1964). Subsequently the farm was subdivided into a 50 m × 50 m grid and each plot assigned one of the previously determined vegetation classes. The exercise was repeated in 1985 using a classification and grid pattern as close as possible to those derived by Poole (1964).

Stock numbers and total liveweight output of lambs and calves were recorded (Wilson 1986). Output of the farm was calculated as the utilized metabolizable energy (UME) output whereby energy from bought-in supplementary feed is deducted from the total energy requirements of all stock on the unit; it represents a measure of the actual output from the grassland (McAdam 1984). Financial output has been calculated on the basis of gross margins (GM) (adjusted for inflation to 1983/84 prices) per suckler cow and per ewe (Wilson 1986) and converted to a per hectare basis.

215

RESULTS AND DISCUSSION

The classes and main species components of the vegetation classification were (i) dense *Calluna vulgaris* (with *Eriophorum vaginatum* and *Scirpus cespitosus*), (ii) *Eriophorum vaginatum* (with *Vaccinium myrtillus, Sphagnum* spp. and *Calluna*), (iii) *Juncus squarrosus* (with *Molinia caerulea* and *Carex* spp.), (iv) *Juncus effusus* (with *Poa pratensis, Agrostis capillaris* and *Festuca rubra*), and (v) *Agrostis, Holcus lanatus, Poa* with *Anthoxanthum odoratum* (nomenclature after Clapham, Tutin & Warburg 1981).

The changes in areas of each class are presented in Table 1 and illustrated in Fig.1. The changes in vegetation on the 'hill' part of the farm have been relatively small. The area of *Calluna*-dominated vegetation has declined (from 44 to 29 per cent of the area of the hill), being largely replaced by weaker *Calluna* cover with *Eriophorum* as the dominant species. A limited amount of lime application and surface drainage has been carried out on the lower slopes of the hill and this has been reflected in the slight increase in area of nutritionally more favourable grasses.

The greatest changes have occurred on the marginal land where the proportion of *J. squarrosus* with *Molinia* has decreased and there has been a corresponding increase in the area of grassland dominated by *A. capillaris, P. pratensis* and *H. lanatus* (with some *Trifolium repens*). This change in the occurrence of *Juncus* and the increase in more productive indigenous grasses has been brought about gradually by regular lime application, surface drainage, *Juncus* control with herbicides, selective application of fertilizers and a greater

TABLE 1. Changes in the proportion of vegetation classes on the hill and marginal sections of the Vogie farm between 1964 and 1985

Vegetation class	Percentage of area in	
	1964	1985
Hill (86 ha)		
Calluna vulgaris dominant	44	29
Eriophorum vaginatum dominant	14	23
Juncus squarrosus, with *Molinia caerulea* and *Carex* spp.	16	14
Juncus effusus, with *Agrostis capillaris* and *Poa pratensis*	26	30
Agrostis/Poa/Holcus lanatus	0	4
Marginal (48 ha)		
Juncus squarrosus, with *Molinia caerulea* and *Carex* spp.	56	10
Juncus effusus, with *Agrostis capillaris* and *Poa pratensis*	40	60
Agrostis capillaris/Poa pratensis/Holcus lanatus	4	30

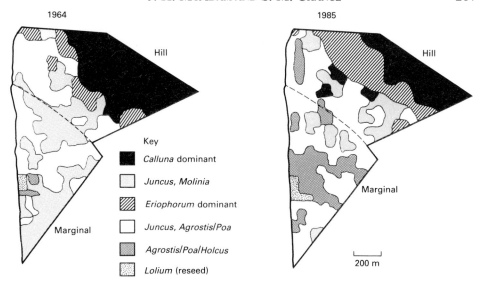

FIG. 1. The distribution of vegetation classes in 1964 and 1985 on the Vogie Farm.

degree of grazing management and control following subdivision (McGaughey & Wilson 1980). Only 2 ha of the farm have been reclaimed and reseeded.

Over the same period, stock numbers (particularly cattle), animal production (Table 2) and UME output (Fig. 2) have increased substantially. The mean UME output of a subsample of Co. Antrim hill farms in 1980 was 13·2 GJ ha^{-1} (McAdam 1983) and output was highly correlated with the proportion of inbye land (McAdam 1984). Hence, with the small amount of inbye land available on the Vogie farm, a UME output of 12·0 GJ ha^{-1} for the same year indicates that grassland was being used relatively efficiently. Although the adjusted gross margin has fluctuated considerably with the profitability of the two enterprises, the financial output has increased faster than the rate of inflation.

From contemporary accounts, the appearance of the farm has not changed appreciably over the 20-year period and this is largely borne out by results of the vegetation mapping, although the decline in heather cover on the upper hill,

TABLE 2. Changes in stocking and output between 1964 and 1985

	1964	1985	% increase
Number of ewes	150	162	8
Number of cows	15	25	67
Lambs produced (t)	2·4	6·8	183
Calves produced (t)	2·6	6·7	158
Gross margin ha^{-1} (£, adjusted)	57	88	54

Changes on a small hill farm

FIG. 2. Utilized metabolizable energy (UME) output (GJ (ha year)$^{-1}$) of the Vogie Farm from 1966 to 1984.

presumably through increased stocking, must give some rise for concern. With increasing pressure on hill farms to avoid major land reclamation and drainage, this monitoring exercise has demonstrated that acceptable levels of improvement in output can be achieved with minimal 'surface type' inputs, while at the same time retaining the natural landscape.

ACKNOWLEDGMENTS

The authors wish to thank the staff at Greenmount College Hill Farm for cooperation and Mr A. Stewart, Greenmount College, for the provision of funding to carry out the exercise.

REFERENCES

Clapham, A.R., Tutin, T.G. & Warburg, E.F. (1981). *Excursion Flora of the British Isles*, 3rd edn. Cambridge University Press, Cambridge.

McAdam, J.H. (1983). *Characteristics of Grassland on Hill Farms in N. Ireland — Physical Features, Botanical Composition and Productivity*. Occasional Publication, Department of Agricultural Botany, Queen's University of Belfast.

McAdam, J.H. (1984). The utilized metabolizable energy output of grassland on hill farms in Northern Ireland. *Grass and Forage Science*, **39**, 129–138.

McGaughey, S.J. & Wilson, J.K. (1980). Increased output from the hills (2). *Agriculture in Northern Ireland*, **55**, 197–200.

Poole, W. (1964). *Botanical report of Greenmount Agricultural College Hill Farm, Glenwherry, Co. Antrim*. Unpublished Report, Department of Agriculture for Northern Ireland, Belfast.

Poore, M.E.D. (1955). The use of phytosociological methods in ecological investigations. III. Practical applications. *Journal of Ecology*, **43**, 606–651.

Wilson, J.K. (1986). The development of a low cost system of management involving the integration of suckler cows and sheep on a hill unit in Co. Antrim. *Hill Land Symposium, 1984* (Ed. M. O'Toole), pp. 248–254. Agricultural Institute, Dublin.

Seasonal patterns of tissue flow and responses of *Molinia* to defoliation

L. TORVELL, T.G. COMMON and S.A. GRANT

Macaulay Land Use Research Institute, Bush Estate, Penicuik, Midlothian EH26 OPY

INTRODUCTION

Molinia caerulea (L.) Moench dominates large areas of poorly drained land in the uplands of Britain. In traditional extensive sheep production systems *Molinia* is usually little grazed (King & Nicholson 1964). When managed to control the accumulation of previously under-utilized dead herbage, *Molinia* communities can provide high-quality feeding value for grazing animals from June to August (Grant *et al.* 1985; Armstrong & Hodgson 1986). In improved two-pasture sheep systems, in which sheep are moved between fenced improved pasture and the open hill, the sheep are mainly on improved pasture at this time of year (Eadie 1971). It is possible that *Molinia* might provide a useful resource for the summer grazing of hill cattle. *Molinia* cover, however, was reported to be severely reduced by heavy grazing (Jones 1967).

The level of grazing that can be sustained while maintaining *Molinia* dominance is currently being investigated in two long-term grazing experiments, both started in 1985. Plots at Cleish, Fife, and at Bell Hill, Roxburghshire, are continuously stocked by cattle, with animal numbers adjusted twice weekly to maintain target *Molinia* lamina lengths corresponding to removal of approximately 33 per cent or 66 per cent of the annual leaf production. Control by lamina length rather than sward height is being used because of the tussocky nature of *Molinia* swards. To aid interpretation of the grazing study, information was collected on the seasonal patterns of *Molinia* growth, and two cutting experiments were set up to measure the more detailed responses of individual *Molinia* tillers and tussocks to the timing, level and frequency of defoliation.

MATERIALS AND METHODS

Rates of leaf extension growth and losses to senescence were measured weekly on marked tillers in undefoliated *Molinia* tussocks. Samples were also harvested each month for analysis of energy (total water-soluble carbohydrate, TWSC) and nutrient (N, P, K) reserve levels in both current and previous season's basal internodes.

The effects of cutting on total leaf production in the current growing season, and on the size and energy reserve levels of the basal internodes, were examined in experiment 1. Treatments removed either 50 per cent or 100 per cent of the length of each lamina from individual *Molinia* tillers, as a single defoliation in

219

either June or July, or twice, in both June and July. In experiment 2, the longer-term responses to defoliation are being followed. *Molinia* tussocks are cut annually to remove either 33 per cent or 66 per cent of lamina, as a single defoliation in June, July or August, or repeatedly in June, July and August. Cutting height is set by measurement of the mean lamina length of three random tillers per tussock prior to each cut. Treatments are being imposed for a minimum of three years before a final harvest after an uninterrupted growing season to assess the longer-term effects on leaf production. Growth is being monitored by annual measurements of leaf extension for two weeks immediately before the first cut in each season. Two out of an original six replicates were harvested at the end of the first year for analysis of energy and nutrient reserve levels of the basal internodes. Both cutting experiments were carried out at two sites; site 1 was at Cleish, Fife, and site 2 at Sourhope, Roxburghshire.

RESULTS

The seasonal pattern of leaf growth and senescence and of TWSC levels in the basal internodes of undefoliated tillers are shown in Fig. 1. *Molinia* has a short but highly productive growing season. The main period of extension growth occurs during June and early July and is largely separated in time from that of senescence, which increases rapidly after the end of August. Senescence of above-ground material is complete by November, leaving the overwintering swollen basal internodes. Energy and nutrient reserves stored in the basal internodes of

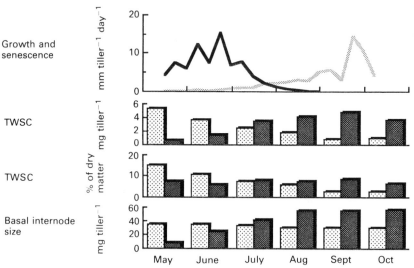

FIG 1. Seasonal patterns of growth in *Molinia*. Symbols are, ——, leaf growth; ▥ , leaf senescence; ▨, current season's basal internodes; ▨ previous season's basal internodes; and TWSC is the total water-soluble carbohydrate.

these previous season's tillers are thought to be important in supporting the growth of new current season's tillers in the following year.

The amount of TWSC in the current season's tiller bases is built up during the summer. The increase is due to change in size of the basal internode rather than its TWSC concentration. TWSC concentration in the previous season's tiller bases declined over the season to low levels by the end of September.

Mineral nutrient concentrations in current bases were highest in May (N 3·0 per cent, P 0·5 per cent, K 2·2 per cent), declined over the summer and levelled off in the autumn. In previous season's bases nutrient concentrates declined progressively over the season.

Leaf production in the current growing season was unaffected by concurrent cutting treatments (Experiment 1, Table 1). TWSC concentration in basal internodes showed no significant effects of cutting treatment, but basal internode size was significantly reduced at the higher level and frequency of defoliation, compared with the uncut control treatment (100 per cent leaf removal in June only, $P < 0.05$; 100 per cent leaf removal in July only, $P < 0.01$; 100 per cent leaf removal in both June and July, $P < 0.001$).

A similar pattern of reduction in tiller base size, but not in concentrations of TWSC, N, P or K in the tiller bases, was found after the first year's cutting in experiment 2; the basal internode size was greatly reduced by repeated 66 per cent leaf removal. Growth was reduced in subsequent seasons following cutting. Repeated defoliation significantly reduced leaf extension rates at both 33 per cent and 66 per cent levels of leaf removal compared with the uncut treatment, the response being greater after two years cutting than after one year (Table 2).

TABLE 1. The effect of defoliation on the total amount of leaf produced in the current growing season and on amounts of energy reserves in the basal internodes. Site 1 is at Cleish, Fife, Site 2 at Sourhope, Roxburghshire. Leaf production data are pooled values from six tillers per tussock and three replicate tussocks, harvested at the end of August. Tiller base samples were harvested in late September from the remainder of each tussock and are from Site 1

Treatment	Amount of leaf (mg tiller^{-1})		Size of tiller base (mg)	Total water-soluble carbohydrate (percentage of dry matter)
	Site 1	Site 2		
Uncut	65·0	71·2	50·8	9·2
50% removal June	69·0	93·0	48·6	10·8
50% removal July	74·6	85·0	41·7	10·9
50% removal June & July	74·4	73·2	46·1	8·2
100% removal June	74·5	75·0	32·9	10·4
100% removal July	83·0	67·2	24·7	6·8
100% removal June & July	85·6	83·8	19·5	7·8
SE of difference between means (12 df)	15·4	20·0	5·4	2·1

TABLE 2. The effect of level, frequency and timing of defoliation on leaf growth (mm tiller^{-1} day^{-1}) measured in early June in the season following cutting treatments. The data are means of three tillers per treatment tussock and four replicate tussocks

	After 1 year of cutting			After 2 years of cutting		
Uncut control	11·2			7·5		
Level of cutting	33%	66%	Mean	33%	66%	Mean
Single cut:						
June only	9·5	9·8	9·7	6·0	7·2	6·6
July only	10·7	8·2	9·4	5·6	5·9	5·8
August only	10·8	10·2	10·5	7·1	5·4	6·2
Repeated cutting:						
June, July & August	8·4	7·0	7·7	5·3	4·0	4·7
Mean	9·8	8·8		6·0	5·6	

SE of difference between means (24 df) for comparison of:

Individual treatments with control	0·99	0·74
Means, level of cut, with control	0·79	0·64
Means, timing and frequency of cut, with control	0·87	0·59

DISCUSSION

The stored reserves of *Molinia* are adequate to mitigate the immediate effects of severe defoliation, as priority is given to the maintenance of leaf expansion at the expense of rebuilding of reserves in the new basal internodes. In consequence, however, the size of the pool of energy and nutrients available to support growth in the following season is reduced. The effects of defoliation are thus deferred in time.

These results suggest that *Molinia* is highly sensitive to defoliation because of a heavy reliance on the recycling of energy and nutrients between successive seasons. More work is needed, however, before the relative importance of energy and nutrients as limiting factors can be assessed. The practical implication of these early results is that *Molinia* dominance is unlikely to be sustained except under very light grazing intensities if grazing is repeated annually.

REFERENCES

Armstrong, R.H. & Hodgson J. (1986). Grazing behaviour and herbage intake in cattle and sheep grazing indigenous plant communities. *Grazing Research at Northern Latitudes* (Ed. by O. Gudmundsson), pp. 211–218. NATO ASI Series A 108, Plenum Press, New York.

Eadie, J. (1971). Hill pastoral resources and sheep production. *Proceedings of the Nutrition Society*, 30, 204–210.

Grant, S.A., Suckling, D.E., Smith, H.K., Torvell, L., Forbes, T.D.A. & Hodgson, J. (1985). Comparative studies of diet selection by sheep and cattle: the hill grasslands. *Journal of Ecology*, 78, 987–1004.

Jones, Ll. I. (1967). *Studies on Hill Land in Wales*. Technical Bulletin No. 2, Welsh Plant Breeding Station, Aberystwyth.

King, J. & Nicholson, I.A. (1964). Grasslands of the forest and sub-alpine zones. *The Vegetation of Scotland* (Ed. by C.H. Burnett), pp. 168–206. Oliver & Boyd, Edinburgh.

Impacts of agriculture on upland birds

C. J. BIBBY

Royal Society for the Protection of Birds, The Lodge, Sandy, Bedfordshire SG19 2DL

SUMMARY

1 Present patterns in moorland vegetation and associated bird communities are shaped predominantly by agriculture.

2 The status of 16 bird species characteristic of upland rough grazings is summarized.

3 Populations of eight of these species are important enough to merit additional status in European conservation law.

4 Of factors influencing numbers and distribution, changes in vegetation arising from (i) grazing and (ii) reseeding have had most impact. Birds of heather-dominated moorland have been most affected.

5 Declines in addition to those attributable to loss of habitat may be due to increased predation from crows (*Corvus corone*) and foxes (*Vulpes vulpes*) related to habitat changes and fragmentation.

6 A graphical model is used to show the importance of the shapes of relationships between birds and habitat loss in understanding the effects of change in different land-use patterns.

INTRODUCTION

The moorland vegetation valued today for qualities of openness, wilderness and wildlife owes its origin squarely to agriculture. Since the extensive removal of tree cover, burning and grazing have combined to prevent its recovery. The extractive nature of hill farming has promoted and maintained an unproductive landscape by methods commonly deplored in many Third World countries. Perhaps the major new trend has been a lessening of reliance on heather for sheep (*Ovis aries*) pasturage. If sheep are wintered elsewhere or fed on the hill, it is no longer important to maintain heather for winter grazing. Sheep numbers have increased because some better land has been improved (e.g. inbye), while traditional moorland management has been practised less widely.

Three other major land uses compete with agriculture in the uplands and thus have alternative impacts on vegetation and birds (Ratcliffe & Thompson 1988). The first of these, forestry, is covered by Thompson, Stroud & Pienkowski (1988). It supports an almost totally different bird fauna from the open land. Second, the use of hills for grouse (*Lagopus lagopus scoticus*) shooting has had a large effect on moorland birds. Its vegetation management is more akin to that associated with the traditional management of hill sheep in attempting to maintain a sustainable yield of heather in its younger and more nutritious stages (Hudson 1986a, 1988). Third, the presently enormous herd of red deer (*Cervus elaphus*) in Scotland represents a less well-regulated grazing force which is only partially managed and exploited as an economic resource. No other land use compares in both scale and influence, nor is likely to. There is also tourism and recreation which have potential to add new income and diversity to rural economies. They can exert effects on soils and vegetation (Watson 1985; Thompson, Galbraith & Horsfield 1987) but are not likely to have extensive primary influences.

In an attempt to assess the relative impact of agriculture and forestry, Table 1 shows area estimates of major vegetation types. The group of bracken (*Pteridium aquilinum*), *Juncus effusus* and moorland grasses represents mainly agricultural impact and covers about twice the area of forestry. Not included here is the additional area converted by reseeding. The extent of *Molinia caerulea*, *Scirpus cespitosus* and *Nardus stricta* is perhaps the worst consequence of heavy grazing in reducing wildlife while being of no great agricultural value (Sydes & Miller 1988). *Molinia* and *Scirpus* alone cover an area of similar magnitude to forestry in the uplands.

MOORLAND BIRDS

Reviews of the upland bird fauna are provided by Ratcliffe (1977) and Fuller (1982); references to key species studies are in Table 2. Many species of the uplands also occur in a wide range of habitats (see Table 1 in Thompson *et al.* 1988). I shall restrict my attention to 16 predominantly sub-montane species (Table 2) which occur mainly in the open uplands used as rough grazings. This list excludes many other birds which could be classed as upland species but with habitats less susceptible to agricultural change (Table 3).

TABLE 1. Estimates of area of major upland vegetation classes in Britain. Data from Bunce, Barr & Whittacker (1981) and R. G. H. Bunce (pers. comm.), and presented in Goodier (1985). Land classes are the main upland types (18–24 and 30–32) from Bunce *et al.* 1981

Vegetation group	Upland land classes		GB total land area	
	Area km^2	% total	Area km^2	% total
Calluna/Vaccinium	14,384	33·2	16,834	25·7
Calluna/Eriophorum	3,098	7·2	3,902	6·0
Pteridium	1,322	3·0	3,789	5·8
Juncus	1,983	4·6	3,711	5·7
Molinia/Scirpus	7,414	17·1	9,276	14·1
Nardus/Agrostis/Festuca	6,614	15·3	13,502	20·6
Forestry	8,485	19·6	14,521	22·1

TABLE 2. British birds predominantly confined to breeding on open moorlands. General references such as Sharrock (1976) and Cramp (1980, 1983, 1985), are used where no specific sources are given (*). Annex I refers to the EEC Directive on the Conservation of Wild Birds

Species	Population (pairs)	Trends	Main habitats	Annex 1	References
Golden eagle	510	stable or small decline	crag and tree nesting, open areas Scotland	+	Dennis et al. 1984 Watson et al. 1987
Red kite	40	increasing slowly	tree nesting, open land hunting, Wales	+	Newton et al. 1981
Hen harrier	500–600	increasing 1950–1970, now decreasing?	heather moors	+	Watson 1977
Peregrine	850	strong recovery from pesticides	crag nesting, aerial	+	Ratcliffe 1984
Merlin	550–650	widespread decline	heather moors	+	Bibby & Nattrass 1986
Red grouse	50,000	declining	heather moors		Hudson 1986a, 1988
Black grouse	20,000	increasing with forestry, overall decline	margins of heather moors		*
Golden plover	30,000? 22,600?	decreasing	open land, short vegetation	+	Ratcliffe 1976 Stroud et al. 1987
Curlew	50,000	increasing	varied moors, grassy margins		*
Greenshank	1500/ 960	stable/threatened in Caithness	northern bogs	+	Nethersole-Thompson & Nethersole-Thompson 1979, Stroud et al. 1987
Dunlin	10,000	some decrease, in south especially	blanket bogs, some coastal		*
Short-eared owl	1000	fluctuates, trends not known	heather moors, young forest for a period	+	*
Raven	5000	some declines	mainly open land with carrion		Marquiss et al. 1978
Ring ouzel	10,000	some declines	rocky areas, heather		*
Wheatear	80,000	declines in lowlands	grassy areas with rocks or walls		*
Whinchat	30,000	decline in lowlands	rough vegetation, especially damp areas or bracken		*

TABLE 3. Examples of upland habitats and birds less susceptible to agricultural change (c.f. those in Table 2)

1	Rivers and lochs	Dipper, osprey, black-throated diver, various ducks (Anatidae)
2	High mountains	Dotterel, ptarmigan, snow bunting
3	Broadleaf woods	Pied flycatcher, redstart, tree pipit
4	Conifer forests	Crossbill, goshawk, siskin
5	Coastal heaths	Arctic skua, whimbrel, arctic tern
6	Very rare birds	Wood sandpiper, Temminck's stint
7	Catholic species	Meadow pipit, skylark, lapwing, buzzard

Moorland bird communities are characteristically species-poor, reflecting the structural simplicity of habitats. Bird densities are low, reflecting the overall low productivity in cold wet areas with poor soils. For several species, such as golden eagle (*Aquila chrysaetos*), hen harrier (*Circus cyaneus*), curlew (*Numenius arquata*) and distinct races of golden plover (*Pluvialis apricaria*) and red grouse (*L. lagopus scotica*), Britain is a stronghold of international significance (Ratcliffe & Thompson 1988). Due to their scarcity and vulnerability, eight of the species in Table 2 are listed on Annex 1 of the EEC Directive on the Conservation of Wild Birds. This legislation requires active conservation measures to be taken, over and above the protection from killing accorded to most birds.

Predators and scavengers

A group of predators is notable among moorland birds. Hen harrier, merlin (*Falco columbarius*) and short-eared owl (*Asio flammeus*) are quite strongly associated with heather. Peregrine (*F. peregrinus*) and golden eagle may be limited locally by availability of crags for nesting but otherwise range widely and take a broad spectrum of prey. For peregrine, which feed solely on birds, the overall abundance of prey might not be much influenced by changes in agricultural practice. The importance of the uplands for raptors may additionally reflect lower levels of persecution than elsewhere in Britain, where several raptors occur at numbers lower than available habitats could probably support. Abundance of carrion associated with extensive sheep grazing is important for ravens (*Corvus corax*), eagles and red kites (*Milvus milvus*) (Dare 1986). This source of food would not be so abundant if the uplands were not managed extensively for agriculture. However, improved sheep husbandry techniques may not suit so many scavenging birds. The breeding success of these species may be determined by abundance of live prey, and thus be susceptible to additional land-use changes such as afforestation (Watson, Langslow & Rae 1987; Thompson *et al.* 1988).

Waders

Most waders (Charadrii) visit the hills for a rather brief breeding season. Their nidifugous chicks are dependent on invertebrates, which can be locally abundant (though often over extremely short spells). Most waders and their foods are associated with wetter areas where they require a combination of cover for nesting and usually more open areas for feeding. Although nests and chicks are very susceptible to both bird and mammal predators, summer migrations to the hills or to the Arctic take many waders to areas with relatively low densities of predators. British uplands support characteristic communities and significant proportions of European populations of the more southerly nesting wader species, especially curlew and golden plover.

Game birds

Both red grouse and black grouse (*Tetrao tetrix*) are herbivores characteristically feeding on heather (*Calluna vulgaris*) and other ericaceous species for much of the year. At least locally, the abundance of their heather-dominated habitat is due to deliberate management for game-shooting in place of domination by sheep-rearing.

Passerines

Of the three passerines listed, wheatear (*Oenanthe oenanthe*) and whinchat (*Saxicola rubetra*) were formerly common in some lowland habitats. I have classed them as moorland birds because they have been virtually eliminated from the lowlands rather than because they are intrinsically upland birds. Ring ouzels (*Turdus torquatus*) have true upland affinities with a preference for steeper and rocky areas, usually with heather; in many southern areas these tend to be less grazed cloughs.

FACTORS AFFECTING BIRD OCCURRENCE AND ABUNDANCE

Data are by no means good enough to quantify temporal trends in numbers of moorland birds except in red grouse (Hudson 1988). There is, however, a clear indication from Table 2 that the birds known to be declining in numbers (especially merlin, red grouse, black grouse, golden plover and dunlin, *Calidris alpina*) are those associated with heather and blanket bog. These declines have been such as to render all of these species very scarce in south-west England and much of Wales.

Birds differ in their habitat requirements and thus in their responses to changes in agriculture. In this section, I outline some of the major factors which influence species richness and abundance.

Geographical location and altitude

Of Britain's upland birds, only the red kite has a southern distribution, being confined to Wales. This species was once widespread, occurring as far north as the
north-east Scottish Highlands as well as more widely in the lowlands, resembling
habitat still occupied on the continent. Its present small range is believed to
reflect the fact that this was the only area where persecution was sufficiently light
to avoid extinction in the late nineteenth century. Greenshank (*Tringa nebularia*)
and golden eagles are virtually confined to Scotland. Suitable habitat for eagles
exists more widely but is scarcer in the less rugged hills of the south; persecution
again probably imposes a limit on breeding range. The other 13 species listed in
Table 2 can occur throughout the British uplands, though several have become
very scarce in more southern areas. If a wider range of 'upland' species had been
considered, a pattern of greater species richness in the central Scottish Highlands
would have been more evident (Hudson 1988). Several species of northern
European distribution occur in Scotland but not elsewhere in Britain (Ratcliffe &
Thompson 1988).

There is no clear trend in variation of individual species abundance with
latitude. There is, however, a tendency for birds to be more numerous in the east
than the west, known from both Scotland and Wales. Distribution maps, such as
those given by Sharrock (1976), show a rather easterly distribution of short-eared
owls on the more productive moors in Scotland, while in contrast the carrion-
feeding raven has a more westerly range. A map of the winter distribution of red
grouse shows not only a relatively easterly range but also that those which do
occur to the west are at low densities (Lack 1986; Hudson 1988). This coincides
with a trend of greater oceanicity to the west, which also interacts with older
harder rocks to produce more peat than mineral soils. Species richness
and abundance tend to decline with altitude. In part this reflects changes in vegetation, especially a decrease in height, but perhaps also a reduction in breeding
season limited by poor weather and shorter flushes of insect abundance.

Land fertility

Other things being equal, higher bird numbers ought to occur in more productive
landscapes. This has been shown for the sparrowhawk (*Accipter nisus*) breeding in
woodland (Newton 1986): better land classes support a greater diversity and
density of small birds taken as prey. The same has been found for the peregrine
(Ratcliffe 1980) and merlin (Bibby & Nattrass 1986). In golden eagles, numbers
appear to be limited by the abundance of carrion in winter, which tends to be
greater in areas of poorer productivity towards the west (Watson *et al.* 1987). This
partly reflects different intensities of sheep management, with fewer animals
being wintered and thus likely to die on the more productive eastern Highland
hills. Breeding success is higher here than in the west, being boosted by more live
prey (such as grouse and lagomorphs), which are associated with better, more fer-

tile ground. This picture is, however, much influenced by the pattern of human persecution, which is itself greater in the east due to active management of sheep flocks and grouse moors. Ravens also tend to be more numerous on poorer land supporting much carrion (Marquiss, Newton & Ratcliffe 1978; Dare 1986).

The herbivorous red grouse varies considerably in density in relation to soil fertility even when the vegetation is little different (Picozzi 1968). Hens are highly selective for quality of heather shoots and female nutrition influences breeding production. Changes in soil fertility thus increase grouse numbers in areas where heather-dominated vegetation prevails. In practice, most agricultural operations which change soil fertility are also likely to change the vegetation either directly (by reseeding) or indirectly (by increased grazing pressure). For most birds, in the long term, such habitat change will override any improvements of soil fertility in remaining areas.

Within grasslands, improved fertility might lead to greater invertebrate abundance and hence favour some birds (Hudson 1988). Hence golden plovers breed at higher densities on some of the more fertile upland soils (Ratcliffe 1976); greater numbers of starlings (*Sturnus vulgaris*) and crows (*Corvus* spp) are supported by improved grasslands in late summer (Kahrom and Edington 1983); golden plovers and curlews feed extensively on grasslands at various stages of improvement. Infestation of tipulids are sometimes a problem in reseeding moorland to grass and may be controlled by insecticides, with as yet unknown, but possibly deleterious, effects on birds.

Vegetation

In natural circumstances, vegetation reflects topography, climate, soil type and drainage. On moorlands, this is strongly overlaid by a number of anthropogenic factors, particularly loss of tree cover and then high grazing intensity (Miles 1988; Ratcliffe & Thompson 1988). Moorland vegetation can sustain grazing without major change at stocking levels compatible with maintaining the original sward (Grant & Maxwell 1988). The main effect of heavy grazing on moorland vegetation is to promote the development of grasses and sedges, such as *Nardus* on poor soils, *Molinia*, *Eriophorum vaginatum* or *E. angustifolium* in wetter areas, or *Agrostis tenuis* and *Festuca* spp. on better drained and less acidic soil (Sydes & Miller 1988). A number of the more characteristic moorland birds exhibit a strong preference for heather-dominated vegetation. Thus red and black grouse, merlin, short-eared owl and hen harrier decline as grass replaces heather. This comprises the largest single impact of agriculture on upland birds.

The conflict of interest between sheep and birds reflects the dilemma of what constitutes the variously used term 'overgrazing'. To birds which favour heather, high grazing levels which convert it to grass are undesirable. It is less clear whether or not overgrazing in the agricultural sense of lowering land or animal productivity is widespread (Mather 1978; Grant & Maxwell 1988).

Decreases in bird populations have been described in the Peak District, England, where sheep numbers trebled between 1930 and 1976 whilst heather-dominated moorland declined by 36 per cent (Anderson & Yalden 1981). Numbers of golden plovers and grouse fell although the documentary evidence for species other than grouse is poor. Curlew and, to a less marked degree, golden plover occur mainly in areas with vegetation varying in height. Without shorter vegetation or conversion to grasses by grazing, it is unlikely that they would breed as widely in homogeneous heather moorland. The curlew is probably the one moorland bird which has increased in numbers as a result of conversion of dwarf-shrub heath to grassland through grazing. A little heather or *Myrica gale* provides nesting cover and more extensive grasslands are exploited for their invertebrates. Lapwings (*Vanellus vanellus*) have also been able to colonize moorland edge pastures and do not occur where there is deep vegetation. Greenshanks also require short vegetation, and the cessation of muirburn can lead to their demise (Nethersole-Thompson & Nethersole-Thompson 1979).

Waders favour wet or poorly drained areas because prey are both more abundant and more accessible to the probers. Moor-gripping for agriculture is rarely sufficiently effective in draining bogs to have much impact on birds, to say nothing of the ground vegetation (Stewart & Lance 1983). On hill margin pastures, however, modern drainage can render areas unsuitable for lapwings and curlews by reducing both food availability and nesting cover provided by rushes (*Juncus effusus*) (Bain 1987). For lapwings, the key adverse factor of pasture improvement is greater predation rate on nests, mainly by gulls (*Larus* spp.), which hunt for earthworms (*Lumbricidae*) but also take eggs (D. Baines, pers. comm.).

Conversion of heather to grass moor has been a major form of habitat loss for merlins in Wales (Bibby 1986). Traditional sites vary in annual occupancy, with the best being occupied in most years. Many sites are now occupied infrequently, and the major factor associated with this appears to be loss of heather relative to grass moor. Pairs in the preferred sites with much heather still tend to rear more young. The nature of this relationship is such that pairs in the poorer sites are unlikely, on average, to rear sufficient young to sustain the population. The causal mechanism underlying preference for heather is unknown (reduced predation risk and more available prey seem likely), but the losses, due mainly to grazing, are clear.

Overall numbers of small passerines do not differ markedly as heather is converted by grazing to grass-dominated moorland. Meadow pipits (*Anthus pratensis*) tend to decline and skylarks (*Alauda arvensis*) and twite (*Acanthis flavirostris*) increase (Moss, Taylor & Easterbee 1979; Knight & Shepherd 1986). Replacement of the one species by the two others reflects preferences for vegetation structure and species (*Molinia*), respectively (Ratcliffe & Thompson 1988).

A distinctive vegetation with abundant bracken (*Pteridium aquilinum*) and scattered surviving trees often occurs on moorland margins. Such areas support higher densities of small passerines and are presently being studied in Wales. While greater bird abundance may in part be related to the better, more freely drained soils, it is more probably related to vegetation type (C. J. Bibby, unpublished). The presence of some woodland, from which bracken typically originates, is an important factor for several species and for bird species diversity. The most characteristic bird of such areas, particularly in southern areas, is the whinchat. The successful development of bracken control measures could have a major impact in such areas. Bracken-covered hill margins also provide important hunting grounds for raptors, especially the merlin, which nests on higher ground but feeds extensively on the moorland margins.

Predation

The role of predators has long been a subject of interest to game biologists attempting to maximize grouse numbers for autumn shooting. In northern England, Hudson (1986a) showed that grouse bags were higher on moors in heavily keepered areas. He suggested that this was due to control of predators, especially foxes, rather than to management of vegetation. In the Peak District, Hudson (1986a) showed that loss of heather alone was insufficient to explain the fall in grouse numbers, which extended to a decline in density on the remaining suitable habitat. The magnitude of this decline was, however, close to that predicted to be attributable to a fall in keepering, the effect of which had been modelled for other areas in northern England. The factor having the greatest effect on year-to-year changes of numbers was chick mortality, followed by over-winter mortality of adults. In a more recent study in Scotland, Hudson (1987) has suggested that predator numbers and grouse mortality rates may be higher than in northern England.

Nesting waders are also very susceptible to predation, although any role of predators in recorded declines of numbers of more southerly populations is not known. Both Bibby (1986) and Newton, Meek & Little (1986) have shown how declining numbers of merlins are associated with poor breeding success, and certainly in Northumbria breeding success has fallen over time. In Wales, but not in Northumbria, nesting success is also greater in heather than on grass-dominated moors. Perhaps increased predation associated with less keeping is one of the factors involved.

The abundance of two major predators, crows and foxes, is likely to be influenced by changes in land use. In Wales, very high numbers of ravens and other carrion-feeding birds are associated with the high and still increasing numbers of sheep (Dare 1986). Carrion crows (*Corvus corone*) probably show a similar trend for they are also substantially dependent on winter carrion (Houston 1977). The abundance of both crows and foxes may additionally be influenced by the spread of forestry providing food and cover and lowering the intensity of human attention to predators (Thompson *et al.* 1988). Neither of these predators is abundant on extensive tracts of semi-natural moorland, where food supplies, especially in winter, are poor. Their numbers, and thus potential for impact on birds, are likely to be substantially influenced by human activity and adjoining land uses. It is therefore possible that the large-scale pattern of future land use could have a significant impact on birds remaining on remnants of suitable habitat. Hudson's (1986a) analysis of grouse numbers in relation to habitat loss in the Peak District might parallel the situation in the uplands of Wales and south-west England, where moorland birds appear to have declined in numbers more strongly than loss of habitat alone would seem to predict.

Disease

The role of disease has been well studied in grouse and two have been shown to be important. Louping-ill is a tick-(*Ixodes ricinus*)-borne virus that affects both sheep and grouse (but apparently not other upland species). It can affect survival of young and adult grouse. This disease is dependent for its maintenance on sheep; ticks without the disease are not known to be harmful. Ticks are particularly associated with bracken, which provides a suitable litter for over-wintering. Hudson (1986a) suggests that decline of grouse on the North York Moors, where bracken has spread considerably, may be due to louping-ill. Louping-ill is thus a disease influenced by two elements (sheep and bracken) of agricultural change.

Grouse are also susceptible to the threadworm *Trichostrongylus tenuis*, which accounts for marked population cycles of grouse found on some moors (Hudson 1986a, b, 1988). Worms pass their immature stages outside grouse and are ingested with heather shoots. High worm burdens may influence both survival of adults and breeding success of females. There may be an interaction between worms and predators, with infected birds being more vulnerable, which raises questions about the extent to which predation mortality is compensated.

Pesticides

Upland raptors were seriously affected by the indirect effects of organochlorine insecticides (Ratcliffe 1970, 1977, 1980). The lower intensity of agriculture in the hills did not provide the same level of pollution as the lowlands, so that, in the case of the sparrowhawk, the nuclei for subsequent recolonization of lowland Britain had survived in upland areas (Newton 1986). Migration of both predators and their prey species was, however, sufficient to contaminate species such as peregrine, even in the remoter upland areas. The winter climate in Scotland is sufficiently mild for many peregrines to remain near their breeding-places. Elsewhere in the world more migratory populations face a greater risk of contamination in more intensively farmed winter quarters and have not made such complete recoveries as those in Britain.

There is no evidence that any British populations of raptors are now sufficiently contaminated to limit numbers; the peregrine has enjoyed a very considerable recovery. But merlins in Orkney and Shetland are not breeding well, and in Orkney numbers have fallen markedly. Nest failures appear to include a high proportion of egg breakages, although organochlorine contamination is not at a level where this would be expected (E. Meek, pers. comm.). Perhaps another pollutant, such as mercury, is responsible (Newton & Haas, in press).

Persecution

Gamekeepers and shepherds have rarely doubted the merit of killing predators. In the heyday of grouse-shooting, raptors were very heavily persecuted irrespective of their possible effects on bags. Many species suffered range reductions and there were several extinctions in Britain (Ratcliffe & Thompson 1988). Red kites, ospreys (*Pandion haliaetus*), white-tailed sea-eagles (*Haliaeetus albicilla*), golden eagles, buzzards (*Buteo buteo*) and hen harriers have never had the chance to fully recover. Although the killing of birds of prey has been illegal for 30 years, their persecution continues to prevent many from achieving much recovery of range or numbers. The species most affected are those which take carrion or eggs (because they are susceptible to poisoning from the non-selective baits put out for crows and foxes; Cadbury 1980). Merlin and peregrine are the upland raptors least influenced by this.

CONCLUSION

Documentation of changes in numbers of upland birds is not very good, and the more gradual changes due to agriculture have not received the same attention as the effects of afforestation. Declines have, however, been noted in eight of the species listed in Table 2. With the exception of raven, these share the common characteristic of preference for, or exclusive occurrence in, heather-dominated moorlands. The conversion of heather to grasslands by grazing and by reseeding constitutes the largest effect of agricultural change on upland birds (Table 1). In the more southerly areas such as Wales and south-western England, many of these birds have suffered declines and local extinctions even where large areas of open ground have not been afforested. This process will be arrested only if grazing levels are lowered to those at which further changes in preferred vegetation types do not occur. In damper areas now dominated by *Molinia*, and of little value to agriculture or birds, it is not clear how long recovery to heather might take, even with continued burning and grazing.

Questions about the impact of crows and foxes on upland birds require further elucidation (Thompson *et al.* 1988). It is suggested that these predators might be supported by forestry and agriculture at greater numbers than would otherwise exist. If Hudson's (1986a) model is correct, this impact is likely to be felt most heavily on small isolated moorland fragments. A similar argument might apply to wide-ranging species, like merlin, for which conversion of heather

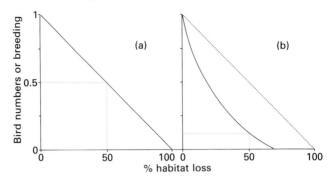

FIG. 1. Two hypothetical relationships between bird numbers (or breeding production) and proportion of habitat lost. In (a) birds are affected linearly in proportion to habitat loss; a 50 per cent decline follows a 50 per cent habitat loss. In (b) the relationship is steeper (dotted line shows (a) for comparison); 50 per cent loss of habitat causes a 90 per cent loss of birds.

to grass generates a progressive loss of habitat suitability. The practical conclusion depends on the shape of the relationship between bird numbers (or breeding success) and habitat loss. If either falls faster than a linear trend, then it would be better to partition large areas of land between those completely converted and those preserved rather than to create an intricate matrix at the average level of conversion. Imagine two large blocks of prime moorland and a demand to convert half their area to an inimical habitat. If the relationship between habitat loss and bird numbers or breeding success were as in Fig. 1a, it would make no difference how it was done; the birds would suffer a 50 per cent loss. If Fig. 1b were a better representation, then the 50 per cent habitat loss, if done as a fine matrix, would cause a 90 per cent loss of birds. If, on the other hand, one block were fully converted and the other were left alone, the loss to birds would only be 50 per cent. Two different patterns of improving the same amount of land could therefore show up to a fivefold difference in numbers of surviving birds (see also Fig. 4 of Thompson *et al.* 1988).

Both for the potential impact of predators and for effects on wide-ranging birds, the scenario of Fig. 1b seems sufficiently plausible to require attention. If it is found to be a good representation, very undesirable consequences could ensue from the current enthusiasm for closer integration between forestry and farming. If moorland birds are more likely to persist on large blocks, a more sympathetic land-use policy should concentrate intensification of forestry and agriculture in some areas, but spare others completely.

ACKNOWLEDGMENTS

I am grateful to Lennox Campbell, Alan Woods, Des Thompson and two referees for comments on previous drafts and to Anita McClune for processing resultant editings.

REFERENCES

Anderson, P. & Yalden, D.W. (1981). Increased sheep numbers and the loss of heather moorland in the peak district, England. *Biological Conservation*, **20**, 195–213.

Bain, C. (1987). *Breeding Wader Habitats in an Upland Area of North Wales (Hiraethog).* Royal Society for the Protection of Birds, Sandy.

Bibby, C.J. (1986). Merlins in Wales: site occupancy and breeding in relation to vegetation. *Journal of Applied Ecology*, **23**, 1–12.

Bibby, C.J. & Nattrass, M. (1986). Breeding status of the merlin in Britain. *British Birds*, **79**, 170–185.

Bunce, R.G.H., Barr, C.J. & Whittacker, H.A. (1981). *Land Classes in Great Britain: Preliminary Description for Users of the Merlewood Method of Land Classification.* Merlewood Research and Development Paper No. 86, Institute of Terrestrial Ecology, Merlewood.

Cadbury, C.J. (1980). *Silent Death: the Destruction of Birds and Mammals through the Deliberate Misuse of Poisons in Britain.* Royal Society for the Protection of Birds, Sandy.

Cramp, S. (1980, 1983 & 1985). *The Birds of the Western Palearctic, Vols 2, 3, 4.* Oxford University Press, Oxford.

Dare, P.J. (1986). Raven, *Corvus corax*, populations in two upland regions of North Wales. *Bird Study*, **33**, 179–189.

Dennis, R.H., Ellis, P.M., Broad, R.A. & Langslow, D.R. (1984). The status of the Golden Eagle in Britain. *British Birds*, **77**, 592–607.

Fuller, R.J. (1982). *Bird Habitats in Britain.* Poyser, Calton.

Goodier, R. (1985). Nature conservation in the hills. *Vegetation Management of Northern Britain* (Ed. by R.B. Murray), pp. 55–70. British Crop Protection Council, London.

Grant, S.A. & Maxwell, T.J. (1988). Hill vegetation and grazing by domesticated herbivores: the biology and definition of management options. *This volume.*

Houston, D. (1977). The effect of hooded crows on hill sheep farming in Argyll, Scotland. *Journal of Applied Ecology*, **14**, 1–15.

Hudson, P.J. (1986a). *Red Grouse, the Biology and Management of a Wild Gamebird.* Game Conservancy Trust, Fordingbridge.

Hudson, P.J. (1986b). The effects of a parasitic nematode on the breeding production of red grouse. *Journal of Animal Ecology*, **55**, 85–92.

Hudson P.J. (1987). Winter problems for scottish grouse. *Game Conservancy Annual Review No. 18*, 146–150.

Hudson, P.J. (1988). Spatial variations, patterns and management options in upland bird communities. *This volume.*

Kahrom, E. & Edington, J.M. (1983). *A Comparison of Bird Populations on Improved and Unimproved Upland Grazing Land at Pwllpeiran E.H.F.* University of Wales, Cardiff.

Knight, A. C. & Shepherd, K.B. (1986). The effects of agricultural improvements on some upland breeding birds in Wales. *Nature in Wales*, **5**, 12–18.

Lack, P. (Compiler) (1986). *The Atlas of Wintering Birds in Britain and Ireland.* Poyser, Calton.

Marquiss, M., Newton, I. & Ratcliffe, D.A. (1978). The decline of the raven *Corvus corax* in relation to afforestation in southern Scotland and northern England. *Journal of Applied Ecology*, **15**, 129–144.

Mather, A.S. (1978). The alleged deterioration in hill grazing in the Scottish highlands. *Biological Conservation*, **14**, 181–186.

Miles, J. (1988). Vegetation and soil change in the uplands. *This volume.*

Moss, D., Taylor, P.N. & Easterbee, N. (1979). The effects on songbird populations of upland afforestation with spruce. *Forestry*, **52**, 124–147.

Nethersole-Thompson, D. & Nethersole-Thompson, M. (1979). *Greenshanks.* Poyser, Berkhamsted.

Newton, I. (1986). *The Sparrowhawk.* Poyser, Calton.

Newton, I., Davis, P.E. & Moss D. (1981). Distribution and breeding of red kites in relation to land-use in Wales. *Journal of Applied Ecology*, **18**, 173–186.

Newton, I., Meek, E. & Little, B. (1986). Population and breeding of Northumbrian merlins. *British Birds*, **79**, 155–170.

Picozzi, N. (1968). Grouse bags in relation to the management and geology of heather moors. *Journal of Applied Ecology*, **5**, 483–488.

Ratcliffe, D. (1962). Breeding density in the peregrine falcon *Falco peregrinus* and raven *Corvus corax*. *Ibis,* **104,** 13–39.

Ratcliffe, D.A. (1970). Changes attributable to pesticides in egg breakage frequency and eggshell thickness in some British birds. *Journal of Applied Ecology,* **17,** 67–107.

Ratcliffe, D.A. (1976). Observations on the breeding of the golden plover in Great Britain. *Bird Study,* **23,** 63–116.

Ratcliffe, D.A. (1977). Upland birds — an outline. *Bird Study,* **24,** 140–158.

Ratcliffe, D.A. (1980). *The Peregrine Falcon.* Poyser, Calton.

Ratcliffe, D.A. (1984). The peregrine breeding population of the UK in 1984. *Bird Study,* **31,** 1–18.

Ratcliffe, D.A. & Thompson, D.B.A. (1988). The British uplands: their ecological character and international significance. *This volume.*

Sharrock, J.T.R. (Compiler) (1976). *The Atlas of Breeding Birds in Britain and Ireland.* Poyser, Berkhamsted.

Stewart, A.J.A. & Lance, A.N. (1983). Moor-draining: a review of impacts on land use. *Journal of Environmental Management,* **17,** 81–89.

Stroud, D.A., Reed, T.M., Pienkowski, M.W. & Lindsay, R.A. (1987). *Birds, Bogs and Forestry: the Peatlands of Caithness and Sutherland.* Nature Conservancy Council, Peterborough.

Sydes, C. & Miller, G.R. (1988). Range management and nature conservation in the British uplands. *This volume.*

Thompson, D.B.A., Galbraith, H. & Horsfield, D.A. (1987). Ecology and resources of Britain's mountain plateaux: land-use issues and conflicts. *Agriculture and Conservation in the Hills and Uplands* (Ed. by M. Bell & R.G.H. Bunce), pp 22–31. Institute of Terrestrial Ecology, Grange-over-Sands.

Thompson, D.B.A., Stroud, D.A. & Pienkowski, M.W. (1988). Afforestation and upland birds: consequences for population ecology. *This volume.*

Watson, A. (1985). Soil erosion and vegetation damage near ski lifts at Cairn Gorm, Scotland. *Biological Conservation,* **33,** 363–381.

Watson, D. (1977). *The Hen Harrier.* Poyser, Berkhamsted.

Watson, J., Langslow, D.R. & Rae, S.R. (1987). *The Impact of Land Use Changes on Golden Eagles in the Scottish Highlands.* Nature Conservancy Council, Peterborough.

Afforestation and upland birds: consequences for population ecology

D. B. A. THOMPSON, D. A. STROUD* AND
M. W. PIENKOWSKI*

*Chief Scientist Directorate, Nature Conservancy Council,
12 Hope Terrace, Edinburgh EH9 2AS, and* Chief Scientist Directorate,
Nature Conservancy Council, Northminster House, Peterborough PE1 1UA*

SUMMARY

1 Almost 1 million hectares of upland grasslands, heaths and bogs have been afforested since 1924. Consequences for upland bird populations are reviewed.

2 The scientific interest of Britain's upland birds is summarized. Palaeoecological information is used to infer for how long present-day bird assemblages may have been widely established in the uplands; there appear to be marked spatio-temporal differences.

3 The 71 species associated with the uplands are considered under four assemblages (predatory and scavenging birds, waders (Charadrii), grouse (Tetraonidae) and small passerines, and freshwater birds) for detailed discussion.

4 At least 34 species are at some risk from afforestation, principally through loss of nesting or feeding range, potentially greater risks of adult, nest or young predation, and apparent increases in water acidity. The distribution of 14 species is being substantially reduced and at least another 8 species are likely to decline through further forestry expansion.

5 New forests are characterized by a greater abundance of common species, which replace the open-ground assemblage composed mainly of less common species. Afforestation may eventually have considerable impacts on ground-nesting birds in remaining unplanted areas, through predation, competition, interference, loss of mate or site fidelity.

6 Graphical models indicate that decline in bird populations is unlikely to be a simple linear function of the amount of habitat afforested; it probably increases supra-proportionately with both the area and quality of ground planted.

INTRODUCTION

The breeding birds of northern and north-west European upland grasslands, heaths, bogs and fellfields are of great interest to ecologists. They breed in a harsh environment and exhibit a spectacular range of social, breeding and migration systems. British upland bird communities are of particular interest to scientists abroad because (i) there is a unique mixture of arctic, boreal, temperate, oceanic and continental species in relatively confined areas, (ii) some of the component species have European strongholds or breed at higher densities than elsewhere, (iii) many of the birds have southern or western outposts in Britain, and three have disjunct distributions (ptarmigan, great skua and twite), and (iv) few other parts of the world have comparable areas of blanket bog, heath and grassland with such high numbers of open-ground nesting birds (Ratcliffe 1989; Ratcliffe & Thompson 1988). In some areas management for grouse, sheep and deer has produced extensive open landscapes which contribute substantially to the last of these interests.

Of the various anthropogenic factors accounting for change in the density or distribution of upland birds (Bibby 1988; Hudson 1988; Ratcliffe & Thompson 1988), recent afforestation has had the greatest impact through the complete transformation of habitat from open ground to typically closed forest. Consequences for landscape value, soils, water quality, vegetation and animals have been widely reviewed (NCC 1986; Ratcliffe 1986; Bainbridge *et al.* 1987; Lindsay 1987; Stroud 1987; Stroud *et al.* 1987; Thompson 1987; Lindsay *et al.* 1988; Gee & Stoner 1988). In this paper we review the known and potential impacts of afforestation on the population ecology of birds breeding or feeding extensively in the uplands. We also discuss variations in the possible timing of establishment by typical open-ground bird communities in the uplands, and summarize statistics on the extent of habitat loss to afforestation.

As often happens in reviewing the literature on habitat loss we have found few examples of good 'control' and 'experimental' site data. Technological advances have facilitated the afforestation of ground previously considered unplantable. Trial plantations incorporating impact studies have not been made, despite extensive changes in land use. When considering effects of habitat loss on the population ecology of just one species we should ideally have information on (i) pre-impact bird population sizes in areas to be planted (experimental) as well as those to be left alone (controls), (ii) the fate of displaced birds, and (iii) changes in population size and productivity through time in planted and unplanted areas. Although time and money have not been available to meet these requirements, we are fortunate, in Britain, in having an upland botanical and ornithological resource that has probably been more comprehensively surveyed and monitored than anywhere else in the world.

UPLAND BIRDS

Up to 71 species may be associated with the Britain uplands: 19 breed almost solely in sub-montane or montane habitats; 19 are opportunists breeding in a range of other habitats as well (see Table 1); 15 use lakes or rivers; and a minimum of 18 have at least a foothold in the uplands. We concern ourselves mainly with those of the sub-montane zone (below 650 m, dropping to 300 m in the north; i.e. above the limits of enclosed farmland and below the potential tree limit; Ratcliffe & Thompson 1988) in areas which are potentially plantable. The

TABLE 1. Seventy-one bird species associated with the British uplands (adapted in part from Ratcliffe 1989) and potential vulnerability to afforestation. The right hand column (*) is somewhat subjective. At least another 22 species associated with woodland or scrub could be added (if upland woods were more widespread and at higher altitudes within the sub-montane zone some of their birds would have significant footholds in the uplands). Rare species (e.g. snowy owl, Nyctea scandiaca, and purple sandpiper, Calidris maritima) have been excluded. The superscripts indicate: 1, the 34 species potentially at some risk from afforestation (see also Ratcliffe 1986; Stroud et al 1987); 2, the 14 species eradicated or substantially reduced in distribution by afforestation (see text; also Ratcliffe 1986; 3, the 8 species likely to decline through further forestry expansion (this is a minimum estimate)

Breed mainly in montane or sub-montane habitat	Opportunistic species with major niches in mountains and moorland	Use upland lakes, rivers and streams	Have at least a foothold in the uplands*
Ptarmigan (Lagopus mutus)	Peregrine falcon (Falco peregrinus)[1]	Black-throated diver (Gavia arctica)[1,3]	Jackdaw (Corvus monedula)
Dotterel (Charadrius morinellus)	Raven (Corvus corax)[1,2]	Red-throated diver (Gavia stellata)[1,3]	Tree pipit (Anthus trivialis)
Snow bunting (Plectrophenax nivalis)	Buzzard (Buteo buteo)[1,2]	Wigeon (Anas penelope)[1]	Grasshopper warbler (Locustella naevia)
Red grouse (Lagopus lagopus scoticus)[1,2]	Kestrel (Falco tinnunculus)	Goosander (Mergus merganser)	Pied wagtail (Motacilla alba)[1]
Golden plover (Pluvialis apricaria)[1,2]	Red kite (Milvus milvus)[1]	Red-breasted merganser (Mergus serrator)[1]	Mistle thrush (Turdus viscivorus)
Dunlin (Calidris alpina)[1,2]	Carrion/Hooded crows (Corvus corone)	Common scoter (Melanitta nigra)[1]	Song thrush (Turdus philomelos)
Twite (Acanthis flavirostris)[1,3]	Meadow pipit (Anthus pratensis)	Teal (Anas crecca)	Oystercatcher (Haematopus ostralegus)
Whimbrel (Numenius phaeopus)	Skylark (Alauda arvensis)[1,2]	Dipper (Cinclus cinclus)[1,3]	Ringed plover (Charadrius hiaticula)
Red-necked phalarope (Phalaropus lobatus)	Wren (Troglodytes troglodytes)	Grey wagtail (Motacilla cinerea)	Herring gull (Larus argentatus)
Arctic skua (Stercorarius parasiticus)[1,3]	Wheatear (Oenanthe oenanthe)[1,2]	Greylag goose (Anser anser)[1]	Lesser black-backed gull (Larus fuscus)
Great skua (Stercorarius skua)[1]	Whinchat (Saxicola rubetra)[1]	Common sandpiper (Actitis hypoleucos)	Great black-backed gull (Larus marinus)
Greenshank (Tringa nebularia)[1,2]	Stonechat (Saxicola torquata)[1,3]	Common gull (Larus canus)	Stock dove (Columba oenas)
Wood sandpiper (Tringa glareola)	Cuckoo (Cuculus canorus)	Mallard (Anas platyrhynchos)	Nightjar (Caprimulgua europaeus)[1]
Temminck's stint (Calidris temminincki)	Lapwing (Vanellus vanellus)[1,2]	Black-headed gull (Larus ridibundus)	Whitethroat (Sylvia communis)
Ring ouzel (Turdus torquatus)[1,2]	Snipe (Gallinago gallinago)[1,2]	Slavonian grebe (Podiceps auritus)	Tawny owl (Strix aluco)
Golden eagle (Aquila chrysaetos)[1,3]	Redshank (Tringa totanus)[1,2]		Willow warbler (Phylloscopus trochilus)
Merlin (Falco columbarius)[1,2]	Curlew (Numenius arquata)[1,2]		Redwing (Turdus iliacus)
Hen harrier (Circus cyaneus)[1]	Black grouse (Tetrao tetrix)		Fieldfare (Turdus pilaris)
Short-eared owl (Asio flammeus)[1]	Chough (Pyrrhocorax pyrrhocorax)[1,3]		

ground is composed predominantly of peatlands, grasslands and dwarf-shrub heaths, and integral lochs, rivers, streams and pool complexes.

Detailed accounts of the upland species are provided elsewhere (Sharrock 1976; Ratcliffe 1977, 1980, 1989; Nethersole-Thompson & Nethersole-Thompson 1979, 1986; Hudson 1986; Stroud *et al.* 1987; Bibby 1988; Ratcliffe & Thompson 1988).

FORESTRY IN UPLAND BRITAIN

Implications of forest absence/clearance for contemporary bird populations

Palaeoecological evidence of long-term and recent changes in extensive forest cover of the British uplands have been well reviewed by Birks (1988). This is helpful in considering for how long the uplands have been used widely by contemporary bird assemblages throughout the Holocene (post-glacial, beginning around 10000 years BP). In Orkney, Shetland, Outer Hebrides, some of the Inner Hebrides and parts of Caithness, however, extensive forest did not exist, and afforestation there is transforming a naturally treeless landscape (references in Birks 1988). In much of remaining northern and western Britain it appears that peat formation overwhelmed large areas of woodland by as early as 9000 BP in the Scottish Highlands (Birks 1975) and, depending on altitude and local topography, between 7500 and 2000 BP elsewhere (McVean & Ratcliffe 1962; Tallis 1964; Birks 1975, 1988; Edwards & Hirons 1982). Present-day plantations are completely different ecologically from the woodlands previously present in some areas.

Deforestation occurred locally from around 5000 BP but extensively and permanently much more recently. Data provided by Birks (1988) suggest that large tracts became treeless through man's activities and so widely available to the birds present today around 3900–3700 BP in the north-west Highlands and eastern Skye; 2600–2100 BP in Wales, England (except the Lake District), northern Sutherland and northern Skye; 1700–1400 BP in the Lake District, Galloway, Knapdale–Ardnamurchan and southern Skye; and 300–400 BP in the Grampians and Cairngorms. These figures point to remarkable spatial variations in potential duration of occupation by contemporary upland bird communities, which are complicated further by location, timing and extent of scrub clearance, burning and grazing, as well as local variations in peat development.

There are thus differences across Britain not only in the community composition and population densities of birds (Sharrock 1976; Ratcliffe 1976, 1980, 1989; Hudson 1986, 1988) but also, apparently, in timing of settlement and establishment. If some of these populations are reproductively closed, it is just possible that individuals in different areas exhibit somewhat specialized adaptations to their particular area. For example, greenshank populations in Speyside may be several thousand years younger than those in north-western Sutherland

and this might contribute to the tendency for Speyside to have individuals significantly less faithful to nest sites (Thompson, Thompson & Nethersole-Thompson, in press; populations were not studied during the same periods). The areas will certainly differ in soil productivity and nutrition, and this may drive corresponding differences in how birds exploit prey or select nest sites. If these speculations are correct, then arguments for preserving the unique spatial variation in Britain's upland birds (Ratcliffe 1977) are all the stronger. Although afforestation in a given district may displace only a fraction of Britain's total population of a species, the loss of birds belonging to the relevant spatio-temporal zone will obviously be more significant.

Present-day forest cover

The forest cover of Britain is now 10·0 per cent (2,265,000 ha) compared with only 5·3 per cent (1,180,000 ha) in 1924 (Forestry Commission 1987). Almost 14 per cent of Scotland is forested. The census (Forestry Commission 1984) for 1979–1982 indicated planting during 1971–1980 was 60 per cent Sitka spruce, 14 per cent lodgepole pine, 5·5 per cent Japanese/hybrid larch, 4·4 per cent Scots pine, 10·7 per cent other conifers and 5·3 per cent broadleaves. The main period of expansion was around 1945–1950 in England and North Wales, and from 1960 onwards in Scotland and South Wales. Britain has experienced a greater expansion of forestry since 1950 than any other European country, but still has the second lowest percentage cover of forest (Ratcliffe & Thompson 1988).

RECENT TRANSFORMATION OF THE UPLANDS

In order to assess the effects of afforestation on upland birds it is necessary, first, to appreciate the extent of open ground being lost. Almost 30 per cent of Britain's surface is upland (Fig. 1a; and Ratcliffe & Thompson 1988). The habitat is valued highly by ecologists and conservationists, yet large areas have been destroyed or transformed recently by agricultural reclamation, drainage and reseeding, and by tree-planting (Fig. 1b). Almost 1 million ha of upland have been planted since 1924 (NCC 1986). Statistics provided by Woods & Cadbury (1987) indicate that the change in moorland during 1946–1981 amounts to reductions of 41 per cent (Wales), 22 per cent (England) and 11 per cent (Scotland). The percentage reduction per annum (during 1976–1981) ran at 1·8 per cent, 1·0 per cent and 1·3 per cent respectively. Sydes (unpublished) and Sydes & Miller (1988) reveal, for instance, that of the 12 per cent of semi-natural moorland transformed post-1945 in six English and Welsh National Parks, 77 per cent have been afforested. In the Less Favoured Areas of England and Wales there has been a 31 per cent loss of agricultural 'rough grassland' during 1949–1981, with 67 per cent due to afforestation. By 1980, Scottish, English and Welsh uplands held forestry plantations

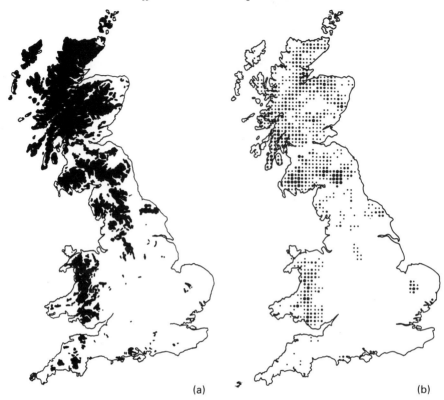

FIG. 1. Distribution of upland habitat and recent afforestation. (a) Rough pasture and moorland are shown shaded dark (adapted from Ratcliffe 1976). (b) Distribution of existing forest (as at 1986) planted since 1920 in every 10 km square of the National Grid: no dot indicates no planting, • less than 10 per cent, • 10–50 per cent, ● greater than 50 per cent (from NCC 1986).

and woodlands amounting to a percentage cover of 12·6 per cent, 7·3 per cent and 11·6 per cent, respectively (Rowan 1986).

Much afforestation has occurred on blanket bog because in agricultural terms it is relatively unproductive and cheap to purchase; land purchase was the only element of commercial afforestation not grant-aided or eligible for tax relief pre-1988 (Bainbridge *et al.* 1987). Extensive areas have been afforested in Wales, the Cheviots and the Southern Uplands of Scotland. Substantial tracts of blanket bog have been planted on the Hebridean islands of Islay, Mull and Skye (Bramwell & Cowie 1983). In south-west Scotland, the Kintyre Peninsula's blanket bog surface has been reduced by over 30 per cent since 1945 because of afforestation (NCC 1986). And, of the 401,375 ha of all pre-afforestation blanket bog in Caithness and Sutherland, 16·5 per cent has been planted, or is programmed for planting, mostly in the last 6 years (figure for up to 1986 given by Stroud *et al.* 1987).

SPECIFIC IMPACTS OF AFFORESTATION ON UPLAND BIRDS

In Fig. 2 we indicate a number of ways in which afforestation can influence upland bird populations. Detailed evidence is now reviewed by considering four separate upland bird assemblages.

Scavenging and predatory birds

The principal impact on scavenging and predatory birds derives from the loss of open-ground feeding habitat. There are three broad categories: (i) sheepwalk, which tends to be grassland-dominated and prevails in the more base-rich areas of upland Britain (e.g. Southern Uplands, Breadalbane Hills); (ii) deer forests composed of mixed grassland and dwarf-shrub heath in more rugged, mountainous areas (e.g. Highlands and Islands); and (iii) grouse moors (McVean & Ratcliffe 1962; Ratcliffe 1977, 1989).

Ravens in many upland areas feed mainly on sheep carrion (Ratcliffe 1962; Newton, Davis & Davis 1982; Dare 1986). Between the early 1960s and 1975

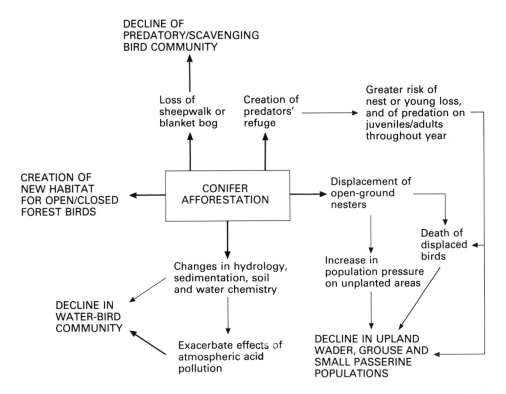

FIG. 2. Examples of how conifer afforestation can influence upland bird populations. Thick lines indicate relationships already established.

there was a 55 per cent decline in occupancy of regular nesting areas, and a 44 per cent reduction in number of breeding pairs, in south Scotland and Northumberland (Marquiss, Newton & Ratcliffe 1978). This was due largely to afforestation post-1940 replacing sheepwalk, and so reducing the availability of sheep carrion (Fig. 3a). Where ravens declined, most territories were vacated within a few years of trees being planted. The percentage of traditional nesting areas occupied was therefore lower in areas where afforestation was most extensive (Fig. 3a). The most heavily afforested areas contained pairs which were less likely to breed, were later breeders and produced fewer young. Good habitat (with alternative sources of prey such as lagomorphs, grouse, other birds and their eggs) was deserted later than poor habitat if both received the same amount of planting (Marquiss *et al.* 1978). Mearns (1983) continued monitoring these ravens, and found that the predicted decline continued and had reached 72 per cent by 1981. In the Lake District, with comparable hill ground but no afforestation, there has been no overall decline in raven numbers since the early 1960s (Marquiss *et al.* 1978; D. A. Ratcliffe, pers. comm.).

Afforestation in mid-Wales, however, has not had any appreciable effect on ravens (Newton *et al.* 1982). Although sheepwalk is obviously being planted it appears that the copious carrion supply in unplanted areas (Dare 1986), much alternative live prey (on which ravens may be more dependent than previously realized, Cross & Davis, unpublished) and a fragmented, uneven-aged cover of forests are compatible, so far, with a sustained high density of breeding ravens. Elsewhere, such as in north-east England, the north Pennines, and north-east Scotland, however, ravens have declined or seem to be disappearing locally because of persecution, changes in sheep husbandry and/or afforestation.

The buzzards of mid-Wales took much more live prey, particularly rabbits (*Oryctolagus cuniculus*), than did ravens, and bred as successfully in heavily afforested localities (up to 40 per cent ground planted within 1 km from nest) as in more open localities (Newton *et al.* 1982). But in Scotland, one Galloway population has declined by 93 per cent ($n = 30$ pairs) between 1946 and 1981 (D. A. Ratcliffe, pers. comm.; Mearns 1983). Although only 50 per cent of the sheepwalk was afforested there (in blanket rather than mosaic pattern), the loss of more productive, agricultural land was locally well above 50 per cent. Red kites are potentially at risk from afforestation of sheepwalk and steep hunting slopes, but may be more dependent on live prey than are ravens (Cramp & Simmons 1980).

The golden eagle is also succumbing to afforestation. Marquiss, Ratcliffe & Roxburgh (1985) studied the fate of four eagle territories in south-west Scotland between the early 1940s and 1979. They found the cessation of breeding in two territories coincided with locally large-scale afforestation, particularly of the most productive ground below 905 m. The eagles fed mainly on large mammal carrion, lagomorphs and large birds (chiefly in summer), and breeding performance was best where the amount of large birds in the spring diet was greatest. It seems that reduced breeding success arose from the loss of somewhat more pro-

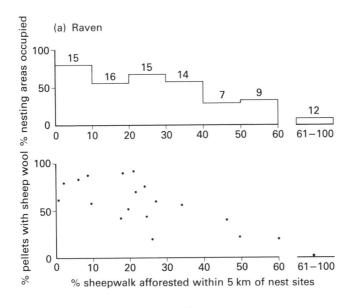

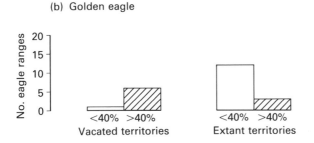

% plantation forestry in middle altitude ground within 4 km of territory centre

Fig. 3. Effects of afforestation on scavenging/predatory birds. (a) Percentage of raven nesting areas occupied (with sample size) and percentage of pellets with sheep wool in relation to extent of sheepwalk afforested within 5 km of nest sites in south Scotland and north England ($P = 0.05$ for top graph, and $P < 0.01$ for sheepwalk afforested within 3 km of nest; $r = -0.668$, $P < 0.01$ for bottom graph). Data analysed from Marquiss et al. (1978). (b) Occupation of golden eagle territories in relation to percentage of range afforested in mid-altitude land (100–300 m) within 4 km of territory centre. Data for mid-Argyll (1956–1985). Vacated territories have significantly more forestry than extant territories ($\chi^2 = 8.31$, $P < 0.01$). Data analysed from Watson et al. (1987).

ductive (e.g. for red grouse) sub-montane ground to trees, forcing eagles to depend more on the less reliable and poorer carrion supplies and live fauna of the higher, unplantable ground. Although eagles in Fennoscandia feed within natural conifer woods on capercaillie (*Tetrao urogallus*) and black grouse (Tjernberg 1981), these forests, which are substantially more open than British plantations, include large bogs and other clear spaces. British eagles rarely forage over plantations (Watson, Langslow & Rae 1987) presumably because they are too bulky to catch live prey amongst dense, young trees and rank vegetation (Marquiss *et al.* 1985), and because they are wary of feeding near areas of cover which could potentially conceal somebody with a gun.

Of course, the breeding ecology of eagles differs markedly between areas, reflecting historical and contemporary differences in land use and available food. Between 1981 and 1985, Watson *et al.* (1987) studied eagles in six parts of Scotland. Across these they found that whereas large amounts of winter carrion accounted for highest breeding densities, good densities of live prey accounted for the best breeding performance per study area. In mid-Argyll, some 140 km north of where Marquiss *et al.* (1985) worked, up to 35 per cent of the eagle's hunting range has been afforested during the past 25 years. Watson *et al.* (1987) found a 30 per cent decline in eagle numbers there and that most of the vacated ranges held more than 40 per cent plantation forestry in the middle-altitude land (100–300 m) within 4 km of the nesting territory centre (Fig. 3b); they suggest that at least 5 out of 15 extant pairs will desert their ranges by 1997. This is because each has more than 30 per cent of the middle-altitude land category afforested, with the amount of tree-planting increasing. Eagle populations in all six areas are being monitored to examined the influences of further changes in land use.

Turning to the predators dependent almost entirely on live prey, we find that merlin, hen harrier and short-eared owl are potentially at risk from afforestation. Merlin, in particular, tend to feed more in open country than in young or native forests (in which they can breed: Newton, Meek & Little 1978,1984; Bibby 1986, 1987). Although the expansion of afforestation in some areas (e.g. Northumbria, Southern Uplands and Wales) has coincided with large declines in population size (Newton *et al.* 1978, 1986; Bibby & Nattrass 1986), a causal link between the two has not been established. No ideal 'controls' exist because several other aspects of the merlin's environment have also changed. First, there is a loss of heather moorland to grassland through overgrazing (Anderson & Yalden 1981; Bibby 1986, 1988; Sydes & Miller 1988). Second, numbers of nest predators, particularly crows and foxes have increased in many areas (e.g. Hudson 1988). Third, there are also local increases in human persecution (Haworth 1987), time-lags in organochlorine pesticide-induced failures in breeding performance (Newton *et al.* 1986), and local reductions in the abundance of farmland passerines available in winter (Bibby 1988). There are also practical difficulties involved in locating all merlin nests within forests (mostly in old crow nests near the edge) so that there may be a slight underestimate in population sizes for afforested areas. Hen harriers and short-eared owls certainly nest in young plantations, but with the formation of thicket forest they leave and tend to feed at the edge of forests or over remaining moorland and marginal agricultural ground. In some parts, there can be eventual population declines despite initial increases in numbers (O'Flynn 1983).

Waders

Detecting any changes in wader numbers in relation to extent of afforestation is complicated by pronounced spatial and year-to-year variations in breeding density and by a tendency for both food supply and risk of predation to impose limits on local population size (Holmes 1970; Evans & Pienkowski 1984; Thompson, Thompson & Nethersole-Thompson 1986; Stroud et al. 1987). Although both afforestation and agricultural intensification have influenced the species composition and distribution of moorland wader communities (Bibby 1988), we concentrate on afforestation per se.

Several long-term studies have monitored changes in wader populations through considerable periods of time. In the central Scottish Highlands, Nethersole-Thompson & Watson (1981) noted the collapse of a greenshank population following afforestation and lack of muirburn in adjacent moorland. In Overscaig, Central Sutherland, there was a net decline in combined numbers of golden plover, dunlin and greenshank in an area near new plantations despite an initial increase in numbers (Stroud et al. 1987). Afforestation of half the Kerloch Moor, Kincardineshire, was followed by a marked decline in the breeding population of golden plovers, possibly because there was an increase in predation of nests of birds breeding on the remaining moorland (R. Parr, pers. comm.). Likewise, in one part of Galloway, A. D. Watson (pers. comm.) has found no dunlin and very few golden plovers on more than 20 km² of open moorland surrounded by much well-established forest; during 1950–60 there were scattered pairs of dunlin and relatively high numbers of golden plover (Ratcliffe 1976). J. B. Halliday (pers. comm.) quantified predation rates on nests of lapwing and curlew on marginal agricultural ground and moorland near conifer plantations in south-east Scotland. Of all nests examined, 35·3 per cent ($n = 34$) of lapwings and 68·2 per cent ($n = 22$) of curlew were robbed by fox or crows. Where lapwing nests were closest together there was a higher risk of predation.

For the Caithness and Sutherland peatlands, Stroud et al. (1987) used statistically significant habitat–bird associations to estimate losses of breeding waders to afforestation. They categorized the habitat into four distinct landform classes (pool complexes with Sphagnum flows; sloping blanket bog with pools; steeper and more broken ground; and montane ground or ground too steep for breeding moorland waders), calculated the average number of dunlin, golden plover and greenshank found on each, and then multiplied up by the total surface area of each class in order to estimate the total numbers of breeding birds. Having examined old, pre-afforestation maps, they estimated that the following number (and percentage) of pairs had been displaced: 912 (19 per cent) golden plover, 791 (17 per cent) dunlin and 130 (17 per cent) greenshank. These losses may have amounted to 4·0 per cent, 7·9 per cent and 13·5 per cent, respectively, of the British populations.

Curlew, redshank, snipe and common sandpiper are typically less abundant upland waders where most afforestation occurs, but are just as likely to be permanently displaced where their nesting habitat is afforested.

Grouse and small passerines

These are considered together because they are important for predatory birds (Ratcliffe 1962, 1980; Newton *et al.* 1982; Watson *et al.* 1987; Bibby 1987; Hudson 1988). Whereas red grouse, ring ouzel, wheatear and skylark are likely to be permanently displaced, others, such as black grouse, meadow pipit, stonechat and whinchat, will occupy open ground within new forests (Moss, Taylor & Easterbee 1979; Ratcliffe 1986). The ring ouzel tends to breed on steep, heathery slopes higher than 250 m (Flegg & Glue 1975) but will nest much lower in the far north and west. In south Scotland, Rankin & Taylor (1985) found that in the afforested uplands only moors larger than 270 ha were likely to retain ring ouzel (as well as merlin and short-eared owl). Poxton (1986) found no ring ouzel nesting in forestry plantations in the Pentland Hills, Scotland.

Well-managed grouse moors can be run very profitably, but losses of grouse due to foxes and crows harboured by nearby plantations can lower the value of estates. In some areas this may in turn contribute to the selling of land for further tree planting (Hudson & Watson 1985; Thompson 1987).

Although few upland passerines breed in mature conifer plantations, there is an overall increase in passerine density and species diversity as plantations age (Moss *et al.* 1979; Newton 1986a, b). However, this does not imply that upland raptors will secure more prey following planting, as they do not appear to hunt in forests. A given upland area with conifer blocks will therefore produce a lower biomass of available prey than similarly sized treeless areas.

Freshwater birds

The final assemblage contains birds of running and open water bodies. Despite numerous adverse impacts of afforestation on soil and water chemistry, hydrology and sedimentation (Gee & Stoner 1988), information on any adverse consequences for breeding birds is patchy. This is partly because extensive conifer afforestation can exacerbate inimical influences of atmospheric acid deposition (Gee & Stoner 1988; see also Fig. 2).

Dippers and grey wagtails have been studied extensively along the River Wye, Wales (see Tyler's 1987 summary). At least three trends are emerging. First, more acidic tributaries have fewer invertebrates (especially Trichoptera and Ephemoptera). Second, dippers breed at highest densities in the less acidic tributaries, and in those with more Trichoptera. And, third, a few tributaries which have increased in acidity, and now hold fewer breeding dippers, drain the catchments with extensive conifer plantations (Ormerod & Edwards 1985; Ormerod, Tyler & Lewis, 1985; Ormerod & Tyler 1986). Numbers appear to be declining because the dippers of acidic streams delay breeding, lay only one clutch, produce small broods, and have nestlings with lower feeding rates and growth rates than those of non-acidic streams (Tyler 1987). Some similar trends are evident in south Scotland (J. Vickery, pers. comm.).

Although fish stocks are reduced by conifer-induced acidification, it appears, at least in Britain, that pisciverous birds have not declined in response to this (Haines 1981; Eriksson 1984). Reasons for this may include the presence of fewer fish that compete for the piscivores' own prey, or the passage of insufficient time for the effect to show. Perhaps birds of rivers and large lochs e.g. goosander, red-breasted merganser, black-throated diver and wigeon are more likely to decline than those of smaller lochans and pools, e.g. red-throated diver and teal, because the latter may feed elsewhere or in local pools with water not drawn from the afforested catchment (see Lindsay *et al.* 1988).

Most of these birds are mobile whilst feeding anyway, and many feed in nearby coastal areas (e.g. red-throated diver and merganser). Likewise, it is unlikely that waders, except perhaps common sandpipers and some greenshanks, will be influenced by such acidified waters. Ormerod & Tyler's (1987) study of Welsh grey waytails provides a good example of mobility and dietary diversification for it shows that wagtails are much less dependent than dippers on upland stream invertebrates and so have not declined on the same streams as dippers. However, Tyler (1987) suggests that streams bordered by broadleaved trees or open moorland have watercourses with more prey for wagtails than those in conifer-afforested ground.

Blanket afforestation can also be associated with a widely fluctuating water table due to changes in the ratio of the rate or intensity of water runoff to precipitation. This may be associated with the flooding of some nests belonging to black- and red-throated divers.

IMPACTS ON GROUND-NESTING BIRDS IN UNPLANTED AREAS

Ground-nesting upland birds in unplanted areas adjacent to or even quite far from the forest edge may also be at risk from afforestation. At least five aspects have to be considered here: (i) predators operating from forests could inflict losses of adults or nests (Fig. 2); (ii) changes in vegetation structure due to cessation of burning and perhaps a decrease in ground wetness (Lindsay *et al.* 1988) could render some areas less attractive in terms of their feeding, nesting or chick-rearing value; (iii) birds may avoid cover which might conceal predators, and so be less likely to nest near forests; (iv) displaced birds may not die but instead settle (or attempt to) on open ground nearby, thus increasing risk of competition and predation for themselves as well as for birds they have joined; and (v) any displaced birds that do attempt to breed may be less successful because they are not familiar with the new ground that they have had to occupy.

The full, adverse consequences of afforestation are therefore extremely complex. The relative area of forest and shape of remaining ground (in terms of predation risk versus forest fragmentation, Andren *et al.* 1985; or ratio of perimeter to surface area, e.g. Thompson & Barnard 1983), quality of ground as utilized by the birds, and population pressures will also influence how populations change through time. And the actual increase in numbers of potential predators will partly depend on the penetrability of new forests, which provide sanctuary for both crows and foxes, as well as the degree of keepering and other land uses. We look at just three of these aspects in a little more detail.

Predation risk and nest or young predation

Although quantitative studies of predation on upland birds are virtually non-existent, scant evidence and work on other habitats point to density-dependent predation. Birds are more at risk where nesting at high densities (Tinbergen, Impekoven & Franck 1967; Hudson 1986; I. Byrkjedal, pers. comm.; J. B. Halliday, pers. comm.; S. Redpath, pers. comm.), mainly because predators search more where prey is abundant, and because of their functional response to prey density (Begon, Harper & Townsend 1986). Some displaced birds may therefore settle in unplanted areas and impose a high risk of predation on both themselves and neighbours' eggs and chicks. Alternatively, the risks may be so great that they refrain from breeding altogether, thus ultimately being lost from the breeding population. Although semi-colonial birds such as lapwings might benefit from communal mobbing (Elliot 1982), this has to be set against the greater detectability of more birds leaving or arriving at the nest (Byrkjedal 1987). Sonerud & Fjeld's (1987) experiment indicates that hooded crows can memorize between years where they predated a nest, and that there should therefore be selection for breeding site dispersal following nest predation. If birds cannot move elsewhere, e.g. because of competition or lack of suitable ground, then they may lose their nests to crows year after year.

   Bird density may well increase further away from the forest edge or trees (Elliot 1982) because the risk of egg and young predation and of adults being taken is lower far away from trees (Whitfield 1985; J. B. Halliday, pers. comm.). Detailed observation of predators and experiments involving the placement of artificial clutches at different densities (Tinbergen *et al.* 1967) and distances from trees are needed to test some of these suggestions (Byrkjedal 1980).

Competition and interference

Available food for most upland birds is spatially patchy (Watson *et al.* 1984; Yalden 1986; Bibby 1987; Coulson 1988; Usher & Gardner 1988). Better areas probably have birds that defend territories more vigorously, lay earlier, produce larger broods with young that grow rapidly, fledge early and reach maturity relatively quickly (Galbraith 1986). These areas should therefore be more productive, sustain more viable populations, and produce more efficient foragers and competitors (Partridge & Green 1985).

   Afforestation of increasingly better-quality habitats will thus eliminate much greater proportions of the breeding populations. More competition may arise following settlement by a few (extra) displaced birds in which case there may be a density-dependent reduction in productivity of the remaining population (Harris 1975; Evans & Pienkowski 1984; Galbraith 1986; Thompson *et al.* 1986). In breeding golden plovers, adult body condition is lowest during the egg-laying period when males continue to be actively territorial, and females are meeting the costs of producing a clutch (Byrkjedal 1985). Competition for space or food because of more population pressure at this stage could therefore reduce the condition of all birds even further, with a subsequently greater risk of adult predation as well as the production of less viable young.

Breeding-site and mate fidelity

Older birds tend to breed more successfully, and this tends to be enhanced by fidelity to breeding sites and/or mates (Oring 1988; Burke & Thompson 1988). There is a marked tendency also for birds to return to breed where they were born (natal philopatry), and this may account for some genetic adaptations to local environmental conditions (Greenwood 1980). Upland birds that lose their nesting habitat to afforestation, and so return to breed in the following year on 'new' ground, may therefore breed less successfully because they have lost or do not encounter last year's mates. Alternatively, they may be handicapped by the lack of knowledge of their former breeding grounds which was vital to their former success. Marked mate and breeding-site fidelity has been found in six waders that breed in the British uplands: greenshanks and redshanks (Thompson *et al.* 1988; Thompson & Hale, in press), dunlin (Soikkeli 1970; Jönsson 1988); golden plover (Parr 1980), curlew (Nethersole-Thompson & Nethersole-Thompson 1986) and common sandpiper (Holland, Robson & Yalden 1982).

Older or more experienced birds breed earlier, and this appears to contribute to their greater success in the population (Oring & Lank 1984; Thompson *et al.* 1986, 1988). They may not breed as early if they are displaced and so lose their familiarity with territories. Furthermore, there is a greater tendency for dunlin and redshank to divorce following breeding failures in the previous year, and in redshanks for previously unsuccessful birds, particularly females, to move farthest to new nest sites (Thompson *et al.* 1988; Thompson & Hale, in press). If displaced birds breed less successfully, and so divorce more frequently, then age- or experience-related benefits of mate compatibility (Coulson & Thomas 1985) will occur less often. This could contribute to a reduction in population viability.

Upland waders, such as greenshank, are extremely faithful to breeding areas, and even to given nest scrapes, and there is strong evidence indicating that exactly the same nests have been used intermittently over 25 years (but not by the same birds!) in at least two parts of Scotland (Thompson *et al.* 1988). This points to a very fine-tuned relationship between habitat (e.g. physiognomy), site fidelity, breeding success and population persistence, which will be undermined following afforestation. An increase in crow numbers could conceivably compel typically site-faithful birds to desert their traditional breeding sites because of repeated predation over several years (Sonerud & Fjeld 1987). Here, we can visualize local increases in dispersal contributing to more population pressure and poorer lifetime breeding performance on remaining unplanted areas.

FURTHER CHANGES IN BIRD POPULATION SIZE

Our graphical model in Figs. 4a, b indicates four possible outcomes for the relationship between afforestation and bird population size. It is most relevant to waders, grouse and small passerines. In A, we have a typical hypothetical frequency distribution of the young:adult ratio (Fig. 4a). Many birds rear few young, but a small proportion are consistently successful and have a high lifetime reproductive success (e.g. Newton 1986a). In B, C and D afforestation displaces the ground nesters, of which a smaller percentage attempt to settle in adjacent suitable habitat. In B, population size in the unplanted area increases through some displaced birds squeezing into smaller territories, and the average young: adult ratio is only slightly reduced (because disproportionately more birds may have to occupy poorer habitats and because of competition; O'Connor 1985). In C, however, immigrants fare badly through loss of local experience, non-

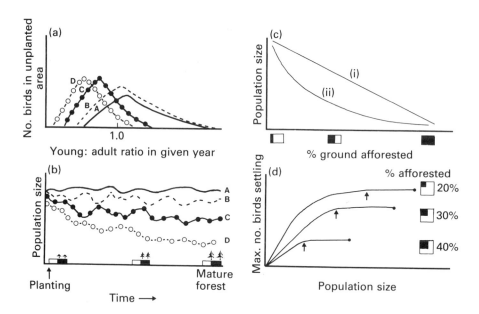

FIG. 4. Hypothetical relationships between forestry expansion and decline in bird population viability in a given area (text gives full details). In (a), giving the frequency distributions of the young:adult ratio, and (b) A is a typical distribution; B has birds squeezing into smaller territories; C has birds suffering interference, nest predation and loss of local knowledge; and D has density-dependent nest predation. In (c) the population-size decline with percentage of ground afforested is (i) linear or (ii) supra-proportional. In (d) the 'carrying capacity' (see arrows) falls off rapidly with a gradual increase in percentage cover of forest.

shrinkage of territories, competition with local residents and high nest and young predation, so the young:adult ratio skews heavily to the low side, but the original occupants do virtually just as well as before. The worst situation is in D, where the settlement of some ousted birds increases overall density to such an extent that there is interference and an overall increase in adult and brood competition, and in the rate of nest, adult and young predation. Here, the 'residents' are only marginally less vulnerable than the incomers, because they have prior knowledge of limiting resources and fight harder to defend these because they have more to lose. As Fig. 4b shows, population D would not be viable and would decline to a level where predators ceased hunting (this seems unlikely because they tend to be generalists) or encountered difficulty locating a few scattered nests. These extremely simple, hypothetical outcomes indicate the likely magnitude of differences between the situations outlined.

The relative amount of ground afforested will also have important repercussions for further changes in population size. In Fig. 4c there are two extremes: (i) the population declines linearly with proportion of ground planted, and (ii) there is a supra-proportionately rapid rate of decline in bird numbers as the area afforested increases. This may be due to the density-dependent processes such as competition or predation or to the greater forest-edge effect arising from mosaic planting. The implication is that small increases in the amount of ground afforested will accompany relatively large decreases in population size, so that even the carrying capacity of a given upland catchment will drop markedly (see arrows in Fig. 4d).

CONCLUSIONS

The paper started with the premise that an internationally distinctive ornithological resource is vulnerable to afforestation. The resource varies spatially not only in species composition and abundance but also, apparently, in duration of occupation throughout the Holocene. The 39 wetland or open-ground nesting species at risk from planting are certainly considerably less abundant than the 35 common species benefiting in the long term from forest expansion (Fig. 5; and Newton 1986b). Overall, at least 34 species associated with the uplands are potentially at risk; 14 are eradicated or substantially reduced in distribution, and at least another 8 species are likely to decline through further forestry expansion (Table 1).

The consequences for birds in unplanted areas have tended to be neglected (except for the large predatory and carrion-eating birds) so we have drawn heavily on relevant fundamental and applied research. We are suggesting that (i) habitat loss will reduce the viability of populations in some of the remaining areas, and (ii) sustained predation, related to more forests, on nests, young and adults, particularly of typically site-faithful species, will contribute to greater dispersal, more competition, density-dependent predation and poorer breeding. Consequently, some seemingly suitable habitat close to plantations may ultimately

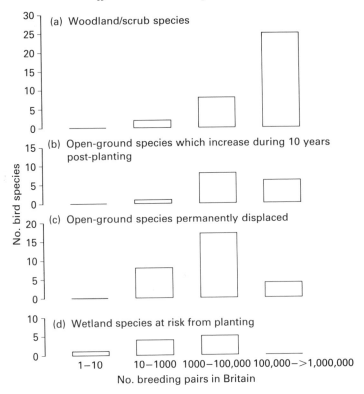

FIG. 5. Relationship between afforestation and species frequency. (a) Woodland or scrub species which colonize and increase in afforested ground (35 species); (b) open-ground nesting species which increase during 10 years post-planting, or breed in restocked ground (15 species); (c) open-ground species permanently displaced by afforestation (29 species); (d) wetland species at risk from afforestation (10 species). Calculated from data in NCC (1986) and Sharrock (1976).

contain far fewer birds than expected. The graphical models (especially Figs 4c, d) have already been applied successfully to provide insights on how overwintering populations of waders on estuaries respond to land reclamation (Goss-Custard 1985); they are now being tested in the uplands.

In Finland there has been detailed monitoring of bird communities in forest-affected bogs for more than 40 years. The work reveals widescale changes in bird community structure and dynamics attributable to changing forest management practices (Helle 1986; Väisänen, Järvinen & Rauhala 1986). There are hardly any comparable long-term studies of bird communities, or even species populations, in the British uplands (except autecological studies by Marquiss *et al.* 1978, 1985; Parr 1980; Ratcliffe 1980; Holland *et al.* 1982; Mearns 1983; Moss & Watson 1985; Hudson 1986; Nethersole-Thompson & Nethersole-Thompson 1986; Watson *et al.* 1978; Thompson *et al.* 1988). A more strategic monitoring programme linked to the planting of trees is clearly desirable. More detailed autecological studies are needed to evaluate the importance of (i) nest/adult predation in population and community functioning (Begon *et al.* 1986), (ii) territoriality and competition, and (iii) the consequences of mate/site fidelity for population ecology. Much more work is needed on the predators to quantify increases in numbers, competition with scavenging/predator birds, and how they operate in relation to plantations.

Meanwhile, with more afforestation there will be increasingly large diminutions in birds characteristic of the uplands. Local populations of some species may go to extinction, and the geographical range and complement of species will diminish further.

ACKNOWLEDGMENTS

We appreciate discussions with Derek Ratcliffe, Desmond Nethersole-Thompson, Adam Watson, Jeff Watson, Terry Burke, Ray Parr, Pat Thompson, Michael Usher, Steve Albon, Colin Bibby, Peter Hudson, Ingvar Byrkjedal, Richard Lindsay, Tim Reed, Derek Langslow, John Goss-Custard, Philip and Pat Whitfield, John Halliday, Hector Galbraith, Chris Thomas, Juliet Vickery, Peter Evans, Steve Redpath, Chris Thomas and Paul Haworth; the NCC and NERC for funding much of the work described here; and Sandra Lackie and Irene Crichton for so kindly word-processing the manuscript.

REFERENCES

Anderson, P. & Yalden, D.W. (1981). Increased sheep numbers and the loss of heather moorland in the Peak District, England. *Biological Conservation, 20*, 195–213.

Andren, H., Angelstam, P., Lindstrom, E. & Widen, P. (1985). Differences in predation pressure in relation to habitat fragmentation: an experiment. *Oikos, 45*, 273–277.

Bainbridge, L.P., Minns, D.W., Housden, S.D. & Lance, A.N. (1987). *Forestry in the Flows of Caithness and Sutherland.* Conservation Topic Paper No. 18, Royal Society for the Protection of Birds, Edinburgh.

Begon, M., Harper, J.L. & Townsend, C.R. (1986). *Ecology: Individuals, Populations and Communities.* Blackwell Scientific Publications, Oxford.

Bibby, C.J. (1986). Merlins in Wales: site occupancy and breeding in relation to vegetation. *Journal of Applied Ecology, 23*, 1–22.

Bibby, C.J. (1987). Foods of breeding merlins (*Falco columbarius*) in Wales. *Bird Study, 34*, 64–70.

Bibby, C.J. (1988). Impacts of agriculture on upland birds. *This volume.*

Bibby, C.J. & Nattrass, M. (1986). Breeding status of the merlin in Britain. *British Birds, 79*, 170–185.

Birks, H.H. (1975). Studies in the vegetation history of Scotland. IV. Pine stumps in Scottish blanket peats. *Royal Society Philosophical Transactions (B), 270*, 181–226.

Birks, H.J.B. (1988). Long-term ecological change in the British uplands. *This volume.*

Bramwell, A.G. & Cowie, G.M. (1983). Forests of the Inner Hebrides — status and habitat. *Proceedings of the Royal Society of Edinburgh, 83B*, 577–597.

Burke, T. & Thompson, D.B.A. (1988). Mating and breeding systems of Western Palaearctic birds. *Sourcebook of Ornithological Data* (Ed. by R.J. O'Connor). Poyser, Calton.

Byrkjedal, I. (1980). Nest predation in relation to snow-cover: a possible factor influencing the start of breeding in shorebirds. *Ornis Scandinavica, 11*, 149–252.

Byrkjedal, I. (1985). Time activity budget for breeding greater golden plovers in Norwegian mountains. *Wilson Bulletin, 97*, 486–501.

Byrkjedal, I. (1987). Anti-predator behaviour and breeding success in greater golden plover and Eurasian dotterel. *Condor,* **89,** 40–47.

Coulson, J.C. (1988). The structure and importance of invertebrate communities on peatlands and moorland, and effects of environmental and management changes. *This volume.*

Coulson, J.C. & Thompson, C. (1985). Differences in the breeding performance of individual kittiwake gulls, *Rissa tridactyla* (L). *Behavioural Ecological Consequences of Adaptive Behaviour* (Ed. by R.M. Sibly & R.H. Smith), pp. 489–503. Blackwell Scientific Publications, Oxford.

Cramp, S. & Simmons, K.E.L. (Eds.) (1980). *The Birds of the Western Palearctic, Vol. 2.* Oxford University Press, Oxford.

Dare, P.J. (1986). Raven, *Corvus corvus,* populations in two upland regions of North Wales. *Bird Study,* 33, 179–189.

Edwards, K.S. & Hirons, R.K. (1982). Date of blanket peat initiation and rates of spread — a problem in research design. *Quaternary Newsletter,* 36, 32–37.

Elliot, R.D. (1982). *Dispersion of lapwing nests in relation to predation and antipredator defence.* Ph.D. thesis, University of Aberdeen.

Eriksson, M.O. (1984). Acidification of lakes: effects on waterbirds in Sweden. *Ambio,* 13, 260–262.

Evans, P.R. & Pienkowski, M.W. (1984). Population dynamics of shorebirds. *Behaviour of Marine Animals, Vol. 5* (Ed. by J. Burger & B.L. Olla), pp. 83–124. Plenum Press, London.

Flegg, J.J.M. & Glue, D.E. (1975). The nesting of the ring ouzel. *Bird Study,* 22, 1–8.

Forestry Commission (1984). *Census of Woodland and Trees 1979–82: Great Britain.* Forestry Commission, Edinburgh.

Forestry Commission (1987). *Forestry Facts and Figures 1986/1987.* Forestry Commission, Edinburgh.

Galbraith, H. (1986). *The influence of habitat on the breeding biology of lapwings* (Vanellus vanellus) *on farmland.* Ph.D. thesis, University of Glasgow.

Gee, A.S. & Stoner, J.H. (1988). The effects of afforestation and acid deposition on the water quality and ecology of upland Wales. *This volume.*

Goss-Custard, J.D. (1985). Foraging behaviour of wading birds and the carrying capacity of estuaries. *Behavioural Ecology: Ecological Consequences of Adaptive Behaviour* (Ed. by R.M. Sibly & R.H. Smith), pp. 169–188. Blackwell Scientific Publications, Oxford.

Greenwood, P.J. (1980). Mating systems, philopatry and dispersal in birds and mammals. *Animal Behaviour,* 28, 1140–1162.

Haines, T.A. (1981). Acid precipitation and its consequences for aquatic ecosystems — a review. *Transactions of the American Fish Society,* 110, 609–707.

Harris, M.P. (1975). Skokholm oystercatchers and the Burry Inlet. *Report of the Skokholm Bird Observatory for 1974,* 17–19.

Haworth, P.H. (1987). *Moorland management and nature conservation in the South Pennines.* Ph.D. thesis, CNAA.

Helle, P. (1986). Bird community dynamics in a boreal forest reserve: the importance of large-scale regional trends. *Annales Zoologici Fennici,* 23, 157–166.

Holland, P.K., Robson, J.E. & Yalden, D.W. (1982). The breeding biology of the common sandpiper (*Actitis hypoleucos*) in the Peak District. *Bird Study,* 29, 99–110.

Holmes, R.T. (1970). Differences in population density, territoriality and food supply of dunlin in arctic and subarctic tundra. *Animal Populations in Relation to their Food Resources* (Ed. by A. Watson), pp. 303–317. Blackwell Scientific Publications, Oxford.

Hudson, P. (1986). *Red Grouse: the Biology and Management of a Wild Gamebird.* Game Conservancy Trust, Fordingbridge.

Hudson, P.J. (1988). Spatial variations, pattern and management options in upland bird communities. *This volume.*

Hudson, P.J. & Watson, A. (1985). The red grouse. *Biologist,* 32, 13–18.

Jönsson, P.E. (1988). *Ecology of the Southern Dunlin,* Calidris alpina schinzii. Ph.D. thesis, Lund University.

Lindsay, R.A. (1987). The great flow — an international responsibility. *New Scientist,* 1542, 45.

Lindsay, R.A., Charman, D.T., Everingham, F., O'Reilly, R.M., Palmer, M.A., Rowell, T.A. &

Stroud, D.A. (1988). *The Flow Country: the Peatlands of Caithness and Sutherland.* Nature Conservancy Council, Peterborough.

McVean, D. & Ratcliffe, D.A. (1962). *Plant Communities of the Scottish Highlands.* Nature Conservancy Monograph No. 1, HMSO, London.

Marquiss, M., Newton, I. & Ratcliffe, D.A. (1978). The decline of the raven, *Corvus corax*, in relation to afforestation in southern Scotland and northern England. *Journal of Applied Ecology*, **15**, 129–144.

Marquiss, M., Ratcliffe, D.A. & Roxburgh, L.R. (1985). The numbers, breeding success and diet of golden eagles in southern Scotland in relation to changes in land use. *Biological Conservation*, **34**, 121–140.

Mearns, R. (1983). The status of ravens in southern Scotland and Northumbria. *Scottish Birds*, **12**, 211–218.

Moss, R. & Watson, A. (1985). Adaptive value of spacing behaviour in population cycles of red grouse and other animals. *Behavioural Ecology: Ecological Consequences of Adaptive Behaviour* (Ed. by R.M. Sibly & R.H. Smith), pp. 275–294. Blackwell Scientific Publications, Oxford.

Moss, D., Taylor, P.N. & Easterbee, N. (1979). The effects on song-bird populations of upland afforestation with spruce. *Forestry*, **52**, 124–147.

NCC (1986). *Nature Conservation and Afforestation in Britain.* Nature Conservancy Council, Peterborough.

Nethersole-Thompson, D. & Nethersole-Thompson, M. (1979). *Greenshanks.* Poyer, Berkhamsted.

Nethersole-Thompson, D. & Nethersole-Thompson, M. (1986). *Waders: their Breeding, Haunts and Watchers.* Poyser, Calton.

Nethersole-Thompson, D. & Watson, A. (1981). *The Cairngorms: their Natural History and Scenery.* Melven, Perth.

Newton, I. (1986a). *The Sparrowhawk.* Poyser, Calton.

Newton, I. (1986b). Principles underlying bird numbers in Scottish woodlands. *Trees and Wildlife in the Scottish Uplands* (Ed. by D. Jenkins), pp. 121–128. Institute of Terrestrial Ecology, Banchory.

Newton, I, Davis, P.E. & Davis, J.E. (1982). Ravens and buzzards in relation to sheep-farming and forestry in Wales. *Journal of Applied Ecology*, **19**, 681–706.

Newton, I., Meek, E.R. & Little, B. (1978). Breeding ecology of the merlin in Northumberland. *British Birds*, **71**, 376–398.

Newton, I., Meek, E.R. & Little, B. (1984). Breeding season foods of the merlin *Falco columbarius* in Northumbria. *Bird Study*, **31**, 49–56.

Newton, I., Meek, E.R. & Little, B. (1986). Population and breeding of Northumbrian merlins. *British Birds*, **79**, 155–170.

O'Connor, R.J. (1985). Behavioural regulation of bird populations: a review of habitat use in relation to migration and residency. *Behavioural Ecology: Ecological Consequences of Adaptive Behaviour* (Ed. by R.M. Sibly & R.H. Smith), pp. 105–142. Blackwell Scientific Publications, Oxford.

O'Flynn, W.J. (1983). Population changes of the hen harrier in Ireland. *Irish Birds*, **2**, 337–343.

Oring, L.W. (1988). Fidelity and philopatry in shorebirds. *Proceedings of International Ornithological Congress, Ottawa*, 19.

Oring, L.W. & Lank, D.B. (1984). Breeding area fidelity, natal philopatry, and the social systems of sandpipers. *Behaviour of Marine Animals, Vol. 5* (Ed. by J. Burger & B.C. Olla), pp. 125–148. Plenum, London.

Ormerod, S.J. & Edwards, R.W. (1985). Stream acidity in some areas of Wales in relation to historical trends in afforestation and the usage of agricultural limestone. *Journal of Environmental Management*, **20**, 189–197.

Ormerod, S.J. & Tyler, S.J. (1986). The diet of dippers *Cinclus cinclus* wintering in the catchment of the River Wye, Wales. *Bird Study*, **33**, 36–45.

Ormerod, S.J. & Tyler, S.J. (1987). Aspects of the breeding ecology of Welsh grey wagtails *Motacilla cinerea*. *Bird Study*, **34**, 43–51.

Ormerod, S.J., Tyler, S.J. & Lewis, J.M. (1985). Is the breeding distribution of dippers influenced by stream acidity? *Bird Study*, **32**, 33–40.

Parr, R. (1980). Population study of the golden plover *Pluvialis apricaria* using marked birds. *Ornis Scandinavica*, **11**, 179–189.

Partridge, L. & Green, P. (1985). Intraspecific feeding specialization and population dynamics. *Behavioural Ecology: Ecological Consequences of Adaptive Behaviour* (Ed. by R.M. Sibly & R.H. Smith), pp. 207–226. Blackwell Scientific Publications, Oxford.

Poxton, I.R. (1986). Breeding ring ouzels in the Pentland Hills. *Scottish Birds*, **14**, 44–48.

Rankin, G.D. & Taylor, I.R. (1985). *Changes within Afforested but Unplanted Ground: Birds.* Nature Conservancy Council, Peterborough.

Ratcliffe, D.A. (1962). Breeding density in the peregrine, *Falco peregrinus*, and the raven, *Corvus corax. Ibis*, **104**, 13–39.

Ratcliffe, D.A. (1976). Observations on the breeding of the golden plover in Great Britain. *Bird Study*, **23**, 63–116.

Ratcliffe, D.A. (Ed.) (1977). *A Nature Conservation Review, Vol 1.* Cambridge University Press, Cambridge.

Ratcliffe, D.A. (1980). *The Peregrine Falcon.* Poyser, Calton.

Ratcliffe, D.A. (1986). The effects of afforestation on the wildlife of open habitats. *Trees and Wildlife in the Scottish Uplands* (Ed. by D. Jenkins), pp. 46–54. Institute of Terrestrial Ecology, Banchory.

Ratcliffe, D.A. (1989). *Upland Birds.* Cambridge University Press, Cambridge.

Ratcliffe, D.A. & Thompson, D.B.A. (1988). The British uplands: their ecological character and international significance. *This volume.*

Rowan, A.A. (1986). The nature of British upland forests in the 1980s. *Trees and Wildlife in the Scottish Uplands* (Ed. by D. Jenkins), pp. 7–13. Institute of Terrestrial Ecology, Banchory.

Sharrock, J.T.R. (1976). *Atlas of Breeding Birds in Britain and Ireland.* Poyser, Berkhamsted.

Soikkeli, M. (1970). Dispersal of dunlin (*Calidris alpina*) in relation to sites of birth and breeding. *Ornis Fennica*, **47**, 1–9.

Sonerud, G.A. & Fjeld, P.E. (1987). Long-term memory in egg predators: an experiment with a hooded crow. *Ornis Scandinavica*, **18**, 323–325.

Stroud, D. (1987). A review of some consequences of open ground afforestation for upland birds. *Forests for Britain: the BANC Report*, pp. 29–34. British Association of Nature Conservationists, London.

Stroud, D.A., Reed, T.M., Pienkowski, M.W. & Lindsay, R.A. (1987). *Birds, Bogs and Forestry: the Peatlands of Caithness and Sutherland.* Nature Conservancy Council, Peterborough.

Sydes, C. & Miller, G.R. (1988). Range management and nature conservation in the British uplands. *This volume.*

Tallis, J.H. (1964). Studies on Southern Pennine peats. III. The behaviour of *Sphagnum. Journal of Ecology*, **52**, 345–353.

Thompson, D.B.A. (1987). Battle of the bog. *New Scientist*, **1542**, 41–44.

Thompson, D.B.A. & Barnard, C.J. (1983). Anti-predator responses in mixed species flocks of lapwings, golden plovers and gulls. *Animal Behaviour*, **31**, 585–593.

Thompson, D.B.A., Thompson, P.S. & Nethersole-Thompson, D. (1986). Timing of breeding and breeding performance in a population of greenshanks, *Tringa nebularia. Journal of Animal Ecology*, **55**, 181–199.

Thompson, D.B.A., Thompson, P.S. & Nethersole-Thompson, D. (1988). Breeding site fidelity and philopatry in greenshanks (*Tringa nebularia*) and redshanks (*T. totanus*). *Proceedings International Ornithological Congress, Ottawa,* **19**.

Thompson, P.S. & Hale, W.G. (in press). Breeding site fidelity and recruitment in a redshank population. *Ibis.*

Tinbergen, N., Impekoven, M. & Franck, D. (1967). An experiment on spacing out as a defence against predation. *Behaviour*, **28**, 307–321.

Tjernberg, M. (1981). Diet of the golden eagle *Aquila chrysaetos* during the breeding season in Sweden. *Holarctic Ecology*, **4**, 12–19.

Tyler, S.J. (1987). River birds and acid water. *RSPB Conservation Review*, **1**, 68–70.

Usher, M.B. & Gardner, S.M. (1988). Animal communities in the uplands: how is naturalness influenced by management? *This volume.*

Väisänen, R.A., Järvinen, O. & Rauhala, P. (1986). How are extensive, human-caused habitat alterations expressed on the scale of local bird populations in boreal forests? *Ornis Scandinavica*, **17**, 282–292.

Watson, A., Moss, R., Rothery, P. & Parr, R. (1984). Demographic causes and predictive models of population fluctuations in red grouse. *Journal of Animal Ecology*, **53**, 639–662.

Watson, J., Langslow, D.R. & Rae, S.R. (1987). *The Impact of Land-use Changes on Golden Eagles* Aquila chrysaetos *in the Scottish Highlands*. Nature Conservancy Council, Peterborough.

Whitfield, D.P. (1985). Raptor predation on wintering waders in south-east Scotland. *Ibis*, **27**, 544–558.

Woods, A. & Cadbury, C.J. (1987). Too many sheep in the hills. *RSPB Conservation Review*, **1**, 65–67.

Yalden, D.W. (1986). Diet, food availability and habitat selections of breeding common sandpipers *Actitis hypoleucos*. *Ibis*, **128**, 23–26.

The impact and ecology of the pine beauty moth in upland pine forests

A. D. WATT AND S. R. LEATHER*

*NERC Institute of Terrestrial Ecology, Edinburgh Research Station, Bush Estate, Pemcuik, Midlothian EH26 OQB and *Forestry Commission, Northern Research Station, Roslin, Midlothian EH25 9SY*

SUMMARY

1 Outbreaks of pine beauty moth (*Panolis flammea*) have occurred in lodgepole pine (*Pinus contorta*) plantations since 1976. These outbreaks can destroy whole forest blocks within 2 years.

2 Large-scale control operations have been required in 6 years between 1977 and 1987.

3 Pine plantations vary considerably in their susceptibility to outbreaks. Factors contributing to this include pine species (Scots pine, *Pinus sylvestris*, has not experienced any outbreaks in Scotland), geographical region (most outbreaks have been in the Highland region of Scotland), soil type (plantations on deep peat are particularly prone to outbreaks) and age of lodgepole pine (so far, 10–27-year-old trees have had outbreaks). More surveys and research are required to confirm relationships between outbreaks and plantation size, tree species and age.

4 Relevant studies of the pine beauty moth's population ecology are summarized in the context of understanding further its impact on upland pine forests.

INTRODUCTION

Afforestation of the British uplands in the second half of the twentieth century has led to a dramatic change in 'typical woodland'. Conifers predominate in upland areas. For example, in Scotland over 80 per cent of all woodland is conifer forest; and Sitka spruce (*Picea sitchensis*) is now by far the most commonly planted tree species in Britain (68 per cent of total area planted 1971–1980) (Rowan 1986). Lodgepole pine (*Pinus contorta*) is the only other tree planted to much extent; it was planted widely in the 1950s and 1960s in the north and east of Scotland because of its ability to grow in wet, acid and infertile soils (Lines 1976). Despite the recent increase in afforestation, however, only 9·4 per cent of upland areas is classed as woodland or forest (Rowan 1986). The pressures to expand forestry much further are considerable and, although the outcome of the political issues surrounding forestry is unpredictable, a further increase in forestry may be both extensive and rapid (Last *et al.* 1986). Threats to forestry, whether they be in the form of pests, diseases, physical hazards or air pollution, must therefore be taken seriously. Moreover, in considering these threats, we must be concerned not just to provide tactical solutions (e.g. chemical pest control) but also to ensure that forestry management is designed to minimize the various threats to forestry.

This paper describes the impact of one insect pest, the pine beauty moth (*Panolis flammea*), on upland forestry in terms of its effect on a single forest (Poulary) as well as its national impact. It then considers briefly the moth's local and national trends in abundance. Research on populations in individual forests is highlighted.

THE IMPACT OF A PINE BEAUTY MOTH OUTBREAK IN POULARY

The Poulary block of Glengarry Forest is just south of Loch Poulary in the Lochaber district of the Highland region of Scotland. Between 1966 and 1978 it was planted for the first time with lodgepole pine and Sitka spruce, predominantly in pure stands but also in some mixtures. In 1986 a severe outbreak of the pine beauty moth occurred in Poulary. No damage had been recorded by this insect prior to 1986; even in 1985, significant defoliation was not observed.

Ground and aerial surveys in September 1986 revealed that (i) 40 ha of lodgepole pine had been completely defoliated and killed by *P. flammea* larvae (Fig. 1), (ii) 80 ha were severely defoliated, and (iii) larvae were present throughout most of the block. Surveys of pupae in autumn 1986 indicated that at least 900 larvae per tree were present in the centre of the forest. Knowing that 15 pupae m^{-2} indicates likely serious defoliation in the following year (Stoakley 1981), we expected 300 ha to be at risk in 1987 (Fig. 1). This was confirmed by sampling eggs in May 1987. Between around 500 to 1000 larvae can completely defoliate a tree; this was exceeded over much of the Poulary block (Fig. 1). A slight increase in the area at risk relative to that predicted from the survey of

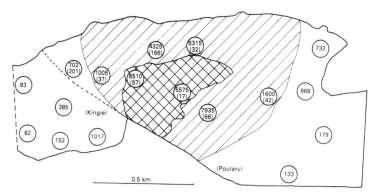

FIG. 1. Parts of Poulary, Glengarry Forest, containing lodgepole pine defoliated and killed by *P. flammea* in 1986 (■), and numbers of pupae exceeding 15 m^{-2} (▨). The number of *P. flammea* eggs laid per tree (with the estimated fecundity in brackets) refer to Poulary in 1987.

pupae was probably due to adult moth dispersal from the 1986 epicentre to other parts of Poulary. This is best demonstrated by calculating the number of eggs laid per female moth in different parts of the forest. The low estimate in the centre of the forest (Fig. 1) suggests that the actual fecundity of *P. flammea* was reduced by the effect of food shortage, or by female moths dispersing elsewhere, or both. The large estimates for eggs per female, particularly in the north and west of the forest do, however, strongly imply substantial adult dispersal of moths from the epicentre to the rest of the forest. Nevertheless, it appears that adult dispersal only served to exacerbate the impact of *P. flammea* on Poulary. Pupal densities outside the dead area in late 1986 indicated that much more defoliation would have occurred regardless of adult moth dispersal. This pattern of outbreak behaviour appears to be typical of *P. flammea* (Stoakley 1979; Watt 1986b; A. D. Watt, S. R. Leather & J. T. Stoakley, unpublished). The initial level and scale of damage can be substantial, but in the following year the area at risk from defoliation rises sharply, and is thus exacerbated, not wholly caused, by dispersal.

In June 1987, Poulary and parts of adjacent blocks were sprayed with Fenitrothion to control the *P. flammea* outbreak. A short delay in spraying, mainly due to poor weather, led to the death of many shoots on some particularly heavily infested trees. No further trees were killed.

THE NATIONAL IMPACT OF THE PINE BEAUTY MOTH

The 1986 *P. flammea* outbreak in Poulary was not an isolated incident. *P. flammea* was first recorded on lodgepole pine in 1973, and only 3 years later the first outbreak occurred at Rimsdale, Sutherland, where 180 ha were defoliated and killed. Between 1977 and 1980 outbreaks occurred at 26 sites. No outbreaks occurred from 1981 to 1983, but 27 occurred between 1984 and 1987 (18 at sites which had already experienced an outbreak). By 1987, over 2 million trees had been killed by *P. flammea* in Scotland. However, many more were saved by the use of chemical insecticides (see Fig. 2).

No other insect has had such a serious impact on established conifer plantations in the United Kingdom. Indeed, a reappraisal of the growth potential of Sitka spruce combined with the destructive impact of *P. flammea* has led to a decline in the planting of lodgepole pine in pure stands in northern Scotland (Stoakley 1986).

SUSCEPTIBILITY OF PINE FORESTS TO PINE BEAUTY MOTH IN SCOTLAND

Differences in species susceptibility

Perhaps the most striking fact about *P. flammea* in Scotland is that no outbreaks have occurred on Scots pine (*Pinus sylvestris*). This has prompted the suggestion that lodgepole pine is the more suitable host plant for *P. flammea* larvae. Indeed,

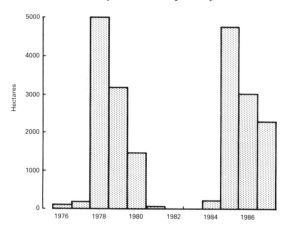

FIG. 2. The total area sprayed with insecticides (Fenitrothion) to control *P. flammea*, or defoliated and killed by *P. flammea* larvae (mainly in 1976 and 1986).

whereas our native pine species has only one serious defoliator (the pine looper moth, *Bupalus piniaria*), lodgepole pine has experienced outbreaks of four defoliating pests (*P. flammea, B. piniaria*, the larch bud moth *Zeiraphera diniana*, and the pine sawfly *Neodiprion sertifer*; Bevan 1987). Apart from *P. flammea*, the only other defoliating insect which has until now been controlled with insecticides on lodgepole pine is *N. sertifer*, a chronic pest against which a viral insecticide has been used in recent years.

Geographical differences

Interestingly, there is evidence to suggest that not all lodgepole pine plantations are susceptible to *P. flammea* outbreaks. All except two outbreaks have occurred in the Highland region, and most of these in Caithness or Sutherland, where the largest concentrations of lodgepole pine occur (Fig. 3). Early on in our experience of *P. flammea* it was noticed that outbreaks were associated with lodgepole pine growing on deep, unflushed peat (Stoakley 1979). This trend has continued and is true not only of the outbreaks which have occurred on the peaty soils of Caithness and Sutherland but also of the 3 most southerly, 'outlying' outbreaks (Fig. 3a). Elchies, in the Moray district of Grampian region, is particularly interesting. Although lodgepole pine grows here on a range of soil types, the outbreak which occurred in 1978 centred on an area of deep, unflushed peat. The association between *P. flammea* outbreaks and deep peat has been the subject of much research, some of which is discussed shortly.

Effects of lodgepole pine age

Outbreaks of *P. flammea* have occurred only on 10- to 27-year-old lodgepole pine. First outbreaks have been recorded on 10- to 25-year-old trees (Fig. 4) and

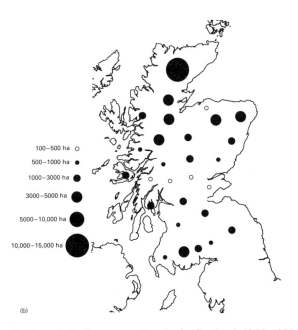

FIG. 3. (a) The distribution of *P. flammea* outbreaks in Scotland 1976–1987. (b) The area of lodgepole pine over 12 years old grown in Forestry Commission Districts in Scotland.

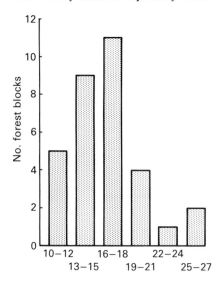

FIG. 4. The age of lodgepole pine in plantations experiencing first outbreaks of *P. flammea.*

second outbreaks on trees 17 to 27 years old. The delay before lodgepole pine is attacked for the first time is probably due mainly to the time taken for *P. flammea* to colonize pine plantations and the rate at which the population increases to damaging levels. The latter is likely to be influenced by weather (Leather, Watt & Barbour 1985; Watt 1986b). In addition, there is evidence that young lodgepole pine trees respond to defoliation by becoming less suitable for *P. flammea* larvae (Leather, Watt & Forrest 1987), whereas older trees appear to show no such response (A. D. Watt, S. R. Leather and G. I. Forrest, unpublished). Despite what is discussed shortly, there is little evidence to suggest that susceptible lodgepole pine plantations will not suffer repeated outbreaks until harvesting.

Mixtures of tree species

All the forest blocks which have experienced *P. flammea* outbreaks contain some conifers in addition to lodgepole pine. North Dalchork Sutherland (2405 ha affected, outbreaks in 1979 and 1984), for example, contains 62 per cent pure lodgepole pine, 4 per cent pure spruce, 11 per cent lodgepole and spruce in mixture, 3 per cent lodgepole and Scots pine in mixture, and 5 per cent pine in other mixtures. Elchies (1448 ha affected, outbreak in 1978) contains 31 per cent pure lodgepole pine, 14 per cent Scots pine, 7 per cent spruce, 3 per cent larch, 25 per cent lodgepole/spruce, 12 per cent lodgepole/Scots pine and 8 per cent pine in other mixtures. It is tempting to suggest that the greater diversity in tree planting in Elchies has, in some way, been responsible for the absence of a second outbreak there (up to 1988). Most forest blocks containing lodgepole pine have a very limited age structure; they were planted over only a few years and seldom contain older trees growing naturally or planted prior to afforestation. North Dalchork, for example, was established between 1959 and 1973; Elchies was planted between 1955 and 1966 but it also contains small areas, amounting to about 5 ha, of Scots pine planted in 1900. Plantations containing lodgepole pine vary greatly in size. The largest plantation where a *P. flammea* outbreak has occurred is North Dalchork (2000 ha) but several plantations under 30 ha, and up to 12 km from other plantations, have also experienced outbreaks.

Although *P. flammea* has been a pest of Scottish forestry since only 1976, some patterns of susceptibility have emerged. Plantations of Scots pine do not appear to be at risk (in contrast to the situation elsewhere in Europe: Berwig 1926; Klimetzek 1979) and although lodgepole pine has experienced many outbreaks these are largely restricted to northern areas and to plantations growing on peaty soils. The importance of plantation structure (size, species composition and age) is unclear and, particularly since this aspect may become more important as forests mature, it should be studied further by more extensive surveys of pine forests. Nevertheless, surveys will indicate only why certain areas are more susceptible to attack. They then have to be supported by detailed population studies, preferably with a strong experimental component.

POPULATION ECOLOGY OF THE PINE BEAUTY MOTH

There have been several studies of the population ecology of *P. flammea* since 1976. These have sought to describe the major factors affecting populations (particularly in susceptible sites), with the eventual intention of predicting when outbreaks will occur, and to discover why particular areas are so susceptible to damage.

Parasites, predators and pathogens

The study of *P. flammea* populations in lodgepole pine plantations has shown that the rate of population increase varies greatly from year to year (Fig. 5). A large number of natural enemies have been shown to have partially affected pine beauty moth numbers (Barbour 1987; Watt & Leather, in press). Several insect parasitoids (braconids, ichneumonids and tachinids) attack *P. flammea*, either emerging from the larval or, more usually, from the pre-pupal and pupal stages. Parasitism can affect 65 per cent of *P. flammea* pre-pupae, but in the populations studied so far it does not act in a straightforwardly density-dependent manner.

Among the range of predators known to attack *P. flammea* are several birds, including coal tits (*Parus ater*), great tits (*Parus major*), fieldfares (*Turdus pilaris*) and ravens (*Corvus corax*) (Watt 1986b). The tit species are larval predators whereas fieldfares and ravens prey on pupae and have both been recorded feeding in large numbers when *P. flammea* is particularly abundant. There are a number

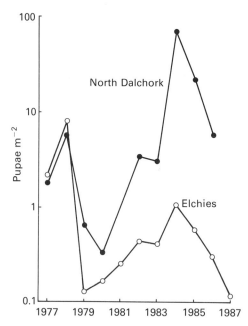

FIG. 5. The number of *P. flammea* pupae in two forests. Control measures were carried out in North Dalchork in 1979, 1985, 1986 and 1987, and in Elchies in 1979.

of other probable predators such as shrews (*Sorex* spp.), ground beetles (carabids), wood ants (*Formica* spp.) and spiders (Araneae), some of which are currently being studied (P. G. Walsh, pers. comm.). The impact of predators on *P. flammea* on lodgepole pine has not been clearly established but estimates of total mortality over winter (excluding parasitism and disease) and comparisons of caged and uncaged populations have given some indication of mortality attributable to predators. The available evidence suggests that predation, like parasitism, does not act in a density-dependent manner.

Two pathogens cause disease in *P. flammea*: a fungus (*Spiralea farinosa*) and a nuclear polyhedrosis virus (*Panolis flammea* NPV). Mortality caused by these increases in a density-dependent manner, but there is no evidence that disease prevents *P. flammea* populations from reaching lethally defoliating levels. Indeed, although there is a wide range of natural enemies attacking *P. flammea*, they are clearly unable to prevent outbreaks occurring in a large number of Scottish sites, where it is difficult to avoid the conclusion that *P. flammea* is limited only by the quantity of its food resource.

Pine foliage and weather

In recent years attention has focused on the quality of the food resources of herbivorous insects in general (Rhoades 1983). The nutritional quality of pine foliage may vary in several ways, for example, within years as pine shoots develop and grow (Watt 1987), and from year to year under the influence of weather (Watt 1986b) or herbivore damage (Leather *et al.* 1987). Although some of this variability can have marked effects on larval survival, its precise role in the variability of *P. flammea* population development remains unclear.

Population studies of the pine beauty moth have also included a preliminary examination of the role of weather. The overwintering pupal stage of *P. flammea* is remarkably resistant to low temperatures (Leather 1984) but temperature has been shown to affect adult emergence, egg laying and egg hatch (S. R. Leather and A. D. Watt, unpublished). Population models have shown that a delay in mating caused by, for example, a period of cold weather, can result in a large reduction in the production of fertile eggs (Leather *et al.* 1985). The consequence of these factors is that there is a positive correlation between fecundity measured in the field (which has been estimated to range from 20 to 120 eggs per female) and temperature throughout March to mid-May (Leather *et al.* 1985; Watt 1986b). Weather may, therefore, hold the key to understanding why *P. flammea* populations behave in an erratic fashion (Fig. 5).

Soil type and host plant species

Research on the susceptibility of certain areas to *P. flammea* outbreaks has concentrated on host plant species and soil type (see earlier). It is tempting to suggest that lodgepole pine is a nutritionally better host for *P. flammea* than Scots pine, and has thus enabled *P. flammea* populations to grow faster on it than on Scots pine. Laboratory studies have shown, however, that the survival of *P. flammea* larvae on lodgepole pine is no higher than on Scots pine, and that shoot growth stage is a more important determinant of survival (Watt 1987). Current work in the field supports this conclusion. It may, therefore, be proposed that lodgepole pine plantations contain a relatively impoverished fauna of natural enemies. Current research suggests that predation on certain stages of *P. flammea* is more significant on Scots pine (Watt 1986b, unpublished), but the effects of a wide range of predators and parasites on the full life cycle of *P. flammea* need to be studied.

The association between outbreaks of *P. flammea* and lodgepole pine growing in deep peat has led to the suggestion that trees growing in deep peat experience water stress which results in their foliage being more nutritious to *P. flammea* larvae than trees growing elsewhere. This is supported by studies carried out at a number of outbreak sites which showed positive correlations between *P. flammea* numbers or damage and (i) peat depth and (ii) tree vigour (Stoakley 1977; Barnett 1983, 1987; Leather and Barbour 1987). However, detailed population studies over a number of years, including population manipulation and predator exclusion, have failed to provide any evidence that the nutritional quality of tree foliage on deep peat is any better than that on other soils (Watt & Leather, in press). The artificial imposition of plant stress in the laboratory has also been shown to have no harmful effect on *P. flammea* (Watt 1986a).

Perhaps the association between *P. flammea* outbreaks and lodgepole pine growing in deep peat is due to the effect of soil type on the survival of the over-wintering pupal stage. However, there is no direct evidence for this, although pupal survival has been shown to be greater in needle litter than in either soil or peat. Since the litter layer is deeper and more persistent on deep peats than on mineral soils, pupal survival is likely to be greater in areas of deep peat (Leather 1984).

An alternative hypothesis currently under investigation (P. G. Walsh, pers. comm.) is that there is a less diverse or less abundant array of natural enemies in areas of lodgepole pine growing in deep peat. Other factors which may predispose lodgepole pine growing in deep peat to attack include the effect of soil type on adult emergence (and subsequent synchronization between egg hatch and plant development) and the effects of topography on adult moth behaviour, e.g. dispersal (Watt 1986b).

DISCUSSION

Taking a wider perspective, it may be argued that the pine beauty moth has high-lighted a major risk inherent in planting exotic conifers in large-scale mono-cultures. However, in any discussion of the risks involved in exotic conifer monocultures we must immediately acknowledge that, in comparison with natural or semi-natural woodland, we are dealing with a relative, not an absolute, comparison of risks. Natural woodland often experiences significant levels of defoliation. For example, in Scotland in 1987 large areas of oak (*Quercus petraea*) and birch (*Betula pendula* and *B. pubescens*) were completely defoliated by *Operophtera brumata* and *Operophtera fagata* respectively (S. R. Leather & A. D. Watt, unpublished) and would have been killed if these trees (unlike pine) did not possess the ability to recover after early defoliation. It seems unlikely that we could ever establish pest-free forestry, but our experience with *P. flammea* has suggested that there may be ways of substantially reducing the risk of pest attack. If, as seems likely, natural enemies prevent outbreaks of *P. flammea* on Scots pine, then we should seek ways of encouraging them into lodgepole pine plantations. An avenue worth exploring is that of increasing plantation diversity, both in terms of species composition and age structure. Indeed, there are interesting indications that some of the more diverse plantations may be becoming less susceptible to attack as they mature.

Although it appears that host suitability is not the factor governing the absence of outbreaks of *P. flammea* from Scots pine, host-plant resistance may prove to be a useful pest management strategy. There is a wide range in the suit-ability of lodgepole pine provenances to *P. flammea* (Leather 1985, 1987) and the use of certain provenances is likely to lead to a reduction in the frequency of *P. flammea* outbreaks (Leather *et al.* 1985; Watt and Leather 1987).

The effects of soil type on the risk of pest outbreaks give cause for concern about the value of planting lodgepole pine over large parts of Scotland. It may be that this problem is insoluble, especially if the importance of soil type lies in its particular suitability for overwintering pupae. If, however, the crucial factor with regard to soil type is that natural enemies are less abundant in areas of deep peat, this situation may alter (at least for some groups of natural enemies) as plantations mature and bring about changes in the soil (see Miles 1988). One factor seems clear: the foliage of lodgepole pine is no more suitable as a food source for *P. flammea* larvae when the trees are grown in deep peat. This does not mean, however, that other insects will not perform better on lodgepole pine growing in stressful conditions. Young *P. flammea* larvae feed on newly expanding shoots and needles and may therefore be less affected by stress-imposed changes in host-plant quality than insects feeding on mature foliage (such as *B. piniaria, N. sertifer*) (White 1974, 1976; Watt 1986a).

The pine beauty moth problem has shown that preventing entry of exotic insects to the UK is no guarantee that pest problems will not occur on exotic conifers. We have yet to see, whether or not other native insects will reach pest status suddenly (like *P. flammea*) or if adaptations to exotic hosts typically occur more gradually. There is an obvious need to monitor the incidence of insects such as the winter moth (*O. brumata*), of only minor pest status on Sitka spruce at present, but which may become more significant in the future. *P. flammea* has also shown that insects can achieve pest status on new (exotic) plants while being no better adapted to them than to their native hosts. This has several implications, including the need to understand fully the role of natural enemies in forest insect population dynamics, and the ways in which forest structure affects their impact. Both are factors of obvious relevance for our understanding of defoliating pests in general.

ACKNOWLEDGMENTS

We thank L. Stanbridge, J. Patterson and A. Ferguson for assistance in the field, J. Patterson for information on the distribution of lodgepole pine and A. Booth, H. F. Evans and J. T. Stoakley for helpful and constructive criticisms of this paper.

REFERENCES

Barbour, D.A. (1987). Pine beauty moth population dynamics: general considerations and the life-table work. *Population Biology and Control of the Pine Beauty Moth* (Panolis flammea) (Ed. by S.R. Leather, J.T. Stoakley & H.F. Evans), pp. 7–13. Forestry Commission, Edinburgh.

Barnett, D.W. (1983). *The nutritional requirements of* Panolis flammea *(D. & S.) larvae.* Ph.D. thesis, University of Edinburgh.

Barnett, D.W. (1987). Pine beauty moth outbreaks: associations with soil type, host nutrient status and tree vigour. *Population Biology and Control of the Pine Beauty Moth* (Panolis flammea). (Ed. by S.R. Leather, J.T. Stoakley & H.F. Evans), pp. 14–20. Forestry Commission, Edinburgh.

Berwig, W. (1926). Die Forleule in Bayern. Historische-statistischc-klimatologische Betrachtung. *Forstwirtschaft Centralblatt*, 70, 165–181.

Bevan, D. (1987). *Forest Insects.* Forestry Commission, Edinburgh.

Klimetzek, D. (1979). *Insekten-Grasschädlinge an Kiefer in Nordbayern und der Pfalz: Analyse und Vergleich 1810–1970.* Institut für Forstzoologie, Freiburg.

Last, F.T., Jeffers, J.N.R., Bunce, R.G.H., Claridge, C.J., Baldwin, M.B. & Cameron, R.J. (1986). Whither forestry? The scheme in AD 2050. *Trees and Wildlife in the Scottish Uplands* (Ed. by D. Jenkins), pp. 20–32. Institute of Terrestrial Ecology, Abbots Ripton, Huntingdon.

Leather, S.R. (1984). Factors affecting pupal survival and eclosion in the pine beauty moth, *Panolis flammea* (D & S). *Oecologia (Berlin), 63,* 75–79.

Leather, S.R. (1985). Oviposition preferences in relation to larval growth rates and survival in the pine beauty moth, *Panolis flammea. Ecological Entomology,* 10, 213–217.

Leather, S.R. (1987). Lodgepole pine provenance and the pine beauty moth. *Population Biology and Control of the Pine Beauty Moth* (Panolis flammea) (Ed. by S.R. Leather, J.T. Stoakley & H.F. Evans), pp. 27–30. Forestry Commission, Edinburgh.

Leather, S.R. & Barbour, D.A. (1987). Associations between soil type, lodgepole pine (*Pinus contorta* Douglas) provenances, and the abundance of the pine beauty moth, *Panolis flammea* (D & S). *Journal of Applied Ecology.* 24, 945–952.

Leather, S.R., Watt, A.D. & Barbour, D.A. (1985). The effect of host plant and delayed mating on the fecundity and lifespan of the pine beauty moth, *Panolis flammea* (Denis and Schiffermuller) (Lepidoptera: Noctuidae): their influence on population dynamics and relevance to pest management. *Bulletin of Entomological Research,* 75, 641–651.

Leather, S.R., Watt, A.D. & Forrest, G.I. (1987). Insect-induced chemical changes in young lodgepole pine (*Pinus contorta*): the effect of previous defoliation on oviposition, growth and survival of the pine beauty moth, *Panolis flammea. Ecological Entomology,* 12, 275–281.

Lines, R. 1976. The development of forestry in Scotland in relation to the use of *Pinus contorta.* Pinus contorta *Provenance Studies* (Ed. by R. Lines), pp. 2–5. Research and Development Paper No. 114, Forestry Commission, Edinburgh.

Miles, J. (1988). Vegetation and soil change in the uplands. *This volume.*

Rhoades, D.F. (1983). Herbivore population and plant chemistry. *Variable Plants and Herbivores in Natural and Managed Systems* (Ed. by R.F. Denno & M.S. McClure), pp. 155–110. Academic Press, London.

Rowan, A.A. (1986). The nature of British upland forests in the 1980s. *Trees and Wildlife in the Scottish Uplands* (Ed. by D. Jenkins), pp. 7–13. Institute of Terrestrial Ecology, Huntingdon.

Stoakley, J.T. (1977). A severe outbreak of the pine beauty moth on lodgepole pine in Sutherland. *Scottish Forestry,* 31, 113–125.

Stoakley, J.T. (1979). *Pine Beauty Moth.* Forest Record No. 20, Forestry Commission, Edinburgh.

Stoakley, J.T. (1981). Control of the pine beauty moth, *Panolis flammea,* by aerial application of fenitrothion. *Aerial Application of Insecticide Against Pine Beauty Moth* (Ed. by A.V. Holden & D. Bevan), pp. 9–14. Occasional Paper No. 11, Forestry Commission, Edinburgh.

Stoakley, J.T. (1986). Protecting the timber resource. *Trees and Wildlife in the Scottish Uplands* (Ed. by D. Jenkins), pp. 108–110. Institute of Terrestrial Ecology, Abbots Ripton, Huntingdon.

Watt, A.D. (1986a). The performance of the pine beauty moth on water-stressed lodgepole pine plants: a laboratory experiment. *Oecologia (Berlin),* 70, 578–579.

Watt, A.D. (1986b). The ecology of the pine beauty moth in commercial woods in Scotland. *Trees and Wildlife in the Scottish Uplands* (Ed. by D. Jenkins), pp. 79–87. Institute of Terrestrial Ecology, Abbots Ripton, Huntingdon.

Watt, A.D. (1987). The effect of shoot growth stage of *Pinus contorta* and *Pinus sylvestris* on the growth and survival of *Panolis flammea* larvae. *Oecologia (Berlin),* 71, 429–433.

Watt, A.D. & Leather, S.R. (1987). Pine beauty moth population dynamics: synthesis, simulation and prediction. *Population Biology and Control of the Pine Beauty Moth* (Panolis flammea) (Ed. by S.R. Leather, J.T. Stoakley & H.F. Evans), pp. 41–45. Forestry Commission, Edinburgh.

Watt, A.D. & Leather, S.R. (in press). The pine beauty moth in Scottish lodgepole pine plantations. *Dynamics of Forest Insect Populations: Patterns, Causes and Management Strategies* (Ed. by A.A. Berryman), pp. Plenum Press, New York.

White, T.C.R. (1974). A hypothesis to explain outbreaks of looper caterpillars, with special reference to populations of *Selidose masuavis* in a plantation of *Pinus radiata* in New Zealand. *Oecologia (Berlin),* 16, 279–301.

White, T.C.R. (1976). Weather, food and plagues of locusts. *Oecologia (Berlin),* 22, 119–134.

The effects of afforestation and acid deposition on the water quality and ecology of upland Wales

A. S. GEE* AND J. H. STONER[†]

*Welsh Water Authority, South Western District, Hawthorn Rise, Haverfordwest, Dyfed SA61 2BH
[†]Welsh Water Authority, Plas y Ffynnon, Cambrian Way, Brecon, Powys LD3 7HP

SUMMARY

1 In Wales, most forests lie in the uplands so that many rivers have headwaters which drain afforested catchments.
2 Recent studies have shown that conifers have their greatest impact on water quality and ecology by exacerbating acidification. They scavenge atmospheric pollutants so that, where soils are poor, streams running through conifer forest are more acidic and/or contain higher aluminium concentrations than those in adjacent moorland.
3 The distribution and abundance of a wide variety of aquatic plant and animal taxa are adversely affected by afforestation of the catchment. Acidity-related water-quality determinants such as pH and aluminium are known to have a direct and harmful physiological effect on several groups, especially salmonid fish. Despite strong empirical relationships between floral assemblages, invertebrates and stream acidity, however, causal mechanisms are still largely unknown for most affected taxa.
4 Whereas certain forest planting regimes and ground preparation techniques can be ecologically beneficial in non-sensitive areas, there is no evidence that these can effectively ameliorate acidification in the base-poor catchments of upland Wales.

INTRODUCTION

Conifer afforestation of upland catchments has increased considerably in recent years, the area under productive forest having almost trebled since the formation of the Forestry Commission in 1919 (Thompson, Stroud & Pienkowski 1988). Afforestation represents one of the major land-use changes in the UK: conifers now cover some 10 per cent of the total land area, and the cover is likely to increase substantially over the next 50 years (Department of Energy 1986). This may result in the afforestation of up to one-third of upland Britain.

Since most Welsh forests lie in the uplands, the majority of the rivers have headwaters draining afforested catchments. These rivers sustain nationally important game fisheries, and are of considerable ecological value. Recent evidence linking afforestation and acidification (Harriman & Morrison 1982; Stoner, Gee and Wade 1984; Fritz et al. 1986) has therefore been received with great concern (NCC 1986).

273

During the last decade, several surveys in Scotland and Wales have demonstrated that streams draining afforested catchments with hard rocks and poor
soils are more acidic and contain elevated aluminium concentrations compared
with those draining similar moorland catchments (Harriman & Morrison 1982;
Stoner *et al.* 1984; Stoner & Gee 1985). As a result, afforested streams contain a
less diverse invertebrate fauna and sparse or non-existent salmonid populations.
Although there is much evidence associating afforestation, acidification and the
absence of certain taxa, there is less information on the mechanisms and
pathways involved. This paper summarizes information on the relationships
between afforestation, acidification and aquatic ecology in Wales, with reference
to catchment studies at Llyn Brianne (Fig. 1) and to some data from other areas.
Only a brief consideration of other effects of afforestation, such as on nutrient,
pesticide and suspended solids inputs, is made as these have been reviewed
recently by NCC (1986) and Ormerod, Mawle & Edwards (1987b).

ACIDIFICATION

Pathways, interactions and deposition quality

In Wales, as in other parts of the British Isles, only a small proportion of the total
precipitation falls directly into surface waters. Most passes over vegetation and
through soils before reaching streams and lakes. The chemical composition of
deposition can thus be altered greatly; conifer afforestation can change the
vegetation of the catchment and can result in particularly pronounced impacts on
surface water quality (United Kingdom Acid Waters Review Group (UKAWRG)
1986). Lee, Tallis & Woodin (1988) and Woodin (1988) provide more details of
acidic deposition in terrestrial habitats in the British uplands.

When the UK Acid Deposition Review Group reported in 1983 (Barrett *et al.*
1983), the quality of atmospheric deposition in Wales was known from only two
sites in the extreme mid-west. Beginning in October 1983, Welsh Water
undertook a 12-month survey of bulk deposition quality throughout Wales.
Although there was much variation in deposition quality in time and space, there
were clear geographical patterns in annual average data for the principal
determinands (Welsh Water, unpublished). Chloride concentrations were highest
near the coast, reflecting the sea-salt effect, but were still high inland. 'Excess'
sulphate concentrations, in contrast, were highest in the east and south of the
country, nearer the industrial centres of England and South Wales. However,
because of the neutralizing effects of magnesium, calcium and ammonium in
these areas, the net, volume-weighted mean pH of bulk deposition was lowest
(pH< 4·5) in the uplands of Mid- and North Wales. Having eliminated the effect
of greater rainfall at higher altitudes by volume-weighting, it is likely that these
data reflect the higher contribution of 'occult' deposition in the uplands, where
mist, fog and low cloud commonly occur and where conifer forests are
widespread (Grace & Unsworth 1988).

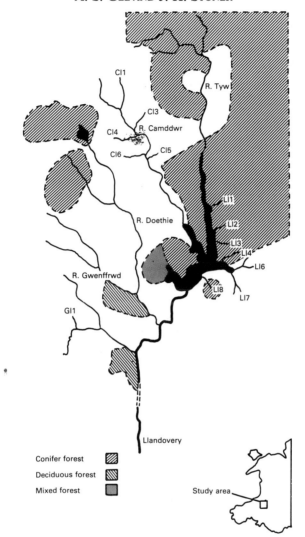

FIG. 1. Streamwater sampling sites at Llyn Brianne.

Throughfall and stemflow

Vegetation can scavenge 'dry' and 'occult' pollutants very effectively. As rainwater passes through a forest canopy, deposited materials are washed from the surfaces of the leaves, together with those anions and cations produced by crown leaching. Further chemical changes occur as the water passes over the surface of the stem leading to an additional lowering of the pH (UKAWRG 1986).

TABLE 1. Acidity and sulphate throughfall under spruce, oak and *Molinia caerulea* compared with ambient precipitation (annual range is given in parentheses)

Vegetation	pH	SO$_4$ (μequiv l^{-1})
25-year-old spruce	4·32 (3·6–5·9)	296 (51–1512)
12-year-old spruce	4·27 (3·7–6·0)	144 (37–1181)
Oak	4·70 (4·1–6·1)	115 (39–444)
Molinia caerulea	4·70 (4·3–5·2)	124 (25–788)
Ambient precipitation	4·60 (3·8–7·0)	54 (10–260)

Conifers are particularly efficient scavengers of atmospheric pollutants and sea salts as illustrated by the results of recent studies in the Llyn Brianne Project (Table 1). Compared with ambient precipitation, H$^+$, SO$_4^{2-}$ and Cl$^-$ concentrations in winter are elevated in throughfall and stemflow from 25-year-old Sitka spruce (*Picea sitchensis*). Concentrations under 12-year-old spruce are intermediate, reflecting an age effect. Although concentrations of SO$_4^{2-}$ and Cl$^-$ under oak (*Quercus robur*) are higher than in precipitation, the production of base cations results in throughfall under oak being generally less acid than precipitation.

Studies at Plynlimon in mid-Wales have shown that evapotranspiration in fully afforested catchments can be as much as 30 per cent of the total water yield compared with 17 per cent for adjacent moorland catchments (Hornung & Newton 1986). Evapotranspiration can therefore concentrate the solution of materials deposited on a conifer forest, further exacerbating the acidification.

Soil chemistry

Surface waters are only likely to be acidified where the soils and underlying geology have insufficient base cations freely available to neutralize the deposited acidity. In Wales, such acid-sensitive regions are restricted predominantly to the uplands of Dyfed and Gwynedd, where the underlying geology comprises shales and mudstones of the Ordovician and Silurian series. The soils in this area are generally base-poor, naturally acidic peats and podzols (Hornung 1986).

In the Llyn Brianne catchments it has been shown recently that, for any given soil type, concentrations of aluminium in soil water are between two and three times higher under 25-year-old conifers than in moorland (Welsh Water 1987). Sulphate and chloride concentrations are also elevated, being around double those found in moorland catchments. Whereas pH levels in acid water do not appear to differ greatly under conifers of different age, aluminium concentrations

TABLE 2. Soil water (B horizon) chemistry under spruce.
Molinia and oak (annual range in parentheses)

Vegetation	pH	Al (μequiv l^{-1})	SO$_4$ (μequiv l^{-1})
25-year-old spruce	4·3 (4·1–4·5)	234 (187–308)	168 (149–237)
12-year-old spruce	4·25 (3·9–4·5)	87 (65–122)	100 (76–122)
Oak	4·5 (3·8–5·0)	63 (52–107)	109 (87–166)
Molinia	4·25 (4·0–4·5)	90 (54–154)	77 (60–111)

are very much higher under 25-year-old trees compared with 12-year-old trees (Table 2). Also, for a given soil type and trees of similar ages, higher concentrations of aluminium are found beneath larch (*Larix decidua*) than beneath spruce.

Surface water chemistry

In Wales and Scotland, streams draining conifer-afforested catchments are more acidic and/or contain higher aluminium concentrations than those in adjacent moorland with similar soils and geologies (Stoner *et al.* 1984; Stoner & Gee 1985; Harriman & Morrison 1982; Harriman & Wells 1985).

Seasonal and episodic variations in stream chemistry occur in poorly buffered streams. For example, in one of the afforested streams draining into Llyn Brianne, pH levels have reached 6·2 in summer and 3·9 in winter. Concurrently, calcium concentrations in the summer reached 130 μequiv l^{-1} (micro equivalents per litre) whilst aluminium concentrations were only 10 μequiv l^{-1}. In winter, minimum calcium and maximum aluminium concentrations of 40 μequiv l^{-1} and 95 μequiv l^{-1} were obtained (Fig. 2). The deployment of continuous monitoring equipment at Llyn Brianne recently has provided data which emphasize the importance of episodic variations in stream chemistry. Forest streams respond rapidly to storm events to the extent that, for example, soluble aluminium concentrations can increase from less than 0·1 to over 1 mg l^{-1} in 2–3 hours.

Where soils have been depleted of base cations, sea salts in rainfall can also result in the acidification of runoff. These sea salts are also scavenged by conifers, thereby exacerbating the acidification process, even in the absence of anthropogenic sulphate. Chloride ions in large concentrations exceed the neutralizing capacity of the soil, and during storm events can release hydrogen and aluminium ions in a way similar to anthropogenically derived sulphate.

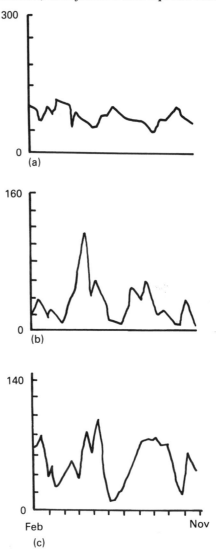

FIG. 2. Seasonal variations in (a) calcium, (b) hydrogen and (c) dissolved aluminium concentrations (µequiv l⁻¹) in Llyn Brianne afforested stream L11.

OTHER CHEMICAL AND PHYSICAL EFFECTS

In upland Wales conifers are thought to influence water quality largely by acidification. Nevertheless, forestry practices *per se* can affect the aquatic environment in a variety of other ways, as recently summarized by Ormerod *et al.* (1987b).

Nutrients

Upland soils are often poor in nutrients, so that afforestation is usually accompanied by the addition of fertilizer. The nitrate, phosphate and potassium added to the soil can leach into drainage waters with noticeable biological consequences, particularly in lakes and reservoirs (Harriman 1978; Brooker 1981). Preparation of the land by ploughing prior to planting can also lead to oxidation of organic material and the release of ammonium, nitrate, phosphorus and sulphate (Hornung & Newton 1986). Forest clearance can also result in eutrophication of runoff waters (Hornung, Stevens & Reynolds 1987).

Pesticides

Herbicides such as Glyphosate®, Asulam and 2-4D are commonly used to suppress the growth of competitive plant species at the planting stage. Insecticides may also be used to control pests. Whereas careful application can minimize the risk of these materials entering watercourses, biologically significant concentrations have been reported in Scottish rivers draining conifer plantations (Morrison & Wells 1981).

Hydrological regime and suspended solids

Conifer afforestation has a pronounced effect on the timing and quantity of run-off water due to changes in the drainage system and evapotranspiration. In newly afforested catchments with extensive artificial drainage, peak discharges have been shown to increase and the duration of storm hydrographs to shorten (Jones 1975; Robinson 1980). In closed canopy forests, evapotranspiration can reduce yield by up to 30 per cent with important consequences for deposited pollutants. Changes in drainage networks and the hydrological regime due to afforestation can result in considerable erosion of soils, which can be exacerbated by poorly engineered road-works. In the Cray catchment, mid-Wales, suspended solid concentrations of over 1100 mg l^{-1} have been recorded (Stretton 1984).

Energy inputs

The amount of light and heat entering a stream has important ecological consequences and can be greatly affected by the proximity and density of forest cover. Smith (1980) has shown that shading by a closed canopy forest can reduce illumination at the stream bed by more than 90 per cent; Ormerod et al. (1987b) recorded that at Llyn Brianne heavily shaded streams have lower summer temperature maxima and higher winter minima than adjacent moorland streams.

Allochthonous material can be an important source of energy for stream invertebrates and fish (Bird & Kaushik 1981). Whilst the decay rate of conifer needles is much less than that of broadleaved species such as oak and alder (Alnus *glutinosa*), conifer litter may be locally important in some upland Welsh streams (Ormerod *et al.* 1987b).

ECOLOGICAL EFFECTS

Primary production

The upland waters of Wales are mostly oligotrophic and are also hydrologically variable. Complex relationships exist between the biological community, water quality and physical conditions. In general, however, afforested lakes and streams have a lower diversity of autotrophs than unafforested systems (UKAWRG 1986).

The strong empirical relationship between diatom community structure and lake acidity allows fossil diatoms in lake sediments to be used to reconstruct the previous pH regime of the lake (Battarbee 1984). Sediment coring studies in mid-Wales have recently shown that the rate of acidification in the afforested Llyn Berwyn accelerated following planting compared with the nearby, unafforested Llyn Hir (Fritz *et al.* 1986; Kreiser *et al.* 1986). Diatoms are scarce in streams draining afforested catchments at Llyn Brianne, possibly because of low pH levels (F. Round, pers. comm.). Whereas studies in North America have shown that the biomass and production of Chlorophyceae (filamentous algae) and Bacillariophyceae (diatoms) increase after tree felling, partial forest clearance in one of the Llyn Brianne catchments (L13) resulted in a predominant increase in filamentous green algae, because the pH remained low, at about 5·2.

In a recent survey of 88 sites on 16 river systems in mid- and North Wales, 47 macrofloral taxa were collected (Ormerod, Wade & Gee 1987c). *Lemanea* spp. and *Fontinalis squamosa* were negatively associated with forest cover (>50 per cent), but this was not attributable to shading since it reflected a catchment rather than a bankside influence. The sites with *Lemanea* and *F. squamosa* had mean pH values of 6·3 and 6·1, respectively, whereas sites dominated by filamentous chlorophytes (*Ulothrix* spp., *Microspora* spp., *Strigioclonium* spp., *Mougesia* spp.) had mean pH values of between 5·5 and 5·8; sites with *S. undulata* had a mean pH of 5·2. Corresponding differences were apparent in other chemical variables such as aluminium concentrations (Ormerod *et al.* 1987c).

Macro-invertebrates

Earlier Welsh studies at Llyn Brianne (Stoner *et al.* 1984) showed that unafforested streams with an average hardness of >8 mg l^{-1} have a reasonably diverse invertebrate fauna, whereas, in contrast, afforested streams with an average hardness of <10 mg l^{-1} have a very restricted fauna. In Wales generally, the diversity of invertebrates is often markedly reduced in streams draining forests (Fig. 3).

During a later survey of 16 river systems (referred to earlier), invertebrate assemblages were classified by TWINSPAN (Hill 1979; Wade, Ormerod & Gee, in

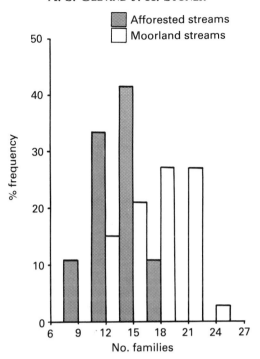

FIG. 3. The percentage frequency distributions of kick-samples with macro-invertebrate family richness from moorland and afforested streams. Data for 72 samples from 25 sites in the Wye and Tywi catchments.

press). The TWINSPAN classification in spring could be related to differences between site groups in terms of pH, calcium, magnesium and aluminium concentrations. Site groups 4 and 5, which tell on the left-hand side of the primary dichotomy (Fig. 4), had mean pH values of >6·4 and 6·1, respectively, in winter. However, site groups 6 and 7, which fell on the right-hand side of Fig. 4, had values of 5·8 and 5·2, respectively. In spring, sites in TWINSPAN group 6 and 7 were more likely to be drawn from catchments with more than 50 per cent forest cover than were site groups 4 and 5 ($\chi^2 = 7\cdot3$, $P < 0\cdot01$).

Seasonal variations in stream chemistry are pronounced in afforested upland streams, and this tends to be reflected in their invertebrate communities. The relationship between forest cover and summer TWINSPAN groups was not significant ($\chi^2 = 2\cdot7$, $P < 0\cdot05$; Wade et al., in press). Patterns amongst invertebrate assemblages in these streams were evident in relation to the classification of macroflora (Ormerod et al. 1987c). For example, 60 per cent of sites with Scapania undulata (but no Lemanea spp.) had fewer than 20 macro-invertebrate taxa, whilst less than 5 per cent of sites with Lemanea were similarly impoverished. Overall, there was a highly significant concordance between site classifications established separately using macroflora and macro-invertebrate fauna

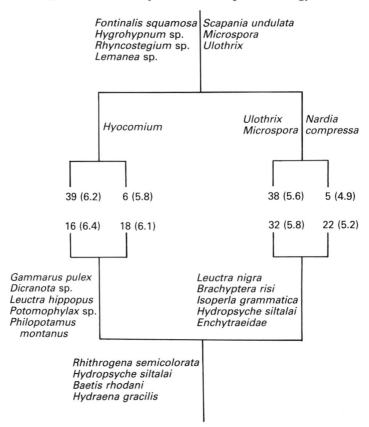

FIG. 4. Concordance between site groupings revealed by independent TWINSPAN of macrofloral and macro-invertebrate assemblages in upland Wales (Group association, level 1, $\chi^2 = 21\cdot2$, $P < 0\cdot001$). The number of sites in each group is shown, with their mean pH in parentheses. Data for a survey of 16 river systems.

(Fig. 4). There was a significant aggregation of *Lemanea* and several inverte-brates known to be scarce in acidic waters. Acid-tolerant taxa (including *S. undulata*, *Nardia compressa*, filamentous chlorophytes and invertebrates) also occurred together.

Fish

Stoner *et al.* (1984) and Stoner & Gee (1985) showed that in mid-Wales trout (*Salmo trutta*) were generally absent from conifer-forested streams and lakes with hardness levels <10 mg l^{-1}. More recent studies at Llyn Brianne have shown that fish are absent from the afforested streams, but densities range from 0 to 0·9 m^{-2} in acidic moorland streams, despite low invertebrate abundance (Welsh Water, unpublished). In a survey of more than 90 upland Welsh streams in 1984, trout abundance correlated positively with pH and negatively with dissolved aluminium concentration (Welsh Water, unpublished). Although there was no statistically significant relationship between trout abundance and the proportion of the catchments afforested, there were significant correlations between pH, aluminium concentrations and percentage afforestation.

Using caged fish, Stoner et al. (1984) showed that in the Brianne streams the mortality of fish was strongly correlated with the concentration of dissolved aluminium. It was also notable that the concentration of aluminium in the gills of fish which died most rapidly was up to 57 times greater than in source fish. By experimentally manipulating water quality in one of those streams the mortality of salmon (Salmo salar) and trout was increased substantially in the presence of aluminium (0.35 mg 1^{-1}) at pH 5.0, but was very low at pH 4.3 with no aluminium (Ormerod et al. 1987a).

In Scotland, trout are also absent from streams draining long-established forests and these streams are more acidic, and contain higher concentrations of aluminium, than those in neighouring non-afforested catchments (Harriman & Morrison 1982). Egglishaw, Gardiner & Foster (1986) have shown recently that there is a strong correlation between Scottish salmon fishery districts having a high proportion of upland conifer forest and those with salmon catches which have declined in the past 30 years.

Other vertebrates

There is little information on the effects of forestry on the distribution of other vertebrates in Welsh upland waters, apart from that resulting from the work of Ormerod, Tyler & Lewis (1985) and Ormerod et al. (1986) on dippers (Cinclus cinclus). The decline of this bird in the River Irfon, a major tributary of the Wye, was accompanied by the afforestation of the catchment and a significant decline in pH. In other afforested streams the scarcity of dippers was associated with low pH (<5.7) and high aluminium concentrations (>0.1 mg 1^{-1}) and was probably due to inadequate food supplies. There is no information at present to suggest that piscivorous birds are affected by acid waters in Britain (Thompson et al. 1988), although it has been suggested that the recolonization of the uplands by some species may be limited by food supply in acid-sensitive regions (S. Tyler, pers. comm.).

Although there are no Welsh data on amphibians in upland forested areas, the work of Cummins (1986), which demonstrated lethal and sub-lethal effects of low pH and high aluminium on frogs (Rana temporaria), suggests that this group may be adversely affected.

As otters (Lutra lutra) are dependent on productive fisheries, it is likely that their distribution and abundance may also be adversely affected by the afforestation of the uplands. Although there are few data currently available, studies in the upper Severn catchment in upland Wales have shown that otters were not resident on streams draining conifer plantations, where pH fell periodically below pH 5.5, whereas a tributary draining open moorland held resident otters (Mason & MacDonald 1987; Woodin 1988).

DISCUSSION

The distribution and abundance of a wide variety of aquatic plant and animal taxa in upland Wales have been shown to be adversely affected by catchment afforestation. In the majority of base-poor waters, significant correlations also exist between ecological status and acidity-related water-quality determinands, especially pH and aluminium. There is, however, a dearth of information in Wales on the possible influence of other forest effects, such as nutrient enrichment, pesticides and suspended solids.

Certain pH levels and aluminium concentrations are known to have a direct and harmful physiological effect on several taxa. In fish, enzyme systems, ion exchange mechanisms, respiration and mucus production have been shown to be affected (Witters & Vanderborght 1987). Whereas hydrogen ion concentrations may need to be substantially elevated to be directly toxic to fish, much lower concentrations are required in the presence of soluble inorganic aluminium. Afforested streams and lakes in upland Wales have been shown to contain seasonal elevations of both these ions to toxic concentrations. Although the growth of trout in Welsh forest streams has been shown to decrease due to the shading effect of trees (Ormerod *et al.* 1987b), there is no evidence that the distribution of fish is affected by the quality or abundance of food.

There is less information on the direct toxic effects of acidity-related water quality conditions on the survival and abundance of invertebrates. Hildrew *et al.* (1984) and Havas & Likens (1985) have suggested that the effect of acidity may be mediated, in part, through changes in the food supply. Certain groups such as mayflies (Ephemeroptera), however, are probably directly affected by low pH and high aluminium concentrations (Bell 1971; Herman & Anderson 1986). Ormerod *et al.* (1987a) have demonstrated, by means of induced episodes of acidity with 0.35 mg m^{-3} aluminium, that the drift of *Baetis rhodani* increases so that its density is reduced in the absence of altered food supplies.

Ormerod *et al.* (1987c) have recently postulated that the striking relationship between floral and invertebrate assemblages in Welsh streams may be a reflection of the latter's dependence on specific food plants and/or plant micro-habitats. The microhabitats provided by *Fontinalis* and *Lemanea* (Hildrew 1978) could influence the density and diversity of invertebrate communities. Forests could influence the composition of the invertebrate community by affecting the supply of allochthonous material and its decomposition. Shredders, for example, could benefit from the reduced rate of detrital breakdown and enhanced fungal biomass at low pH (Hildrew *et al.* 1984). Because cellulose-rich food material such as filamentous algae can be eaten by many invertebrate taxa under certain conditions, Ormerod *et al.* (1987c) have suggested that these animals may not be food-limited unless the trophic pathways (e.g. carbon leaching, microbial conditions and assimilation) are acidity-dependent.

Floral taxa such as *Scapania* and *Nardia* have been shown to be tolerant of high aluminium concentrations at low pH levels (Caines, Watt & Wells 1985) and are characteristic of acid forest streams in Wales (Ormerod *et al.* 1987c).

Lemanea, however, which is found only in less acid waters, has shorter, finer and more gelatinous filaments below pH 5·5.

Although there are very strong empirical relationships between floral assemblages, invertebrates and stream acidity, causal mechanisms are still largely unknown. Yet there is a great deal of information for upland Wales relating conifer afforestation to the acid conditions known to reduce faunal diversity. Poorly buffered upland waters are being continually acidified by acid deposition which is substantially accelerated by conifer afforestation with the release of toxic aluminium from soils. In non-sensitive areas the formation of buffer strips and bankside zones of broadleaved trees has been shown to be ecologically beneficial (Mills 1980; Smith 1980). In acid-sensitive areas, i.e. most of upland Wales, however, no effective treatments have yet been proven (Woodin 1988). Nevertheless, recent applications for forestry grants referred to Welsh Water by the Forestry Commission include many suggestions for ameliorating some of the above inimical effects on water quality. These include buffer strips, alternative ploughing techniques, mixture of species (including hardwoods) and limited bankside liming. There is no evidence yet, however, that these measures will have the desired effect. There is, therefore, an urgent need for more research on the efficacy of such measures, similar to those currently being undertaken in the Llyn Brianne Project (Welsh Water 1987).

ACKNOWLEDGMENTS

We are grateful for the assistance of the many field and laboratory staff whose work is summarized here, and to Dr S. J. Ormerod for commenting on the manuscript. Current work at Llyn Brianne is part-funded by the Department of the Environment and the Welsh Office.

REFERENCES

Barrett, C.F. *et al.* **(1983).** *Acid Deposition in the United Kingdom.* Warren Spring Laboratory, Stevenage.

Battarbee, R.W. (1984). Diatom analysis and the acidification of lakes. *Philosophical Transactions of the Royal Society of London, Series B,* **305,** 451–477.

Bell, H.L. (1971). Effects of low pH on the survival and emergence of aquatic insects. *Water Research,* **5,** 313–319.

Bird, G.A. & Kaushik, N.K. (1981). Coarse particulate organic matter in streams. *Perspectives in Running Water Ecology* (Ed. by M.A. Lock & D.D. Williams), pp. 41–68. Plenum, New York.

Brooker, M.P. (1981). The impact of impoundments on the downstream fisheries and general ecology of rivers. *Advances in Applied Biology,* **6,** 91–151.

Caines, L.A., Watt, A.W. & Wells, D.E. (1985). The uptake and release of some trace metals by aquatic bryophytes in acidified waters in Scotland. *Environmental Pollution (Series B),* **10,** 1–18.

Cummins, C.P. (1986). Effects of aluminium and low pH on growth and development in *Rana temporaria* tadpoles. *Oecologia,* **69,** 248–252.

Department of Energy (1986). *Energy Forestry in Britain: Environmental Issues.* Consultation document prepared by Environmental Resources Limited for the Department of Energy (ETSU) (October), London.

Egglishaw, H.J. Gardiner, R. & Foster, J. (1986). Salmon catch decline and forestry in Scotland. *Scottish Geographical Magazine*, **102**. 57–61.

Fritz, A. et al. (1986). *Palaeoecological evaluation of the Recent Acidification of Welsh Lakes. 1. Llyn Hir, Dyfed.* Department of Geography, University College, London.

Grace, J. & Unsworth, M.H. (1988). Climate and micro-climate of the uplands. *This volume.*

Harriman, R. (1978). Nutrient leading from fertilized forest watersheds in Scotland. *Journal of Applied Ecology*, **15**, 933–942.

Harriman, R. & Morrison, B.R.S. (1982). Ecology of streams draining forested and non-forested catchments in an area of central Scotland subject to acid precipitation. *Hydrobiologia*, **88**, 251–263.

Harriman, R. & Wells, D.E. (1985). Causes and effects of surface water acidification in Scotland. *Journal of Water Pollution Control*, **84**. 215–224.

Havas, M. & Likens, G.E. (1985). Toxicity of aluminium and hydrogen ions to *Daphnia catawba*, *Holopedium gibberum*, *Chaeoborus puntipennis* and *Chironomus anthrocinus* from Mirro Lake, New Hampshire. *Canadian Journal of Zoology*, **63**, 1114–1119.

Herman, N.J. & Anderson, K.G. (1986). Aluminium impact on respiration of lotic mayflies at low pH. *Water, Air and Soil Pollution*, **30**, 703–709.

Hildrew, A.G. (1978). Ecological aspects of life history in some net-spinning Trichoptera. *Proceedings of the 2nd International Symposium on Trichoptera, 1977*, pp. 269–281. Junk, The Hague.

Hildrew, A.G. Townsend, C.R. Francis, J. & Finch, K. (1984). Cellulolytic decomposition in streams of contrasting pH and its relationship with invertebrate community structure. *Freshwater Biology*, **14**, 323–328.

Hill, M.O. (1979). *TWINSPAN — a FORTRAN Program for Arranging Multivariate Data in an Ordered Two-way Table by Classification of the Individuals and Attributes.* Cology and Systematics, Cornell University, Ithaca, New York.

Hornung, M. (1986). *Llyn Brianne Project: Catchment Characteristics SubGroup. First Progress Report.* Institute of Terrestrial Ecology, Bangor.

Hornung, M. & Newton, M.D. (1986). Upland afforestation: influences on stream hydrology and chemistry. *Soil Use Management*, **2**, 61–65.

Hornung, M. & Stevens, P.A. & Reynolds, B. (1987). The effects of forestry on soils, soil water and surface water chemistry. *Environmental Aspects of Plantation Forestry in Wales* (Ed. by J.E.G. Good), pp. 25–36. Institute of Terrestrial Ecology, Bangor.

Jones, A. (1975). *Rainfall, run-off and erosion in the upper Tywi catchment.* PhD thesis, University of Wales.

Kreiser, A. et al. (1986). *Palaeoecological Evaluation of the Recent Acidification of Welsh Lakes. 2. Llyn Berwyn, Dyfed.* Department of Geography, University College, London.

Lee, J.A., Tallis, J.H. & Woodin, S.J. (1988). Acidic deposition and British upland vegetation. *This volume.*

Mason, C.F. & MacDonald, S.M. (1987). Acidification and otter (*Lutra lutra*) distribution on a British river. *Mammalia*, **51**, 81–87.

Mills, D.H.O. (1980). *The Management of Forest Streams.* Forestry Commission leaflet No. 78, HMSO, London.

Morrison, B.R.S. & Wells, D.E. (1981). The fate of Fenitrothion in a stream environment and its effect on the fauna, following aerial spraying of a Scottish forest. *Science of the Total Environment*, **19**, 223–252.

NCC (1986). *Nature Conservation and Afforestation in Britain.* Nature Conservancy Council Peterborough.

Ormerod, S.J. Tyler, S.J. & Lewis, J.M.L. (1985). Is the breeding distribution of dippers influenced by stream acidity? *Bird Study*, **32**, 33–39.

Ormerod, S.J., Allinson, N., Hudson, D. & Tyler, S.J. (1986). The distribution of breeding dippers (*Cinclus cinclus* L., Aves) in relation to stream acidity in upland Wales. *Freshwater Biology*, **16**, 501–507.

Ormerod, S.J., Boole, P., McCahon, P., Weatherley, N.S., Pascoe, D. & Edwards, R.W. (1987a). Short-term experimental acidification of a Welsh stream: comparing the biological effects of hydrogen ions and aluminium. *Freshwater Biology*, **17**, 341–356.

Ormerod, S.J., Mawle, G.W. & Edwards, R.W. (1987b). The influence of forest on aquatic fauna. *Environmental Aspects of Plantation Forestry in Wales* (Ed. by J.E.G. Good), pp. 37–49. Institute of Terrestrial Ecology, Bangor.

Ormerod, S.J., Wade, K.R. & Gee, A.S. (1987c). Macro-floral assemblages in upland Welsh streams in relation to acidity, and their impotance to invertebrates. *Freshwater Biology*, **18**, 545–557.

Robinson, M. (1980). *The Effect of Pre-afforestation Drainage on the Streamflow and Water Quality of a Small Upland Catchment*. Report No. 73, Institute of Hydrology, Wallingford.

Smith, B.D. (1980). The effects of afforestation on the trout of a small stream in southern Scotland. *Fish Management*, **11**, 39–57.

Stoner, J.H. & Gee, A.S. (1985). Effects of forestry on water quality and fish in Welsh rivers and lakes. *Journal of the Institute of Water Engineers and Scientists*, **39**, 27–45.

Stoner, J.H., Gee, A.S. & Wade, K.R. (1984). The effects of acidification on the ecology of streams in the upper Tywi catchment in West Wales. *Environmental Pollution, A*, **36**, 125–157.

Stretton, C. (1984). Water quality and forestry — a conflict of interests: Cray reservoir, a case study. *Journal of the Institute of Water Engineers and Scientists*, **38**, 323–330.

Thompson, D.B.A., Stroud, D.A. & Pienkowski, M.W. (1988). Afforestation and upland birds: consequences for population ecology. *This volume.*

United Kingdom Acid Waters Review Group (UKAWRG) (1986). *Acidity in United Kingdom Fresh Waters*. Interim Report, Department of the Environment, London.

Wade, K.R., Ormerod, S.J. & Gee, A.S. (in press). Classification and ordination of macroinvertebrate assemblages to predict stream acidity in upland Wales. *Hydrobiologia.*

Welsh Water (1987). *Llyn Brianne Acid Waters Project. An Investigation into the Effects of Afforestation and Land Management on Stream Acidity*. First Technical Summary Report, April, Welsh Water, Llanelli.

Witters, H. & Vanderborght, O. (1987). Ecophysiology of acid stress in aquatic organisms. *Annales de la Société Royale Zoologique de Belgique*, **Supplement 1**, 1–472.

Woodin, S.J. (1988). Acidic deposition and upland conservation: an overview and the way ahead. *This volume.*

Effects of clearfelling on surface water quality and site nutrient status

P. A. STEVENS, J. K. ADAMSON*, M. A. ANDERSON[†] AND M. HORNUNG*

*Institute of Terrestrial Ecology, Bangor Research Station, Penrhos Road, Bangor, Gwynedd LL57 2LQ, *Institute of Terrestrial Ecology, Merlewood Research Station, Grange-over-Sands, Cumbria LA11 6JU and [†]Forestry Commission, Alice Holt Lodge, Wrecclesham, nr. Farnham, Surrey GU10 4LH*

INTRODUCTION

Clearfelling can result in greatly increased concentrations of major nutrients in streamwaters (Vitousek & Melillo 1979). For example, post-felling nitrate concentrations were 50 times greater than pre-felling levels at Hubbard Brook in the US (Likens, Bormann & Johnson 1969). Such losses, coupled with nutrient removal in timber at harvest, have led to concern over long-term site nutrient depletion, particularly where intensive methods are employed (Anderson 1985).

In Britain, large areas of upland conifer plantations are reaching felling age; this paper describes the effects on streamwater chemistry of felling Sitka spruce (*Picea sitchensis* (Bong.) Carr.) plantations at Kershope Forest, Cumbria, and Beddgelert Forest, North Wales. In addition, estimates of nutrient inputs and losses through a 50-year crop rotation, including the harvest phase, are presented for Beddgelert Forest. The potential for site nutrient depletion is then discussed.

SITES AND METHODS

The Kershope Forest site lies on thick glacial till at an altitude of 350 m. The site is gently sloping and the soils are predominantly stagnohumic gleys (Avery 1980). Two 2 ha plots are bounded by drainage ditches which cut into impermeable till, thus creating artificial catchments. One plot was felled in 1983, leaving one plot as an unfelled control. Samples of drainage ditch water were collected weekly. The results presented are the discharge-weighted means for the ditches draining these plots. Ditch sample concentrations were statistically analysed by analysis of variance of the 52 values for each ion in each of the control and felled plots.

At Beddgelert Forest, 62% of a 1·4 ha forested catchment was felled in September 1984. The catchment lies at 330 to 500 m altitude and is dominated by ferric stagnopodzol soils (Avery 1980). The stream at Beddgelert was gauged and weekly streamwater samples were analysed from September 1982 to September 1986. Streamwater ion concentrations were calculated and statistically analysed in the same manner as those for Kershope. During the same period, estimates of nutrients in incoming precipitation were made, using bulk precipitation collectors. Nutrient outputs in the stream pre- and post-felling were calculated, using total discharge from the catchment and discharge-weighted mean streamwater nutrient concentrations. Additional nutrient outputs at harvest from the felled catchment were estimated by using an unfelled control catchment as a reference. Further details of the two sites can be found in Hornung *et al.* 1986.

At Beddgelert, stratified random samples of 'crown' and 'stem' (>7 cm diameter) material from 32 trees were weighed, chipped and chemically analysed. The sampling scheme was designed to allow estimation of the total nutrient removal from the site by 'conventional' and 'whole tree' harvesting techniques; in the former, only the stem is removed from the site, leaving a carpet of brash on the ground; in the latter, all above-ground parts of the tree are removed from the site. Also at Beddgelert, 24 soil profile pits were sampled; samples were analysed for total and extractable nutrients. Chemical analysis of all samples was made, using standard methods (Allen *et al.* 1974).

RESULTS

Drainage water chemistry

At both Kershope and Beddgelert Forests, felling resulted in a number of changes to ditch or stream water chemistry (Table 1). At Kershope, significant increases in concentration of PO_4–P, NO_3–N and NH_4–N reflected release from decaying felling debris and possibly increased rates of decomposition and nitrification in the forest floor. Potassium concentration did not increase significantly. The discharge-weighted mean K concentration of the felled catchment was greatly influenced by a single summer storm with high discharge and concentration. At Kershope, concentrations of airborne elements (Na, Ca, Mg, Cl and SO_4–S) decreased significantly. Mean pH (as H) decreased to a small but significant degree, but Al concentration remained comparatively constant. A detailed description of the effects of felling on Kershope water chemistry is to be found in Adamson *et al.* (1987).

Felling of 62 per cent of the catchment at Beddgelert resulted in similar changes to those noted at Kershope. The main differences at Beddgelert were significant increases in K and Al concentration after felling and Ca, PO_4–P and NH_4–N which showed little change. Phosphate and NH_4–N were rarely detected in the Beddgelert stream — adsorption of P by the mineral soil and active soil nitrification occur at this site.

Site nutrient depletion

The amounts of nutrients accumulated in the crop are substantial and at harvest represent a major drain on the site nutrient capital, especially if whole-tree harvest is undertaken (Table 2). Nutrient inputs from precipitation are adequate over 50 years to provide the N, K and Ca for the crop, but not the P. However, since drainage water losses of N and K are similar to inputs over 50 years, then the trees presumably exploit the available soil nutrient stocks. Very little P is lost

TABLE 1. Discharge-weighted mean concentrations and maximum concentrations for drains and streams at Kershope and Beddgelert Forests. Results for Kershope are for 1984, the first year after felling of the experimental plot. Results for Beddgelert are for a single stream: pre-felling from October 1982 to October 1984, post-felling from October 1984 to 1986. Concentrations are mg l^{-1} except pH. Asterisks describe the significance levels of F-tests on results from analysis of variance on weekly sample concentrations — see text (*P < 0.05; **P < 0.01; ***P < 0.001)

| | Kershope | | | | Beddgelert | | | |
| | Mean | | Maximum | | Mean | | Maximum | |
	Control (unfelled)	Felled	Control (unfelled)	Felled	Pre-felling	Post-felling	Pre-felling	Post-felling
pH	3.9	3.8***	4.5	4.1	4.3	4.5	—	—
Na	11.9	7.0***	17.8	10.8	7.7	5.9***	12.0	10.5
K	0.90	2.10	2.50	3.30	0.22	0.63***	0.77	1.28
Ca	6.8	4.9***	10.8	7.2	0.8	0.8	3.1	1.4
Mg	1.8	1.3***	2.5	1.8	0.9	0.8***	1.6	1.2
Fe	0.40	0.80***	0.70	1.20	<0.02	<0.02	0.06	0.03
Al	1.4	1.5	2.6	1.8	0.9	1.2***	1.2	1.8
PO$_4$–P	0.005	0.026***	0.017	0.058	<0.005	<0.005	0.01	0.01
NO$_3$–N	2.0	4.5***	4.0	12.0	0.6	1.1***	1.8	2.0
NH$_4$–N	0.2	0.9*	0.3	1.6	<0.1	<0.1	0.3	0.2
Cl	23.2	11.8***	37.0	24.0	13.4	10.1***	22.0	18.0
SO$_4$–S	7.3	4.0***	9.5	6.9	2.2	2.0***	5.7	5.0

TABLE 2. Nutrient balance sheet at Beddgelert Forest for a 50-year Sitka spruce crop rotation. Units are kg ha^{-1}. Phosphate P concentrations in the stream were below detection limit. n.d. = not determined. Additional losses from the site in stream at harvest are assumed to be the same for both types of felling practice

	N	P	K	Ca
Nutrient input in bulk precipitation — 50 years	520	7·5	223	375
Nutrient output in stream — 50 years	678	0	215	1,035
Nutrient accumulation in crop — all above-ground material	428	43·5	144	279
Nutrient accumulation in crop — stem only	128	12·3	38	151
Additional loss for site in stream at harvest	126	0	19	0
Overall loss from site — whole-tree harvest	712	36	155	939
Overall loss from site — conventional fell	412	4·8	49	811
Extractable nutrients in soil to rooting depth (base of Bs horizon)	n.d.	24·6	94	112
Total nutrients in soil in rooting depth (base of Bs horizon)	9,400	2,940	59,700	1,390

in streams at this site (this nutrient is normally deficient in upland forests and is tightly conserved), and the large Ca outputs are explained by bedrock weathering beyond the reach of most roots.

DISCUSSION

The effect of felling on ditch and streamwater solute concentrations was generally insignificant in both ecological and water quality terms. The main exception was the maximum NO_3–N concentration after felling at Kershope, which exceeded World Health Authority guidelines for water quality. However, the overall effect downstream would be limited as dilution from unfelled areas would rapidly occur. Similarly, the increased but small concentrations detected at Kershope after felling are unlikely to lead to algal blooms in downstream lakes or reservoirs. Of the remaining ions, only Al occurred in concentrations high enough to be of significant ecological and water quality importance. Streams draining upland conifer plantations have been found to contain higher concentrations of Al than adjoining non-forested streams (Harriman & Morrison 1982; Gee & Stoner 1988). Clearfelling might be expected to result in reduced Al concentrations. There is no evidence of a reduction at Kershope and, at Beddgelert, Al concentration increased significantly after felling. Certainly, any anticipated reduction in Al in streams draining upland conifer forests must be delayed until at least two years after felling.

The reduction in concentrations of Na, Mg, Cl and SO_4–S after felling at both sites is explained by the removal of airborne material captured by the spruce canopy and subsequently leached through the soil.

The potential for site nutrient depletion at Beddgelert would appear to be greatly increased by more intensive harvesting practice. Phosphorus is clearly being depleted by both conventional and whole-tree harvest techniques, but is normally applied as a fertilizer at planting and is not considered to be a problem. Losses of K and Ca at harvest are also a matter for concern. Extractable soil K to rooting depth is only double the overall loss from the site with conventional felling, and less than that with whole-tree harvest. Calcium removal at harvest (even with conventional felling) is greater than the extractable Ca in the soil to rooting depth.

If the trees utilize extractable nutrients in the soil, rather than nutrients in incoming precipitation (a little of each is a more likely situation), then nutrients must be released from the 'total' soil stock to replenish the 'extractable' fraction. If the rate of release is inadequate, then, in the long term, the trees will be forced to rely upon the throughput of nutrients in precipitation.

There appears to be no lack of available N at Beddgelert — the losses to streamwater recorded at this site are unusually large and the very limited forest floor accumulation indicates mineralization of organic material.

In the long term, K and Ca depletion by both conventional and whole-tree harvest will take place at Beddgelert unless soil weathering rates are adequate to replenish the soil-extractable stocks. If weathering rates are inadequate, then trees must improve their efficiency of utilization of nutrients in incoming rainfall, or fertilizer must be applied.

ACKNOWLEDGMENTS

We thank S. Hughes, the ITE Chemical Service at Merlewood and the Forestry Commission Analytical Service for undertaking the chemical analysis. This work was partly funded by the Department of the Environment.

REFERENCES

Adamson, J.K., Hornung, M., Pyatt, D.G. & Anderson, A.R. (1987). Changes in solute chemistry of drainage waters following the clearfelling of a Sitka spruce plantation. *Forestry*, **60**, 165–177.

Allen, S.E., Grimshaw, H.M., Parkinson, J.A. & Quarmby, C. (1974). *Chemical Analysis of Ecological Materials*. Blackwell Scientific Publications, Oxford.

Anderson, M.A. (1985). The impact of whole-tree harvesting in British forests. *Quarterly Journal of Forestry*, **79**, 33–39.

Avery, B.W. (1980). *Soil Classification for England and Wales*. Soil Survey Technical Monograph No. 14, Soil Survey of England and Wales, Harpenden.

Gee, A.S. & Stoner, J.H. (1988). The effects of afforestation and acid deposition on the water quality and ecology of upland Wales. *This volume.*

Harriman, R. & Morrison, B.R.S. (1982). Ecology of streams draining forested and non-forested catchments in an area of central Scotland subject to acid precipitation. *Hydrobiologia*, **88**, 251–263.

Hornung, M., Adamson, J.K., Reynolds, B. & Stevens, P.A. (1986). Influence of mineral weathering and catchment hydrology on drainage water chemistry in three upland sites in England and Wales. *Journal of the Geological Society, London*, **143**, 627–634.

Likens, G.E., Bormann, F.H. & Johnson, N.M. (1969). Nitrification: importance to nutrient losses from a cutover forested ecosystem. *Science*, **163**, 1205–1208.

Vitousek, P.M. & Melillo, J.M. (1979). Nitrate losses from disturbed forests: patterns and mechanisms. *Forest Science*, **25**, 605–619.

Are there any economics of upland forestry ?

C. PRICE

Department of Forestry and Wood Science, University College of North Wales,
Bangor, Gwynedd LL57 2UW

SUMMARY

1 Economics, as a behavioural science, can successfully explain the pattern of investment in afforestation in terms of low rates of return but large fiscal incentives. As an evaluative methodology, economics encounters formidable problems in assessing the changes at site brought about by afforestation, in predicting probable secular trends and in tracing the economic ripples generated by afforestation.

2 While timber productivity and quantitative effects on water yield can be well predicted, water quality, recreation, landscape, wildlife and employment impacts are intractable even to sophisticated evaluative techniques. Projected trends in timber markets and appropriate discount rates over the course of a rotation are also very speculative.

3 There is poor information on the value of funds in alternative investments, while the welfare effects of price changes and natural resource depletion have invariably been underestimated.

4 The appropriate response to so much uncertainty about the economic value of forestry is caution, which suggests only a slow expansion of forestry in the uplands.

5 The 'need' for more timber is not absolute, and the case for afforestation should depend also on whether proposals are environmentally acceptable.

INTRODUCTION

The discipline of economics has two aspects. Positive economics is a behavioural science, whose methods and objectives in some ways resemble those of ecology. Normative economics, by contrast, actively seeks to make value judgements about the deployment of resources, and is more related to ethics and politics. As a behavioural science, economics ought to help us understand the way in which human individuals and organizations respond to price stimuli: producers to the lure of profit, consumers to the maximization of satisfaction within the constraints of income. Its achievements are measured by the extent to which it generates testable hypotheses, yields statistically significant relationships and facilitates accurate predictions.

The private perspective

Economics can claim some modest success in its account of the past pattern of forestry activity in the uplands. The history is generally one of progressive deforestation through firing, grazing and felling (Birks 1988). While self-sufficiency continued as an imperative, active management of remaining forests continued.

With the industrial revolution and the increase of international trade, however, planting and replanting could be regarded as just one possible means of achieving a timber supply, to be tested as an investment by the same criteria as and in competition with a range of industrial investments. The outcome was that forestry, with its extraordinarily long production cycle, offered a comparatively low rate of return. Planting persisted largely for amenity reasons and as an adjunct to farm activities rather than as a commercial timber enterprise. In the uplands, where the economic pressures for deforestation were greater, and the rewards from planting low-productivity sites particularly small, forestry reached its lowest ebb.

The rise of state forestry

The First World War, having devastated the remaining forest area, temporarily renewed the importance of self-sufficiency, but this was an objective of national concern not reflected in private profits. From 1919 until 1957, therefore, the great bulk of afforestation fell to the state sector, which was nominally freed of the requirement to achieve an economic return on its investment. At the same time, the other imperative — improved self-sufficiency in food — confined forestry expansion to the least productive agricultural land. Silviculture flourished in the British uplands as never before.

But in 1957, following the publication of the Zuckerman Report (1957), the concept of profitable investment came to the fore in state forestry. The technique of discounted cash flow analysis (which had in fact been invented more than a century earlier by a German forester, Martin Faustmann) showed the long-delayed returns from forestry in a poor light. Methodical application of the technique led to changes in ideal rotation length, thinning regime and species choice, which to this day distinguish British silvicultural practice from its continental counterparts.

Fiscal encouragement

In the mean time, private activity had remained at a low level (planting only 18 per cent of the national total during 1950–51, rising to 35 per cent by 1956–57), reflecting a continuing judgement that forestry was a poor investment.

The remarkable surge of interest in private afforestation since about 1960 is also quite explicable in economic terms. With the foundation of the welfare state, high marginal rates of tax became established practice under both Labour and Conservative governments. The unique tax provisions for forestry, however, allow an investor to set the cost of planting against tax liabilities from other sources (an allowance which has sometimes been worth as much as 98 per cent of planting costs), and yet to enjoy virtually tax-free income from the crops when they mature, under schedule B tax (see details in Hart 1987). The post-tax rate of return on planting has sometimes been as high as 14 per cent, comparing very favourably with post-tax rates in industrial investment. It was only necessary for this tax avoidance potential of forestry to be discovered and for the requisite managerial and institutional arrangements to be put in place.

Private afforestation equalled state activity by the early 1970s, and currently forms 81 per cent of all new planting. It seems that its expansion has gathered enough momentum to survive the significant fall in tax rates over the last decade. The tax arrangements are, however, only of much interest to those with high incomes and rates of tax. For the average upland farmer, paying a lower rate of tax, the fiscal incentives are worth much less, and have done little to promote farm forestry.

The structure of tax incentives explains too why the private sector has occasionally been prepared to forgo grant aid by planting land which has not been approved for afforestation by the Forestry Commission. The grant reduces the net cost of planting eligible for tax relief, and hence even now may be worth only 40 per cent of its face value, or about £100 per hectare. This sum is often less than the premium price on land with, or likely to obtain, the Forestry Commission's clearance for planting. For those not mindful of their public image, the financial incentives are thus, if anything, towards forestry on land which society would prefer to remain unplanted!

Response to new grants

The test of a good theory is its ability to predict, and the recent informally canvassed Ministry of Agriculture and Fisheries (MAFF) proposals for farm forestry grants give forest economists the opportunity to make some pronouncement. Four factors have to be considered here: (i) the effect of high rates of discount on the present value of delayed returns, (ii) the lack of scale economies in small farm forest enterprises, (iii) the dearth of forestry expertise on farms, and (iv) the perception that forestry has detrimental effects on the productivity of a farm enterprise (Thomas & Maclean 1984). Bearing these in mind, the expectation is that uptake of existing grants will be small, even where land no longer has any agricultural use. The new scheme aims to give annual income to farmers while the tree crop grows to commercial size, but grant levels are low in Disadvantaged and Severely Disadvantaged Areas — essentially the uplands. There is a widespread feeling that 36,000 ha of planting in three years, mooted as a target by MAFF, is unlikely to be achieved unless the forestry grants, net of financial outlay on planting, are comparable with whatever short-term agricultural revenue and subsidy could be obtained from the same piece of land. Discounting makes the distant revenues from forestry all but irrelevant as an incentive.

ECONOMICS AS EVALUATIVE METHODOLOGY

Whatever success economics achieves in describing the world as it exists, the focus of interest in the economics of upland forestry has long been upon what it can tell us about afforestation as it should be. Over the last 25 years there has been a plethora of economic evaluations. The majority have reflected the perspective of the profit-maximizing private individual, although often transposing it into the social context. The conclusion is that, in the absence of fiscal subvention, the rate of return on plantations normally lies below the rate which individuals, or society, might expect by investing elsewhere, and that it would hence be in the public interest to reduce or even halt afforestation (Ramblers' Association 1971; Treasury 1972; Bowers 1982; Grove 1983; National Audit Office 1986). Some authors reinforce this case by referring to negative environmental effects of massive afforestation (Nature Conservancy Council 1986; Gee & Stoner 1988; Thompson, Stroud & Pienkowski 1988).

Other studies appeal to positive social and environmental effects of afforestation which are held capable of raising the social return on forestry to an acceptable level, even if it is intrinsically unattractive as a private venture (Wolfe & Caborn 1973; Centre for Agricultural Strategy 1980; Denne, Bown & Abel 1986).

Little has come to light which was not known to forest economists decades previously (Faustmann 1849 in Gane & Linnard 1968; Hiley 1930; Zuckerman 1957; Walker 1960). Some authors show a penchant for over-simplification or selectivity which borders on either the naïve or the conspiratorial. If there is one thing that both categories of study tend to share, it is an air of certainty about their conclusions and the supporting facts: their crusading spirit does not admit any doubts.

But if we aim to discover what (if anything) economics can tell us about the value of forestry to the nation, rather than to use economics to bolster our prejudices, we have to ask three questions open-mindedly. We have to ask these for a specific site, and not seek generalizations before we understand the particulars:

(i) What would happen at the site if an extra hectare were to be afforested?
(ii) What will change in the world, irrespective of whether or not the extra hectare is planted?
(iii) What would change in the world as a result of what happened to that hectare?

There is an all-too-frequent assumption that the answer to the first question is obvious, and that the second and third questions can be ignored. We must, moreover, clarify the ultimate objective. In what follows, this is taken to be maximization of economic well-being, rather than of profit or Gross National Product (GNP). Prices, where they exist, are often a reasonable index of what is considered to be valuable. Where they are not, economists resort to indirect techniques of estimating what people would be willing to pay, if a market existed.

Predicting on-site change

Productivity

The key to forestry's poor financial performance in the northern temperate uplands is a relatively slow growth rate, leading to protracted rotations, small-dimensioned (low value) products, high susceptibility to climatic and biotic hazards and, in consequence of all, usually a negative net present value at the prevailing 5 per cent discount rate. By contrast, in the tropics and parts of the southern temperate zone, where growth rates may be five times the mean for the UK, rates of return on afforestation are quite ample enough to attract private investment without subsidy. However, even within the uplands, growth rates vary significantly. In 1977, 1·3 per cent of state plantations were expected to achieve a 5 per cent rate of return (Forestry Commission 1977), while for unconventional short-rotation regimes on favourable sites, 10 per cent has been claimed possible (Moore & Wilson 1970).

The first evaluative step, therefore, is not an economic appraisal, but a physical prediction. Many attempts have been made to relate performance to site factors. Given its influence on soil and climate, it is perhaps not surprising that elevation emerges as the most pervasively successful predictor of productivity (Mayhead 1973; Mayhead & Broad 1978; Price & Dale 1982), typically accounting for 50 per cent of the variance in productivity within a forest area. Elevation is also a key factor in determining whether windthrow risk will allow the trees to be grown to commercially desirable dimensions. Other important site factors can be identified locally (Pyatt, Harrison & Ford 1969). Foresters on the whole would have little difficulty in identifying sites in the upland zone where, say, a 5 per cent rate of return could be expected on plantations. These would typically be low-lying sheltered sites with deep, well-drained soils — sites which have hitherto tended to remain in farming. Under the new agricultural provisions, however, they may well become available to forestry. If attainment of a 5 per cent return at current prices were the only requirement, then forest economists in partnership with field foresters could successfully point the way forward. But even on the site itself there are more complex effects to measure.

Hydrological effects

Perhaps the most readily quantified non-market effects of forests are those on water runoff. The evidence is fairly unequivocal (Calder & Newson 1979) that afforestation reduces the yield from grassland sites in the uplands. Qualitative effects have attracted more recent attention and are presently less susceptible to prediction (Gee & Stoner 1988).

Economic appraisals of even the gross volume yield effects are few (Collett 1970; Treasury 1972; Barrow, Hinsley & Price 1986). Interestingly, elevation appears as a key variable here too, with financial costs of lost runoff rising for the less productive, high-lying sites. Again, it seems that the line could be identified which divides sites where an overall 5 per cent rate of return would or would not be expected. In principle, evaluation of qualitative effects awaits only reliable physical prediction. Once effective corrective treatments for excess sediment loads, acidity and unsatisfactory base status have been worked out, they are economically as easy to cost as the normal commercial operations of forestry. And, to the extent that the result of afforestation is, say, loss of lucrative sport fishing, methods have been developed to evaluate this in economic terms (Radford 1985). Indeed, such methods have also been developed for the re-creation value of the forests. The equity issue — who ought to pay these costs? — lies at the edge of economics. None the less, the principle that the land user who causes the malign effect should pay seems well agreed in theory.

Recreation

Since the publication of Clawson's classic travel cost method of recreation evaluation (Clawson 1959), recreation economics has been a burgeoning disci-pline. Applications to British forests appear in Mutch (1968), Treasury (1972) and Christensen (1985). The method draws parallels with markets for other commodities. Their value is indicated by what consumers are willing to pay for them, except that in the case of informal outdoor recreation payment is for transport and accommodation rather than for a final product. Unfortunately, the elegance of the method and its ability to render a plausible cash answer have tended to disguise its practical and logical problems, e.g. valuation of time (Radford 1985), influence of substitute sites (D. S. Connolly & C. Price, unpubl.), effect of multi-site visits (Christensen, Humphreys & Price 1985) and an extreme sensitivity of the value obtained to the means used for fitting a demand curve to the data (Christensen 1985).

Some appraisals have imputed a generalized cash value for forest recreation, almost independent of location or forest type (Treasury 1972). Even the more site-specific studies demonstrate just willingness to pay for recreation at the site itself. They fail to attribute value (positive or negative) to the vegetation cover, and therefore indicate nothing about the merit or demerit of replacing, say, open moorland by plantations. This leads to a need to evaluate environmental quality on a cash scale.

Landscape

In theory the value attributable to each aspect of environmental quality (including presence/absence of forest cover, species, age of plantations) is implicit in variation of willingness to travel to different sites. Given a sufficiency of data, the separate contribution of each aspect should be revealed by multiple regression analysis. Many studies have indeed successfully explained variability in observers' responses to landscapes in terms of objectively measured features (e.g. Coventry–Solihull–Warwickshire 1971). However, such studies normally assume that the effects of individual features are strictly additive, and that dependence on any cardinally measurable features is linear in form. Neither theory nor observation supports these assumptions (Price 1978). Hence, the increment of value ascribed to a given change in one feature in one set of circumstances cannot properly act as predictor for the value of an equal change made in other circumstances. There can be no legitimate generalized cash equivalent for the addition of one hectare of forest to the landscape.

Conservation

Despite the brave efforts of Helliwell (1973), Everett (1979) and Rodme (1980), the case for wildlife evaluation is pretty much the same. There are additional problems of the uncertainty and irreversibility of some ecological changes, and the value, if any, to be ascribed to the very existence of ecosystems, species and individuals (Price 1985, 1987b). When even the conceptual validity of some of these items is controversial, how does one realistically hope to find agreement on a cash value?

Employment and rural depopulation

Since the twin processes of rationalization and mechanization of operations cut the labour intensity of commercial forestry by a factor of 5 in 20 years, it has been a debatable point whether or not the rural employment objective of forestry has much practical importance. The evidence is that the relative job opportunities in forestry and farming vary widely within the uplands (Holmes & Inglis 1977; Johnson & Price 1987), but that (like the timber crop) forestry jobs are long in maturing . We have also to ask whether upland depopulation is not partly a matter of people voting with their feet against the idyll of the rural lifestyle; whether mechanization simply means people make machines to harvest trees, instead of harvesting trees directly; and whether, in this case, forestry has any peculiar difference from other industries as a means of creating jobs. The unemployment problem is, after all, now a pervasive rather than a local one. Given the doubts about the efficacy of forestry in job creation, it is questionable whether it is worth applying sophisticated economic techniques to derive a 'social' cost of labour, particularly if such techniques are not used in evaluating alternative uses of investment funds.

Review

Given the imposing problems listed above, one might easily regard the economic study of many non-market effects of forests as no more than a diverting intellectual exercise. It is not that methods of appraisal have not improved, but rather that perception of apparently insuperable difficulties has advanced faster than perception of solutions. These economic evaluations can be helpful only if they are used as a means of calibrating our judgement. But, when we allow scope for discretionary interpretation of quantitative data, we have truly discovered the slippery slope into the abyss which divides objective science from machiavellian politics!

Predicting the world context

Economists are wont to deride attempts to predict the price of any commodity 50 or so years in the future. Yet it is clear that such a prediction for timber is un-avoidable if any useful economic conclusions at all are to be drawn for forestry. The difference between stable prices and those rising in real terms by 3·5 per cent per year is the difference between most upland plantations being unprofitable and most of them being profitable (Price & Dale 1982).

The simplest predictive expedient is to extrapolate past price trends. Some analysts detect a 1–2 per cent rise in real prices for standing timber over the last 100 years (even this is disputed). On the other hand, over the last 20 years prices to British growers have fluctuated irresolutely. In any case, extrapolation requires the continuation of the causal trends, a condition unlikely to be observed. The past 100 years of timber supply involved pushing back frontiers of natural forests; the next 50 will show increasing reliance on managed forests with increasingly binding constraints on supply.

It is therefore more satisfactory to attempt to project the underlying determinants of price:supply, demand and the elasticities (responsiveness) of price to changes in these variables. Here again the long-term nature of forestry exacerbates the prediction problem. Over the next 50 years there are expected to be acute localized shortfalls of supply, especially in relation to fuel-wood requirements. This demand, however, is largely from a non-monetized subsis-tence sector which is not directly competing in world markets. The global picture is expected to be of demand overtaking supplies at current prices, although by a relatively small margin, as population and economic activity expand, and plantations and managed forests increasingly supersede exploitation of natural resources. The degree of shortfall depends on the course of increasing demand, itself the subject of varying predictions of global and regional growth rates. On the supply side, studies for the USA and Europe project only 43 and 30 years ahead, respectively. The supply situation is much harder to predict for the tropics, where even current timber inventories are unknown, and where stock impoverishment and deforestation lie substantially outside the control of state forest services.

Many official studies avoid committing themselves to a view of elasticities, despite their crucial role. Price elasticity of demand is a measure of the responsiveness of quantity purchased to changes in price. Near-zero elasticities imply that large price increases are needed to counteract the effect of a small shortfall of supply. Estimated elasticities for timber products range from -0.7 to $+3.0$; values as close to zero as -0.01 have been recorded. On the basis of such values and FAO's predicted shortfall of timber supply in 2025 (quoted in Centre for Agricultural Strategy 1980), we can confidently, but unhelpfully, predict that the price of timber in that year will lie in the range from a little below present prices to a 2.6×10^{10}-fold increase!

My own guess (it is no more than that) is an elasticity of around -0.5, implying a 54 per cent increase in price to the consumer by 2025, and perhaps a doubling in real prices by the end of a forest rotation initiated now. What that implies for prices in the forest depends sensitively on assumptions made about future harvesting and processing costs. But, if supply and demand do not follow their assumed most probable trend, the answer could again be very different.

The social rate of discount

It is, however, not just the scale of future revenues that determines the profitability of forestry, but their timing. The high rate of discount acting on the long production cycle seals the financial fate of upland forestry. Despite the monolithic image of the test discount rate set by government, the appropriate value, and even the conceptual meaning, of the social discount rate has been a matter of furious debate among economists for many decades.

The basis for the social discount rate is seen either as the rate of return on available investment funds, or as a reflection of society's apparent preference for consumption earlier rather than later in time. In either case a delayed revenue or benefit is deemed less valuable than its equivalent made available now. Far from being a self-evident truth, this supposition is actually quite extraordinary, and it is particularly extraordinary that many conservationists have made it the focus of their case against afforestation in the uplands (e.g. Grove 1983). That values should be systematically diminished, simply because of their long-term futurity, seems the very antithesis of conservation philosophy.

No doubt there are values that change systematically over time, through changing taste and changing technology (Price 1978), but it is absurd to believe that these factors apply equally to all products. As for the preference of consumers for benefits now rather than in the future, it will come to be balanced in due course by a retrospective preference for benefits now rather than in the past. It is immediacy rather than earliness *per se* that consumers prefer. As for future generations, discounting makes their interests of no consequence compared with the interests of this generation — but it is unlikely that future generations would share this view! The only validation for the discounting of forestry's revenues would be an expectation that per capita consumption of forest products and their close substitutes would increase, and additional units of consumption would hence have progressively more trivial value. This is quite at odds with any scenario that predicts increased scarcity of forest products. Indeed, the two issues of discounting and of future supply and demand for timber not only are inseparable, but are in fact one issue. Discounting, as I see it, has no entitlement to an independent existence.

Seen from this viewpoint, competition from investments which are alternatives to forestry is to be represented in economic evaluation, and not by their rate of return, but by the appropriately weighted sum of benefits they are able to generate over their lifetime. The function of weighting is again simply to reflect the expected change in abundance of the products of the investment.

Moving the earth without a firm place to stand

The world beyond the site is itself affected by the decision to invest in forestry. Forestry investment prevents physical and financial resources being deployed elsewhere. And in due course its stream of products influences production in a wider context.

Forgone investment opportunities

Once we abandon the convention of accepting the prevailing market rate of return on investment as a discount rate, the value of funds devoted to investment differs from (and will normally be greater than) the face value of those funds. The opportunity cost of forestry investment can thus be constructed only with a detailed knowledge of how the funds would have been deployed. Would some part of them have been consumed immediately? If part had been invested, at what rate of return? The market rate of return is by no means a clearly defined, unique value, and it has been argued (Forestry Commission 1979; Smith 1979) that historical and current rates of return are overstated by the test discount rate. What environmental externalities (positive and negative), and how many jobs, would have been created by this alternative investment? What portion of the revenues from the investment would have been reinvested during the life of a forestry investment? Partly because the social discount rate has been taken to embody all these factors, they have been remarkably ill-researched, with even past investment behaviour by governments and private individuals being poorly understood.

Even in retrospect, with all costs and revenues of the forest rotation known, it is hard to judge whether a forestry investment has been a better option than an alternative deployment of finance. The task of predicting relative values is thus a daunting one. There are defensible assumptions under which the premium on investment funds is negative, and others under which it is indefinitely large.

Physical resource use

Dropping the presumption in favour of a high discount rate also makes it impossible to ignore the issue of non-renewable resource depletion. Commercial forestry, particularly on the rough terrain and impoverished soils of the uplands, is an increasing user of such resources, whose social cost may, under plausible conditions, have no definable upper limit (Price 1984). Like an indefinite premium on investment funds, this simply precludes any sensible decision rule.

Pecuniary externalities

One of the first lessons that students of economics learn is that a change in supply of any commodity causes a change in its price, all else being equal. This is expected to induce a response from other suppliers of the commodity, with the market finally settling to a new total quantity supplied and a new price for the product. Yet in practice these induced responses are almost universally ignored, the justification being that, under the optimal conditions of general market equilibrium, a small shift in resource allocation and income distribution has no effect on total social welfare. Yet if one thing can be agreed about the world, even about its economic condition, it is that optimality and equilibrium are far from being attained. One local but topical example is the forest's reduced water runoff, which causes more fossil fuel to be burnt (and more acid precipitation to be formed) to replace lost hydroelectric power. Conventionally the means of adjusting electricity supply is assumed to be a matter of indifference, but only because incremental pollution effects are not costed.

A more widespread response to halting upland afforestation may be seen in terms of greater price incentives to supply competing products, e.g. by clearing tropical high forest for pulpwood plantations and thereby not only displacing rich ecosystems, but depriving forest-living populations of their traditional firewood resources. It may be difficult to conceive that decisions about one hectare of British land could possibly have such far-reaching repercussions. But, if the world functions at all in the way that economists suggest, then the size of these abstruse externalities is a constant percentage of the direct effects of afforestation. This is irrespective of the scale of planting, and may indeed constitute the major portion of all the resultant welfare changes (Price 1987a). However, to balance the picture, similar indirect advantages can be claimed for continuing with low-output agriculture on sites suitable for afforestation (Price 1987b). It is comforting, but illusory, to think that there is nothing we can do in Britain about the ecological, fuel-wood and other crises of distant lands.

CONCLUSIONS

Even in the appraisal of physical effects of afforestation, felling and replanting, we are still very tentatively feeling our way. In relation to the long-term economy of natural resources, there is scope for an enormous range of assumptions. On the future for upland forestry, I am neither optimistic nor pessimistic, but agnostic. The more I study the problem, the larger the number of dimensions I find added to my agnosticism. The less the world economy is perceived as governed by stable and self-seeking optima, the more necessary it becomes to take overt account of the inscrutable interrelatedness of everything. In these circumstances, those who claim to know the value of another hectare of forest seem simply to be expressing a personal need for security and certainty, something which reality, unfortunately, cannot offer. There is almost nothing that can be said with confidence about the economic values of different forestry options. While economic evaluation of long-term ecological effects of commercial afforestation may be the last intellectual frontier for the economist to cross, it is by no means the only frontier.

Coping with risk and uncertainty

If there are any normative economics of upland afforestation, therefore, they are the economics of risk, uncertainty and the unknown. Some of the physical risks of forestry are sufficiently understood to be quantifiable and thus susceptible to treatment by the standard investment appraisal methods (e.g. windthrow). Political risks such as a drastic revision of capital taxation are risks to the individual, but should not greatly influence society's evaluation of forestry. But the future of prices, the relative values of natural and financial resources over the span of a rotation length, the nature of economic interaction, and some of the ecological consequences of unalloyed commercial forestry all remain in the realm of uncertainty. One might go as far as to estimate a probability distribution of eventual different timber prices, but this would be based on substantial guesswork.

Interestingly, the forms of advice proffered by the economist in the face of uncertainty resemble those of the layperson: either proceed with caution, or try to pretend that the problem does not exist. The latter strategy is achieved by raising the discount rate (Bromwich 1976), thus giving even less weight to the uncertain future. The logic of this ostrich outlook is obscure but it is, unfortunately, a widespread one. The alternative strategy is quite opposite in its implications, favouring those actions which give emphasis to long-term security (Wald 1950). But in this context caution indicates different policies to the timber grower and to the upland conservationist, to private forester and to government.

Economics or what else?

My own inclination would be to impose upon upland ecosystems as little of the stress of commercial forestry as is compatible with meeting the nation's needs. But absolute need is a concept that economists go to great lengths to avoid: it is a concept open to conflicting interpretation, and economists have no special role in defining it!

Perhaps it can be agreed that neither the landscape nor the ecosystems of the uplands need any more Sitka spruce (*Picea sitchensis*) which has never been more dominant in the planting programme. Nor do the forests in the regions where afforestation is rapid show any signs of bursting at the seams under recreation pressure. While the need for employment in rural areas is manifest, the effectiveness of commercial afforestation in supplying it, particularly in the short term, is in some doubt. With a possible decline in agricultural activity in the uplands, farm forestry may enhance overall labour intensity, especially to the extent that it is technologically backward. If it relies on the expertise of roving contractors the implications for truly rural employment are doubtful.

Despite recent preoccupation with multi-purpose forestry, therefore, it seems that the main justification for upland afforestation should continue to be the need for timber. The rather outdated statistic of 8 per cent self-sufficiency in timber is often quoted for Britain. But it cannot be too often repeated that timber output is now growing rapidly, against a background of relatively static consumption. The forests already planted are sufficient, once in sustained production, to meet nearly 50 per cent of our presently defined needs. Looking at the wasteful current use of timber products, it is difficult to believe that a considerable reduction in consumption would greatly inconvenience us.

My view, then, is that the economic evidence supports a mild pressure for extending national planting, but that this pressure should not override substantial environmental or social objections at the local level. Water authorities, conservationists and connoisseurs of landscape need to sharpen their concepts of the acceptability of commercial forestry in very specific localities, and to give a clear set of priorities for land where afforestation should preferably be avoided. There is then a case for discussion, because the imperatives for forest expansion are not absolute: they bear comparison with the imperatives established by other causes. Perhaps many people felt they knew that anyway, without the aid of economics, but at least they should be reassured that the economics of upland forestry cannot legitimately confute their instinct. Although my main message is that claims by economists that they have established a value (positive or negative) for upland forestry should be treated with scepticism, that is not to say that economics has nothing to tell us. The economic case for afforestation certainly grows weaker with greater altitude; many aspects of the environmental case against forestry grow stronger.

Behavioural economics again

The economist's remaining task is to suggest a package of fiscal measures in response to which landowners might be expected to pursue the optimal form of forestry. The uncertainties of determining that optimum make it difficult to present suggestions with any finesse. Yet it is clear that current measures, depending more on individual income than on suitability of site, lack the sensitivity to steer afforestation to those sites where it is most likely to prove acceptable. Differential and discretionary grants in theory provide the mechanism.

Unfortunately, grants are of little value under the present tax regime. In the mean time afforestation has become part of the everyday currency of tax avoidance, despite tax rate changes which have made it much less lucrative. Not even the possible scrapping of schedule B taxation would deter investment. The exemption of growing timber from capital gains tax and the option of sale in mid-rotation to a charitable institution would leave the tax advantages essentially intact until such time as a really radical upheaval occurs in the general taxation of capital gains. Without such a change, the economist can do no more than explain the problem, and commend the application of directer controls.

REFERENCES

Barrow, P., Hinsley, A.P. & Price, C. (1986). The effect of afforestation on hydroelectricity generation: a quantitative assessment. *Land Use Policy*, 3, 141–151.

Birks, H.J.B. (1988). Long-term ecological change in the British uplands. *This volume.*

Bowers, J.K. (1982). Is afforestation economic? *Ecos*, 3, 4–7.

Bromwich, M. (1976). *The Economics of Capital Budgeting.* Penguin, Harmondsworth, Middlesex.

Calder, I.R. & Newson, M.D. (1979). Land-use and upland water resources in Britain — a strategic look. *Water Resources Bulletin*, 15, 1628–1639.

Centre for Agricultural Strategy (1980). *Strategy for the UK Forest Industry.* Centre for Agricultural Strategy, University of Reading.

Christensen, J.B. (1985). *An economic approach to assessing the value of recreation with special reference to forest areas.* Ph.D. thesis, Department of Forestry and Wood Science, University College of North Wales, Bangor.

Christensen, J.B., Humphreys, S.K. & Price, C. (1985). A revised Clawson method: one part-solution to multidimensional disaggregation problems in recreation evaluation. *Journal of Environmental Management*, 20, 333–346.

Clawson, M. (1959). *Methods of Measuring the Demand for and Value of Outdoor Recreation.* Reprint 10, Resources for the Future, Washington.

Collett, M.E.W. (1970). External costs arising from the effects of forests upon streamflow in Britain. *Forestry*, 43, 87–93.

Coventry–Solihull–Warwickshire (1971). *A Strategy for the Sub-region, Supplementary Report 5: Countryside.* Coventry–Solihull–Warwickshire Sub-Regional Study Group, Coventry.

Denne, T., Bown, M.J.D. & Abel, J.A. (1986). *Forestry: Britain's Growing Resource.* UK Centre for Economic and Environmental Development, London.

Everett, R.D. (1979). The monetary value of the recreational benefits of wildlife. *Journal of Environmental Management*, 8, 203–213.

Forestry Commission (1977). *The Wood Production Outlook in Britain.* Forestry Commission, Edinburgh.

Forestry Commission (1979). *Comparison of Rates of Return in Forestry and Industry.* Planning and Economics Paper 34, Forestry Commission, Edinburgh.

Gane, M. & Linnard, W. (1968). *Martin Faustmann and the Evolution of Discounted Cash Flow.* Institute Paper 42, Commonwealth Forestry Institute, Oxford.

Gee, A.S. & Stoner, J.H. (1988). The effects of afforestation and acid deposition on the water quality and ecology of upland Wales. *This volume.*

Grove, R. (1983). *The Future for Forestry.* British Association of Nature Conservationists, Cambridge.

Hart, C.E. (1987). *Taxation of Woodlands.* Hart, Chenies, Coleford, Glos.

Helliwell, D.R. (1973). Priorities and values in nature conservation. *Journal of Environmental Management*, 1, 85–127.

Hiley, W.E. (1930). *The Economics of Forestry.* Clarendon Press, Oxford.

Holmes, W.D. & Inglis, C.J. (1977). *Employment in Private Forestry in Scotland in 1974.* Department of Forestry, University of Aberdeen.

Johnson, J.A. & Price, C. (1987). Afforestation, employment and depopulation in the Snowdonia National Park. *Journal of Rural Studies*, 3, 195–205.

Mayhead, G.J. (1973). The effect of altitude above sea-level on the yield class of Sitka spruce. *Scottish Forestry*, 27, 231–237.

Mayhead, G.J. & Broad, K. (1978). Site and the productivity of Sitka spruce in Southern Britain. *Quarterly Journal of Forestry*, **72**, 143–150.

Moore, D.G. & Wilson, B. (1970). Sitka for ourselves: the 25 year rotation. *Quarterly Journal of Forestry*, **64**, 104–112.

Mutch, W.E.S. (1968). *Public Recreation in National Forests: a Factual Study.* Booklet 21, Forestry Commission, Edinburgh.

National Audit Office (1986). *Review of Forestry Commission Objectives and Achievements.* HMSO, London.

Nature Conservancy Council (1986). *Afforestation and Nature Conservation.* NCC, Peterborough.

Price, C. (1978). *Landscape Economics.* Macmillan, London.

Price, C. (1984). Project appraisal and planning for over-developed countries: (I) The costing of nonrenewable resources. *Environmental Management*, **8**, 221–232.

Price, C. (1985). Economics, natural justice and the conservation of marine mammals. *Proceedings of the Conference on Conservation of Marine Resources and Marine Parks, Cochin, India, January 1985.*

Price, C. (1987a). Upland land use: towards the elusive balance. *Agriculture and Conservation in the Hills and Uplands* (Ed. by M. Bell & R.G.H. Bunce), pp. 156–159, Institute of Terrestrial Ecology, Merlewood.

Price, C. (1987b). *Does Shadow Pricing Go On For Ever?* Monograph, Department of Forestry and Wood Science, UCNW, Bangor.

Price, C. & Dale, I.D. (1982). Price predictions and economically afforestable area. *Journal of Agricultural Economics*, **33**, 13–23.

Pyatt, D.G., Harrison, D. & Ford, A.S. (1969). *Guide to Site Types in Forests of North and Mid-Wales.* Forest Record 69, Forestry Commission, Edinburgh.

Radford, A.R. (1985). *The recreational value of salmon fisheries in England and Wales.* M.Phil. thesis, Portsmouth Polytechnic.

Ramblers' Association (1971). *Forestry: Time to Rethink.* Ramblers' Association, London.

Rodme, N.J. (1980). *Cost–benefit analysis: the social evaluation of national nature reserves.* Ph.D. thesis, University of Cambridge.

Smith, C. (1979). The similarity of opportunity cost and social discount rate when measured in real terms. *Journal of Agricultural Economics*, **30**, 63–66.

Thomas, T.H. & Maclean, S.T. (1984). Farmers' attitudes to forestry. *Welsh Studies in Agricultural Economics*, **August 1984**, 45–58.

Thompson, D.B.A., Stroud, D.A. & Pienkowski, M.W. (1988). Afforestation and upland birds: consequences for population ecology. *This volume.*

Treasury (1972). *Forestry in Great Britain: an Interdepartmental Cost/Benefit Study.* HMSO, London.

Wald, A. (1950). *Statistical Decision Functions.* Wiley, New York.

Walker, K.R. (1960). The Forestry Commission and the use of hill land. *Scottish Journal of Political Economy*, **7**, 14–35.

Wolfe, J.N. & Caborn, J.M. (1973). *Some Considerations Regarding Forestry Policy in Great Britain.* Forestry Committee for Great Britain, Edinburgh.

Zuckerman, S. (1957). *Forestry, Agriculture and Marginal Land.* HMSO, London.

POSTSCRIPT

The UK budget of March 1988 did make radical change in the taxation of forestry. The Chancellor avoided the cosmetic step of merely scrapping schedule B tax. The replacement of tax rebates by grants has not greatly affected the rate of return from afforestation, but it has brought the possibility of directer controls on the land to be afforested.

MANAGEMENT OF ECOLOGICAL CHANGE

Nature conservation in Norway

O. GJAEREVOLL

Botanisk Avdeling, Universitetet i Trondheim, 7000 Trondheim, Norway

SUMMARY

1 The Norwegian conservation legislation of 1954 is described. This contained proposals for 16 national parks, all of which have now been established. National parks are vigorously protected.

2 The 1970 legislation created a new type of protected environment, the 'protected landscape'. The protection afforded to these sites resembles that of national parks in many other countries.

3 There is an extensive consultation phase before any conservation plan is implemented. Examples quoted are of plans for forests, mires, and cliffs used by nesting sea birds.

4 In 1986 a new plan proposed an additional 27 national parks and 14 protected landscapes, together with extensions to existing parks. When this plan is implemented, approximately 12 per cent of the Norwegian land area will be protected.

5 Svalbard is a special case due to both the Arctic and the endemic elements of its fauna and flora. Approximately 56 per cent of the land area is protected.

6 Norway is a major contributor to the conservation of western Europe's environmental resource.

INTRODUCTION

Norway, including Svalbard, has some of the most magnificent scenery in Europe. This magnificent scenery spans central European deciduous forest, great coastal mountain ranges and Arctic tundra, embracing skerries, fjords, glaciers, mountains, mires and forests. Norway has not, however, been a pioneer nation in nature conservation. It was generally held that, in a country with a small population and large expanses of wilderness, nature protected itself. It was felt there was, therefore, no need to preserve particular areas. Attitudes are changing, however, and since the Second World War there have been some important developments.

CONSERVATION LEGISLATION AND PROTECTED AREAS

During the first 50 years of this century, the Norwegian wilderness was considerably reduced. An important contributory factor was the extensive development of the country's water-power resources, leading to major encroachments even in the remotest wilderness areas. There was also an increasing demand from walkers, climbers and skiers for areas with undisturbed nature. The first section of a Conservation Act, passed in 1954, reads:

> The King shall determine that areas of land or water shall be protected when this is considered desirable for scientific or historical reasons, or because of the beauty or distinctive character of the area.

This new law also instituted a new body, the National Council for Nature Conservation, with remits in both advisory and initiatory capacities in nature conservation. The first major task facing the Council was the preparation of a nationwide plan for national parks. It was presented to the government in 1964 and contained proposals for 16 parks, embodying an area of 6300 km², involving only state-owned ground. It received parliamentary approval in 1967 and is now fully implemented. The first two parks, Rondane and Börgefjell, were established before 1964.

In Norway, the usual procedure with a plan is to send it out to all relevant parties for discussion prior to the government preparing a report for parliament. Once parliament has decided on the plan in principle, the government can then deal with the individual projects within the plan prior to their establishment by Royal Decree.

The 1954 Conservation Act soon became obsolete and in 1970 it was replaced by the Nature Conservation Act. This new law allowed for three types of nature protection: national parks, protected landscapes and nature reserves. The category 'protected landscape' is used for areas which have already been disturbed in some way, but where important conservation assets remain. On the coast, for example, it is particularly difficult to find fair-sized areas that have not suffered some encroachment. If certain aspects of Norway's very varied and magnificent scenery are to be protected, some partially disturbed areas will have to be included in the series of conserved sites. So far, 45 protected landscapes have been established. Areas of protected landscape are also used as buffer zones, bordering a national park to separate it from exploited areas. Attempts are made to avoid further technical encroachment in areas of protected landscape, but some commercial activity is allowed, such as timber felling, farming of summer pastures and fish farming. Mining, quarrying and large developments connected with the tourist industry are not allowed. The regulations controlling the protected landscapes resemble those relating to national parks in many other countries.

Norway has very strict legislation governing national parks. No encroachment, such as buildings, road construction, mining, quarrying, gravel extraction, river-system regulation or power-line construction, is allowed. Motorized traffic on either land or water is forbidden. Hikers' organizations have permission to mark routes for walkers discreetly. Hunting of small game and fishing are usually allowed. Ordinary grazing is allowed, and the Lapps are able to use the national parks for reindeer (*Rangifer tarandus*) farming as they have from time immemorial.

The national parks contain many different types of scenery, but they fail to cover the full extent of variation within the country. Norway spans 13° of latitude from *c.* 58°N to beyond 71°N; in addition there is Svalbard which reaches approximately 81°N. The geology varies greatly, ranging from hard, acid rocks to soft, readily weathered shales, and the topography varies considerably over short distances. Even with the moderating influence of the Gulf Stream, significant climatic variation remains. Since south-westerly winds bring most precipitation, the mountain and fjord districts in the west of Norway are characterized by high precipitation whereas areas east of the mountains, lying in the rain shadow, may have typically continental conditions with an annual average of 250 mm. These variations produce a large number of environments, creating a fantastic patchwork of landscapes and biota. Many of these fall outside national parks, but they can be given protection as nature reserves under Clause 8 of the Nature Conservation Act, which states:

> Areas of undisturbed or almost undisturbed nature, or which constitute nature types of scientific or educational importance, or have a unique character, may be preserved as nature reserves.

Nature reserves enjoy the strictest type of protection. No interference of any kind is allowed and even access to some bird reserves is forbidden during the nesting season. Some reserves are obvious cultural landscapes and in these it is necessary to employ specialized management techniques. Nature trails are provided in some reserves, both for school use and to enlighten the general public.

CONSERVATION PLANNING

Several national plans have been implemented with the aim of establishing a network of nature reserves. Both the planning and the scientific investigations are done from a nationwide perspective. In the early days, the work was carried out by people from universities, but later the Norwegian Institute for Ecological Research was established to deal with all kinds of ecological investigations, including conservation planning. As this symposium deals with ecological change in the uplands, incorporating concepts of both landscape and ecology, it should be stressed that the many phytosociological studies carried out during this century have been of fundamental importance for the planning and evaluation of nature reserves.

Before establishing nature reserves, there is an investigation phase, during which a number of localities are taken into consideration and their conservation quality evaluated. It seeks to evaluate the international, national, regional or local value of the locality. Responsibility for further practical work is then assumed by the county conservation office, one of which is located in each of Norway's 18 administrative counties. The office deals with the hearings, negotiations with landowners, and discussions of the plan with other county bodies such as agriculture and forestry administrations. The result is presented in a county plan; an example is the nesting cliff conservation plan for the county of Finnmark

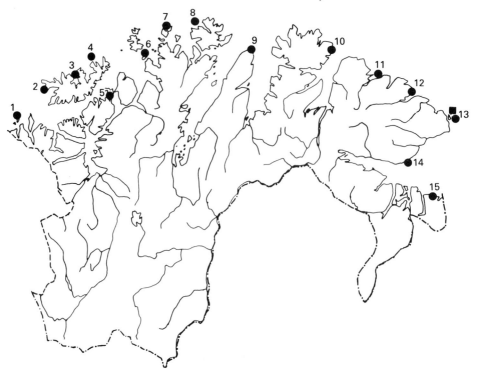

FIG. 1. The nesting cliff conservation plan for the county of Finnmark, Norway. Numbers indicate
the following bird nesting cliffs: 1, Loppa; 2, Andotten; 3, Storgalten; 4, Lille Kamøya; 5, Eidvägen;
6, Reinøykalven; 7, Hjelmsøya; 8, Gjesvaerstappan; 9, Svaerholtklubben; 10, Omgangstauran;
11, Kongsøya; 12, Syltefjorstauran; 13, Hornøya/Reinøya; 14, Ekkerøya; and 15, Kobbholmfjord.

(Fig. 1). The plan is then considered by the Ministry of the Environment and
finally confirmed by a Royal Decree.

One of these plans deals with nemoral (temperate deciduous) forests.
Deciduous trees characteristic of the nemoral forests of central Europe play a
secondary role in Norway but, since they represent the northernmost outposts of
these types of vegetation, areas carrying these trees are of great importance. They
are found in eastern and southern Norway and along the coast northwards to the
Arctic Circle, chiefly where soil and climate are favourable. Many of these
biotypes, which are very rich in numbers of plant and animal species, are in
vulnerable locations owing to building or forestry, and can easily be converted
into valuable spruce forest, resulting in a completely different and poorer
ecosystem.

Another plan, for mires (bogs and fens), is being implemented. Because of the
great variation in soil conditions, precipitation and evaporation, as well as the
altitudinal, longitudinal and latitudinal gradients, mires show extreme diversity
in development. They therefore form the most varied type of environment in

Norway and are in a class of their own as regards ecological studies. The mire plan is meeting strong opposition from economic interests; rich fens in particular can be easily converted into forest and farmland. A national plan for protecting wetlands, primarily as bird sanctuaries, is closely linked to the mire plan. However, there are also many conflicts with economic interests. Estuaries, for example, are often found near towns and harbours, and shallow lakes can be drained for agricultural purposes.

At present a plan for coniferous forests is under consideration. Although the Directorate for State Forests has so far established 54 virgin forest reserves throughout the country, this is insufficient to have all of the biogeographical regions represented. Private forests must necessarily come into consideration, and for obvious reasons the coniferous forest plan is very controversial. The most recently initiated plan concerns deposits essential to an understanding of the Quaternary geological history of Scandinavia — moraines, drumlins, eskers, raised shorelines, sand-dunes, etc. Each of these national plans involves lengthy preparation, involving scientific investigations, negotiations with landowners and public hearings. The plans are now starting to be implemented and about 700 reserves have so for been established. This number should be more than doubled when the work is complete!

The development of Norway's water-power resources is of particular interest. The conflict between nature conservation and the utilization of the water-power resources has been acute. This, more than any other conflict, has demonstrated how difficult it is to compare the value of nature with the economic values of other forms of land use. The conflict is now partly settled. The Storting (parliament) has decided to safeguard about 200 watercourses with a water-power potential of about 2×10^{10} kWh (or about 12–13 per cent of the total potential in Norway).

The 1964 national park plan did not mark the completion of the programme for parks and reserves. In 1982 the Ministry asked the National Council for Nature Conservation to prepare a new countrywide plan for national parks and other large protected areas. In view of the decreasing amount of relatively undisturbed ground, everyone was aware that the preparation of such a plan was a matter of urgency. Whereas the present national parks in Norway mainly consist of ground owned by the state, it was decided that the new plan should not be influenced by questions of land ownership but instead be based on quality and diversity. The plan, which was presented in June 1986, proposed 27 new national parks, 14 new areas of protected landscape and 3 large reserves; in some cases a combination was recommended. The plan also recommended extending 9 existing national parks.

As Norway is a mountainous country, alpine scenery is strongly represented in the new plan, but several of the new protected areas are proposed in coastal and forest districts. A biogeographical division of the Nordic countries into 60 regions, presented by the Nordic Ministerial Council in 1977, helps to achieve maximum representativeness when protection plans are being prepared. Norway extends over 27 of these regions and all except the most densely forested areas of southern and south-eastern Norway are represented in the new plan (Fig. 2). If, following the usual procedures, the plan is adopted in full, there will be 43 Norwegian national parks (excluding those in Svalbard). The national parks, areas of protected landscape and nature reserves will then cover an area of about 37,000 km^2, approximately 12 per cent of Norway's land area.

CONSERVATION IN SVALBARD

Norway's responsibility as regards nature conservation in Svalbard is clearly expressed in the international Svalbard Treaty of 1920. Article 2 states that Norway shall be free to adopt measures to ensure the preservation and, if necessary, re-establishment of the fauna and flora of the archipelago and of its territorial waters. The land areas of the Arctic are among the last few surviving intact ecosystems in the world. Their economic exploitation would be a step towards making the world more monotonous and far less interesting biologically. Islands and archipelagos have always played a very important role in biological research.

In Svalbard there are three species of land mammals: polar bear (*Thalassarctos maritimus*), reindeer (*R. tarandus*) and arctic fox *(Alopex lagopus)*. The polar bear is totally protected, and it is estimated that between 200 and 400 cubs are born each year around the coasts of Svalbard. Special interest is attached to the Svalbard reindeer (*R. t. platyrhynchos*), a unique race related to the Peary reindeer (*R. t. pearyi*) in the Canadian Arctic. When Norway became responsible for Svalbard in 1925, the reindeer stock had been strongly depleted, but protection measures, which were introduced immediately, resulted in the population increasing to the present estimate of between 10,000 and 12,000 animals. The bioresources of Svalbard were overexploited for centuries, both on land and especially off shore. When the walrus (*Odobenus rosmarus*) finally became totally protected in 1952, the population, which once numbered tens of thousands, had been practically exterminated. The stock has only recently begun to show signs of recovery.

Part of the Gulf Stream flowing along the west coast gives Svalbard the most ameliorative climate in the Arctic and therefore a flora which, considering the latitude, is both luxuriant and rich in species. The high primary production in the relatively warm sea off the west coast results in some of the largest concentrations of sea birds in the world.

In 1968 the Norwegian Polar Research Institute started preparing a draft plan for national parks and nature reserves in Svalbard. The matter was urgent because oil exploration was already in progress and hundreds of claims were located in areas that were given high priority by conservationists. The conflict was settled in 1973 when three national parks, two large nature reserves and fifteen bird sanctuaries were established. The protected areas amount to about 35,000 km^2, approximately 56 per cent of the total land area of Svalbard (Fig. 3).

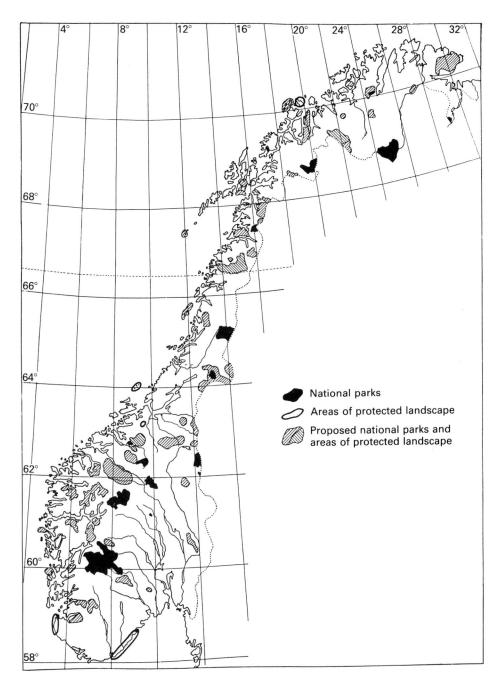

FIG. 2. The existing and proposed national parks and areas of protected landscape in Norway.

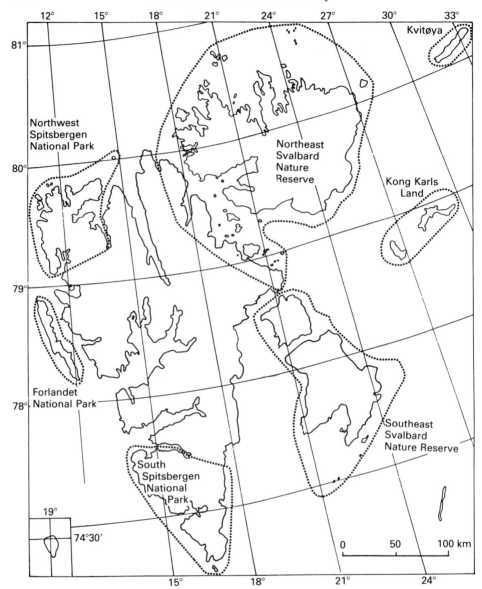

FIG. 3. National parks and nature reserves in the Svalbard area.

It should be stressed that it is, in principle, just as valuable to protect the Arctic bioresources as it is to exploit them for economic purposes.

CONCLUSION

The work of safeguarding Norwegian wildlife and wilderness is carried out mainly from a national stance, although Norway is aware of its broader responsibilities. In western Europe, Norway is one of the few countries with considerable areas of natural environment remaining intact. The oceanic areas are exemplary. The safeguarding of nature in Norway is thus a major contribution to the preservation of western Europe's environmental resource. Norway's efforts are also of significance from a British point of view: the natural environment is a common heritage, sharing many affinities.

Range management and nature conservation in the British uplands

C. SYDES AND G. R. MILLER

Nature Conservancy Council, 12 Hope Terrace, Edinburgh EH9 2AS and Institute of Terrestrial Ecology, Hill of Brathens, Banchory AB3 4BY

SUMMARY

1 Grazing and burning upland vegetation has caused woodlands to be replaced by dwarf-shrub heaths and, ultimately, by grasslands. There have been associated changes in fauna and in soils.

2 In the eighteenth century, year-round grazing by sheep replaced a pastoral system in which a variety of herbivores grazed in summer only. Management of red grouse and red deer populations for sporting use developed in the nineteenth century.

3 Since the 1940s grazing pressure on rangeland has probably intensified on upland vegetation. Red deer numbers in Scotland and sheep stocks, particularly in England and Wales, have increased.

4 During the past 40 years graminoids have extensively replaced dwarf shrubs as the dominant vegetation in parts of northern England.

5 Management on SSSIs must prevent further losses of dwarf-shrub heath. This could be done by reducing the density of herbivores but might be achieved by other changes in management practice.

6 Regeneration of trees and shrubs needs to be reinstated in the remaining woodland and scrub, and consideration to be given to extending the area of this vegetation.

7 Most upland wildlife is not protected by SSSIs and depends on sympathetic management of the remainder of the uplands. However, management for nature conservation may be similar to that required for the long-term survival of present grazing and sporting land use.

INTRODUCTION

The aims of this paper are to (i) review past and current management practices used to exploit British upland biota, (ii) identify the consequent successional trends in vegetation, (iii) examine future management options, and (iv) consider what consequences these might have for nature conservation and for other land uses. Much of the discussion will focus on the status and management of vegetation dominated by heather (*Calluna vulgaris* (L.) Hull). Heather moorland is an integral part of the British upland landscape, where it is currently exploited for agriculture, sport, recreation and tourism. However, it is also important for nature conservation. Heather moorland is a vegetation type peculiar to the

323

Atlantic seaboard of Europe and it supports a unique community of upland animals (Ratcliffe & Thompson 1988).

RANGE MANAGEMENT PRACTICES

Much of the area we are considering lies below the upper altitudinal limit of forest, some 650 m above sea-level in north-east Britain. The original forest has been extensively replaced by structurally less complex vegetation, still composed of native species but physiognomically resembling communities of the low-alpine zone. The dates for replacement of forest suggested for different parts of Britain by the studies of pollen preserved in peat and sediments (Birks 1988) indicate the impact of man's activities in felling, burning and grazing upland vegetation. The relative inaccessibility of many upland areas and the scale of the changes suggest that pastoral management was the main factor bringing about these changes.

Farming systems

Early place-names and later documentary evidence indicate that the uplands were exploited by transhumance. In summer, farmers lived in temporary hill settlements, encouraging their stock to roam over the unimproved vegetation; in winter the stock was herded on to lower, more sheltered pastures. Records suggest that the hill vegetation was grazed by a mixture of sheep, goats, cattle and some horses. Large commercial sheep flocks were maintained on monastic estates in Wales, northern England and the south of Scotland in medieval times but the present dominance of the sheep in upland Britain was a much later development. One of the earliest recorded changes in pastoral management of the uplands is the increasing use of store cattle for trade with the lowlands and with England; this trade was already considerable in Wales by the fifteenth century and had reached the Highlands by the seventeenth century. However, the remainder of the domestic herbivore population remained mixed and was probably mainly used to meet local needs. Enclosure in Wales in the early eighteenth century led to the establishment of sheep farms; transhumance, the trade in cattle and the mixed herbivore population were completely swept away. The same process began in Scotland during the late eighteenth century and was completed by about 1830.

Game shooting

In Scotland and northern England the management of many estates from about 1860 onwards became geared to the production of large stocks of red deer (*Cervus elaphus* (L.)) and red grouse (*Lagopus lagopus scoticus* (Latham)) for shooting. As a result, numbers of deer and grouse probably increased considerably during the latter half of the nineteenth century. They remain an important source of income on many upland estates.

Fire

It is likely that forest and scrub were extensively burned even in the uplands. Carbon particles are abundant in the upper layers of peat and freshwater sediments although fires started by natural causes may also contribute to this. Fires might have been lit to drive game, to destroy habitat for large predators such as the wolf (*Canis lupus* L.), and to create open space for foraging by domestic livestock. Presumably they would have been uncontrolled, extensive and of haphazard occurrence in both location and time.

By the nineteenth century, range vegetation was being burned regularly, especially where heather or *Molinia caerulea* (L.) Moench were predominant. These and many other range species are relatively indigestible to ruminants (Kay & Staines 1981). Most species are particularly unpalatable in winter and therefore the vegetation cannot support sufficient stock throughout the year to utilize the primary production fully in summer. Surplus production accumulates as wood or litter but can be burned off to encourage or expose young growth. Upland shepherds are believed to have operated a rough-and-ready rotation, burning large blocks of ground annually. Although the degree of control over the timing, location, extent and intensity of these fires may have been minimal, the planned use of fire began to be considered seriously during the late nineteenth century with the desire to produce large numbers of red grouse for shooting. The birds have more exacting requirements than do many other range herbivores, and burning must be strictly controlled to maximize benefits (Miller & Watson 1974). The importance of grouse, deer and sheep to the upland economy has stimulated research on the principles and practice of burning (Muirburn Working Party 1977).

Stocking rates

Overall stocking rate, or herbivore density, is a crude and imperfect measure of grazing pressure on vegetation. Crude, because it makes no allowances for diurnal and seasonal changes in herbivore distribution and takes no account of local concentrations of stock on small areas of good grazing or shelter; imperfect, because it is usually impossible to obtain data on the numbers of animals actually pastured on upland grazings. Only indirect evidence about possible changes in stocking rates is available; conclusions must be based on inference and are therefore only tentative. However, no other guide to grazing pressure is available.

Two parameters determine overall stocking rates — the total amount of land available and the numbers of herbivores using it. There has been a substantial loss of area of range grazings to seeded grassland and to afforestation since the 1940s. This loss has been a particular issue in southern Britain because the rangeland that remains there is both small and fragmented. However, it is a national phenomenon. During 1946–81, the greatest proportional loss of range, about 40 per cent, was in Wales, whereas the greatest absolute loss, nearly 5×10^5

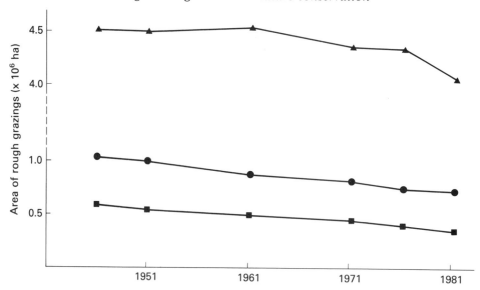

FIG. 1. Area of freehold rough grazings in Scotland (triangles), England (circles) and Wales (squares), 1946–81 (source: RSPB 1984).

ha, was in Scotland (Fig. 1). Overall, some 1.1×10^6 ha of freehold grazings have disappeared throughout the UK during this period.

The annual agricultural census shows that in 1966 there were some 29 million sheep in Great Britain, about the same number recorded a century earlier (MAFF 1968). However, there was a marked redistribution of sheep during this interval. In 1875, the overall density of sheep on agricultural land in the lowland counties of south and east Britain was similar to, or even greater than, that elsewhere. Densities have since declined in the south-east but have risen in northern and western counties (Fig. 2), where upland grazings are extensive but decreasing in area (Fig. 1). This redistribution may be related to increased production from sown grassland in the north and west, either by increased use of fertilizers or by improvement from rough pasture. However, much improved pasture in the uplands is closed to stock for part of the summer to provide grass for winter feed. Thus, many of these animals are turned on to semi-natural vegetation at this time.

Large increases in sheep numbers have been noted from many upland parishes in different parts of England and Wales since the 1950s (Anderson & Yalden 1981; Ball *et al.* 1982; Miller, Miles & Heal 1984). Taking together all the English and Welsh Less Favoured Areas (EEC designation for upland parishes), sheep numbers rose by 87 per cent from 1951 to 1981 (Table 1). Coupled with a 9 per cent loss in total agricultural area, this represents a doubling of the overall sheep stocking rate. Data from Scotland are less readily available. Albon & Clutton-Brock (1988) found little overall change in numbers of hill sheep in the Highland region since the 1940s, while RSPB (1984) recorded only a small

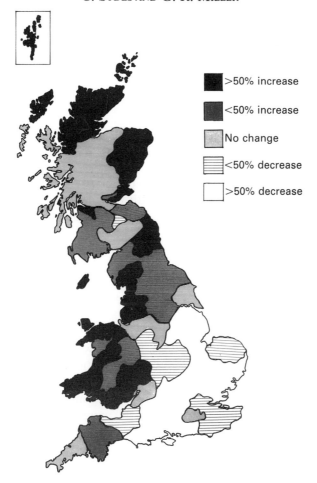

>50% increase

<50% increase

No change

<50% decrease

>50% decrease

Fig. 2. Changes in the density of sheep on agricultural land, 1875–1966 (source: MAFF 1968).

increase in the total Scottish sheep population during the same period. However, the loss of 0.5×10^5 ha of upland grazings may have resulted in some local increases of stocking rate.

Fluctuations in sheep numbers in the uplands have probably occurred for centuries in response to economic factors, to epidemics and to anomalous weather events such as the 1946–47 winter. Since the war, however, livestock trends have been increasingly influenced by government policies implemented through subsidy payments. Upland farmers may have been able to keep more sheep by converting semi-natural vegetation into seeded grassland, by more intensive management of existing enclosed pastures, and by improved veterinary practices. The actual extent to which stocking rates on the open hill have changed must remain debatable. However, it seems reasonable to suspect that the

TABLE 1. Numbers of sheep recorded from EEC Less Favoured Areas in
England and Wales, 1951–81 (source: RSPB 1984)

Year	No. of sheep ($\times 10^6$)	Total agricultural area (10^6 ha)	No. of sheep ha^{-1}
1951	5·20	1·80	2·9
1961	6·85	1·74	3·9
1971	7·49	1·67	4·5
1981	9·75	1·63	6·0

substantial expansion in the size of the sheep population in northern England and
Wales during the last 40 years, coupled with the concomitant shrinkage in the
area of rangeland, has resulted in greater grazing pressures for at least part of the
year.

Data on trends in red deer numbers are also sparse but, again, there is
evidence of increasing densities in Scotland. No reliable estimates of population
size were available until the 1960s, when the Red Deer Commission began
systematic counts. First, these have established that deer densities vary widely
throughout Scotland, from 0·5 to 31 deer km^{-2} (Stewart 1985). Presumably these
variations relate to habitat characteristics, possibly including competition from
hill sheep (Albon & Clutton-Brock 1988). Second, when compared with earlier
guesses about population size, the recent censuses indicate an increase in the total
population size from about 1950 (Fig. 3). Gains have been greatest in eastern
Scotland, for example 154 per cent in the east Grampians between 1966 and 1986
(Staines & Ratcliffe 1987). The main cause of the present abundance of deer is the

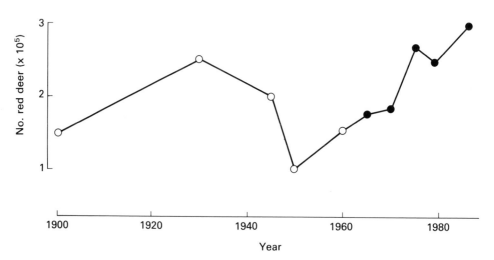

FIG. 3. Total numbers of red deer in Scotland, 1900–1986: open circles are rough estimates, solid
circles represent reliable censuses (source: Staines & Ratcliffe 1987).

deliberate underculling of hinds. In addition, some red deer are sustained through winter by food supplements and, more significantly, many now live in the new conifer plantations, where they are difficult to shoot.

SOME ECOLOGICAL CONSEQUENCES OF RANGE MANAGEMENT

Vegetation

In general, burning and grazing pressures are inversely correlated with the structural complexity of range vegetation (Fig. 4). Young trees, and shrubs of any age, are vulnerable because their apical meristems are exposed to and killed by fire and by repeated browsing. Some dwarf shrubs, notably heather, are at least partially resistant to defoliation and have the ability to regenerate from the base of a cut or burnt stem. However, even they succumb eventually to severe or repeated burning and grazing (Welch 1984). Graminoids, on the other hand, grow from basal meristems which are protected because of their location close to or below the soil surface. It is not surprising, therefore, that increasing exploitation of the uplands by domestic herbivores has gone hand in hand with a progressive replacement of trees and shrubs by dwarf shrubs and, ultimately, by graminoids. Tall woody vegetation is now found on only a minute proportion of the uplands.

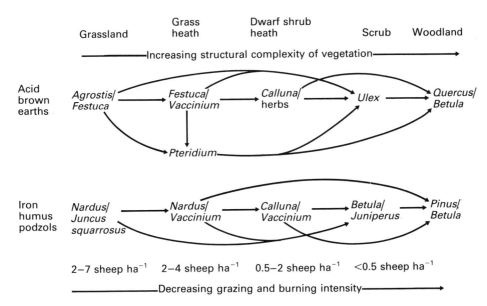

FIG. 4. Postulated successional trends amongst some upland plant communities of acid, freely drained soils in response to decreasing grazing and burning pressures (sources: Ball *et al.* 1982; Miles 1985).

TABLE 2. Estimated loss of heath-dominated upland vegetation in Cumbria 1940s to 1970s (source: NCC 1987)

	Area (km^2) in 1940s	Area (km^2) in 1970s	Area lost km^2	%
Dwarf-shrub heath	299	89	210	65
Blanket bog	617	541	76	12

Overall habitat diversity may well have increased for a period as heathland and grassland expanded at the expense of woodland but this trend may have been reversed as woodland and scrub were reduced to their present extent. A reduction in the area of heathlands dominated by dwarf shrubs would lead to further uniformity as the vegetation becomes dominated mainly by graminoids.

Present upland vegetation is by no means stable and there is accumulating evidence of the recent replacement of heather by graminoids. Thus, in the Peak District, Anderson & Yalden (1981) have linked a 36 per cent loss of heather moorland over 66 years to increasing sheep stocks. At Exmoor, Miller *et al.* (1984) have related an encroachment by grasses and bracken (*Pteridium aquilinum* (L.) Kuhn) around the fringes of heather moor to heavy grazing and haphazard burning. On a broader scale, analysis of a large sample of aerial photographs of Cumbria indicates a loss of nearly 300 km^2 of heather-dominant vegetation since the 1940s (Table 2). A small proportion of this loss was to improved grassland and forestry, but most (267 km^2, 93 per cent) was replaced by unimproved vegetation dominated by graminoids. The Nature Conservancy Council's National Countryside Monitoring Scheme will document such gross changes in vegetation throughout Britain.

Enough is known from historical records, monitoring studies, management experiments and field experience of the relationships amongst plant communities to predict the general direction and nature of vegetation changes in the uplands (Fig. 4). Intensive grazing or frequently repeated burning tends to produce short grassy swards and, ultimately, bare ground and erosion. Conversely, a relaxation of pressure may allow the development of scrub and woodland. These upland successions are not yet fully understood but are thought to be reversible and to follow several alternative pathways (Miles 1988). Dwarf-shrub heaths occupy a pivotal position amidst this network of variation (Fig. 4) and also play an important role in characterizing upland landscapes. Ball *et al.* (1982), postulating a 50 per cent increase in the overall density of domestic livestock in the uplands, predicted that the most conspicuous impact would be a 66 per cent reduction in the area occupied by dwarf-shrub heath. On the other hand, a halving of current stocking rates might double the area of heath as well as allowing scrub and woodland to regenerate.

Little is known about the rates at which these successional changes might proceed but clearly they would be widely variable. Much depends on environmental condition — climate, soil fertility, drainage — but also on the availability of colonists and on the characteristics of the existing dominant. For example, low-growing bents (*Agrostis* spp.) and sheep's fescue (*Festuca ovina* L.) can be overcome by heather within 10–15 years of the removal of sheep (Jones 1967). On the other hand, the rhizomatous growth and the tall, dense fronds of bracken can enable it to obstruct succession for many decades. Similarly, grazing and burning can rapidly promote the dominance of *Trichophorum cespitosum* (L.) Hartman, *M. caerulea* and *Eriophorum vaginatum* L. over heather on wet acid soils in western Britain. Once established, however, their densely tussocky growth form makes these species resistant to further succession even when the management changes.

Fauna

The decline of upland woods means that many formerly widespread and abundant animal species are now either extinct or restricted in numbers and range. Wolf, beaver (*Castor fiber* L.) and brown bear (*Ursus arctos* L.) have all gone; others such as wildcat (*Felis sylvestris* Scheber) and pine marten (*Martes martes* (L.)), once widespread in lowland woods, found refuge in the uplands. The adaptation of red deer to a treeless habitat may have been partly contrived by man's destruction of predators and partly by deliberate underculling. Many other species have, of course, thrived on open rangeland. For example, mountain hare (*Lepus timidus* L.), golden plover (*Pluvialis apricaria* (L.)) and raven (*Corvus corax* L.) have probably all extended their distribution and increased in numbers. A particularly important beneficiary has been the red grouse.

Losses of heather as a result of intensive burning and grazing inevitably cause decreases in the population size of red grouse. Indeed, analyses of grouse bag data (Williams 1974) provide statistical evidence of an overall decline in grouse stocks between the turn of the century and the 1940s, particularly in England, Wales and western Scotland. This may have serious consequences for predators such as the golden eagle (*Aquila chrysaetos*). Although eagles can survive on carrion on the graminoid-dominant sheepwalks of western Scotland, their breeding performance is best wherever heather remains dominant and live prey are still abundant (Watson, Langslow & Rae 1987).

The replacement of dwarf shrubs by the shorter, less structurally complex graminoids alters the balance between other bird species; skylark (*Alauda arvensis* L.) and meadow pipit (*Anthus pratensis* (L.)) might be expected to increase, whereas waders such as golden plover and dunlin (*Caldiris alpina* (L.)) might decrease (Reed 1985). In addition, the lack of cover may limit species richness because there is a smaller variety of niches. A comparative paucity of bird species on the central grass moors of Exmoor has been attributed partly to the absence of dwarf shrubs (Miller *et al.* 1984).

Soils

Appreciable changes in soil conditions almost certainly followed from forest clearance. First, the loss of tree canopy will have reduced the interception and evaporation of rain and net precipitation on to upland soils will have increased. Increased soil wetness encourages the accumulation of surface organic matter and the development of mor conditions. Second, there have probably been changes in the soil's chemical and biological properties as a direct consequence of vegetational change. Miles (1985) has reviewed such plant–soil reactions. Ericoids and the grasses, *Nardus stricta* L. and *Deschampsia flexuosa* (L.) Trin., seem to cause acidification and podzolization. On the other hand, birch (*Betula* spp.), bracken and bent (*Agrostis* spp.) and fescue (*Festuca* spp.) grasses produce a mull type of humus and are thought to promote depodzolization. Clearly, the establishment of great tracts of heathland and of some types of grassland has encouraged the development of mor humus with consequent reductions in exchangeable cations and pH.

At one time, great concern was expressed about the possible losses of nutrients from upland ecosystems as a result of burning and the cropping of livestock. Several attempts were made to draw up a nutrient balance sheet, comparing possible losses due to burning, grazing, leaching and drainage with accessions from rainfall, dust and weathering. Some of these parameters are difficult to measure but the general conclusion (Muirburn Working Party 1977) is that, with one clear exception, nutrient gains from rainfall exceed the possible losses due to burning and animal production. The exception is phosphorus. There appears to be a continuing depletion of this nutrient, which is generally scarce in upland soils and plays a vital role in the nutrition of all organisms. More research is needed here, if only because these conclusions relate mainly to eastern Scotland. In western Scotland, where fires are often uncontrolled, very large, and burned at short intervals, other nutrients may be in deficit (Hobbs & Gimingam 1987).

RANGE MANAGEMENT FOR NATURE CONSERVATION

Management methods

The vegetation of existing upland Sites of Special Scientific Interest (SSSIs) is predominantly dwarf-shrub heath, ombrogenous mire (both having abundant heather) and grassland (Table 3). Other vegetation types, including woodland, occupy less than 10 per cent of the total area. The main priority at these sites is to safeguard the existing vegetation and fauna, which justify their designation as SSSIs. As much of the vegetation has been created by repeated burning and continual grazing, these management practices must continue. However, close control is essential.

The use of fire as a management tool is now well understood and the possibility of unwanted successional changes, nutrient losses and erosion can be minimized. The main obstacles to good burning management are poor weather

TABLE 3. Available data on the area occupied by different vegetation types on upland SSSIs

	England (19 sites)		Scotland (25 sites)	
	ha	%	ha	%
Woodland	92	< 1	2,700	2
Scrub	112	< 1	4	< 1
Dwarf-shrub heath	11,563	24	30,313	22
Grassland	19,734	41	24,010	17
Herb and fern communities	1,056	2	597	< 1
Moss heath	51	< 1	7,485	5
Ombrogenous mire	14,445	30	70,193	51
Soligenous mire	1,286	3	853	< 1
Springs	3	< 1	7	< 1
Total	48,342		136,162	

and high labour costs. These can be overcome, given the commitment and money to manage effectively.

Control over grazing pressure is less easy. On small blocks, up to *c.* 100 ha, fencing or careful shepherding can adjust stocking rates and grazing intensities to desired levels. However, such direct control is much more difficult on tracts of heathland, grassland and blanket bog. Here, the manipulation of grazing pressures must rely on management of the vegetation, by burning and on crude adjustments to the overall stocking rate. Ideally, vegetation management by grazing should be founded on a sound understanding of what influences diet selection by herbivores. Some information about seasonal grazing patterns on different plant communities is available for sheep (Hunter 1962) and for red deer (Charles, McCowan & East 1977). However, much more needs to be known about preferences for individual plant species and communities. In particular, how is selection affected by herbivore interactions, stocking rates, the availability of different plant communities and their juxtaposition? The successful manipulation of vegetation patterns to fulfil specified nature conservation and animal production objectives depends in part on finding answers to these questions.

Management issues

Two particular issues need to be addressed by upland conservationists in managing SSSIs. One concerns the loss of heather cover on heathland and blanket bog. The second relates to the present and future status of upland woods.

Heather-dominant vegetation

Where dwarf shrubs are being lost the simplest remedy may be to reduce the density of sheep on the range. A programme of regular burning combined with stocking densities well below commercial levels might appear to be the easiest way to maintain a dwarf-shrub heath for nature conservation. Low densities of red deer, a native herbivore, may be more appropriate than sheep. However, management on SSSIs can rarely be defined without considering the requirements of the landholder. Therefore it is usually necessary to define a stocking rate of sheep that will maintain heather and is close to a commercially acceptable density. A generally defined density of herbivores is unlikely to provide the best compromise as this will vary considerably with the potential of the area to support herbivores. The management model (Grant & Maxwell 1988) that maximizes animal production from upland areas with heather provides a more realistic approach. Conservation managers can use this to indicate maximum stocking rates but must take into account the greater structural range of heather required for wildlife conservation.

The diminution of heather cover might be checked in many cases, perhaps most, by changes in management of the stock to spread the grazing animals more evenly over the range. On many estates, shepherding is now minimal and burning is erratic. The increase in winter feeding may be even more significant as it tends to concentrate the sheep in limited areas at this crucial period when there is little evergreen food apart from heather. More intensive shepherding, a more equitable distribution of burnt patches in both time and space, and the rotation of winter feeding points would all encourage stock to make more complete use of the ground.

It must be recognized, however, that management-induced uniform tracts of heather moor are not the desired objective. To maintain a full range of structural diversity within the heather-dominant vegetation, some patches will be burned on a very long rotation or even not burned at all so that they become uneven aged with a wide range of associated species. Other vegetation types, including grasslands of all types, will be maintained as part of the vegetation mosaic. Whether dwarf shrubs should be allowed to spread at the expense of some grassland will depend on the local balance of types and objectives.

Woodland

Woodland and scrub occupy a mere 2–3 per cent of upland SSSIs (Table 3). The structurally complex vegetation is generally rich in associated animal species and many of these relicts are of great importance to nature conservation. However, most are on the verge of extinction, some patches are so small and isolated as to be liable to be lost to a runaway fire or to some other catastrophe. The majority will simply vanish as soon as existing trees die because grazing continually checks regeneration. Only the largest support a wide range of animals.

Given good seed sources and suitable niches for germination, tree and shrub invasion can proceed rapidly where grazing is absent or minimal. Isolated examples of good regeneration occur throughout the uplands. The exact relationship between stocking rate and tree establishment is not well understood and probably varies widely from site to site. In general, fewer than 0·5 sheep ha^{-1} might be necessary to allow regeneration, given the present paucity of seed trees in the uplands. Ground treatment — burning or removal of the A_0 horizon — may sometimes be needed to initiate or to sustain regeneration.

Where conditions can be created to allow regeneration of existing upland woodlands, it must be considered whether the relict woodlands are large enough to protect the range of species that would be expected to exist there. An expansion could permit them to act as better reservoirs. However, given that present SSSIs were chosen to safeguard their present diversity of wildlife, it is necessary to determine what proportion of secondary woodland could be tolerated. A limited expansion of woodland is unlikely to result in significant losses from open habitats generally. However, one current prognosis for the uplands suggests that sheep stocking rates may decline, so allowing scrub and woodland development (Ball *et al.* 1982). If this happens it will be necessary to determine more precisely what proportion of secondary woodland can be tolerated and at what point regenerating trees will have to be destroyed to maximize benefits for nature conservation.

The upland mosaic

The management needed to maintain and diversify upland wildlife is generally at odds with government support for monoculture systems, whether these be forestry or sheep-grazing (Mutch 1984). On an SSSI this conflict may be resolved through voluntary management agreements with the landholder. However, only about 10 per cent of the uplands are protected by SSSI designation. Therefore, the bulk of upland wildlife depends on the management policies adopted on the remainder of ground. However, conservation management need not conflict with an extensive farming system designed to sustain a modest level of production in the long term. Over-intensification can cause undesirable vegetation successions, which, in turn, cause an eventual loss of production (Arnalds & Rittenhouse 1986). Indeed, an apparent decline in lambing percentages in the western Scottish Highlands from 1900 to 1970 (Mather 1978) suggests that this may have already happened there.

The creation and maintenance of an appropriate balance between grassland, well-burnt heather moor and upland woods may help to sustain or even to improve future animal production. Bent–fescue grassland makes a major contribution to the hill sheep's diet, especially during summer (Hunter 1962). Other grassland types, however, are less valuable. Heather not only provides year-round food and cover to red grouse but is also a valued winter food for red deer and sheep, especially where deciduous graminoids predominate. Maintaining heather can resolve the conflict between sporting and grazing interests, ensuring the survival of these two complementary sources of income from moorland. The mosaic burning pattern which boosts grouse numbers (Miller, Watson & Jenkins 1970) can also be beneficial to deer (Miller & Watson 1974).

The possible benefits to large herbivores of a modest expansion of upland woods have not been well researched. However, scattered copses would provide winter shelter for both red deer and sheep. The same shelter also extends the growing season of ground vegetation in spring and autumn, providing nutritious herbage at periods crucial to lambing success. In addition, the vegetation under birch may be of generally superior nutritional value to that on open range, owing to the mull-forming litter deposited annually by the trees. These possibilities all need investigation. Such a mosaic of vegetation would be inherently less stable than uniform tracts of heather or grass, simply because a greater variety of propagules and of germination niches would be available. It would therefore require careful monitoring and management to maintain a state of dynamic equilibrium. The degree and frequency of intervention would depend to a large extent on soil conditions.

The case for multiple land-use systems designed to sustain increased upland productivity has been argued for many years, for example by McVean & Lockie (1969). It has been suggested that multiple land use, although theoretically desirable, has economic disadvantage in terms of management efficiency (Mutch 1984). This is particularly true where sporting and grazing rights are held separately. None the less, some landowners are committed to the idea (Gordon-Duff-Pennington 1985). Two major aims of range management for nature conservation — the maintenance of heather moor and the development of small woods — might be usefully incorporated into the management systems of most upland landholders. Nature conservation, field sports and hill farming could all benefit.

REFERENCES

Albon, S.D. & Clutton-Brock, T.H. (1988). Climate and the population dynamics of red deer in Scotland. *This volume.*

Anderson, P. & Yalden, D.W. (1981). Increased sheep numbers and the loss of heather moorland in the Peak District, England. *Biological Conservation,* **20**, 195–213.

Arnalds, A. & Rittenhouse, L.R. (1986). Stocking rates for northern rangelands. *Grazing Research at Northern Latitudes* (Ed. by O. Gudmundsson), pp. 335–345. Plenum Press, New York.

Ball, D.F., Dales, J., Sheail, J. & Heal, O.W. (1982). *Vegetation Change in Upland Landscapes.* Institute of Terrestrial Ecology, Cambridge.

Birks, H.J.B. (1988). Long-term ecological change in the British uplands. *This volume.*

Charles, W.N., McCowan, D. & East, K. (1977). Selection of upland swards by red deer (*Cervus elaphus* L.) on Rhum. *Journal of Applied Ecology,* **14**, 55–64.

Gordon-Duff-Pennington, P. (1985). Economic aspects of vegetation management. *Vegetation Management in Northern Britain* (Ed. by R.B. Murray), pp. 123–128. BCPC Publications, Croydon.

Grant, S.A. & Maxwell, T.J. (1988). Hill vegetation and grazing by domesticated herbivores: the biology and definition of management options. *This volume.*

Hobbs, R.J. & Gimingham, C.H. (1987). Vegetation, fire and herbivore interactions in heathland. *Advances in Ecological Research,* **16**, 87–173.

Hunter, R.F. (1962). Hill sheep and their pasture: a study of sheep grazing in south-east Scotland. *Journal of Ecology,* **50**, 651–680.

Jones, L.I. (1967). Studies on hill land in Wales. *Welsh Plant Breeding Station Technical Bulletin,* No. 2.

Kay, R.N.B. & Staines, B.W. (1981). The nutrition of red deer (*Cervus elaphus*). *Nutrition Abstracts Review (B)*, **51**, 601–622.

McVean, D.N. & Lockie, J.D. (1969). *Ecology and Land Use in Upland Scotland*. Edinburgh University Press, Edinburgh.

MAFF (1968). *A Century of Agricultural Statistics: Great Britain 1866–1966*. HMSO, London.

Mather, A.S. (1978). The alleged deterioration in hill grazings in the Scottish Highlands. *Biological Conservation*, **14**, 181–195.

Miles, J. (1985). The pedogenic effects of different species and vegetation types and the implications of succession. *Journal of Soil Science*, **36**, 571–584.

Miles, J. (1988). Vegetation and soil changes in the uplands. *This volume*.

Miller, G.R. & Watson, A. (1974). Some effects of fire on vertebrate herbivores in the Scottish Highlands. *Proceedings of the Annual Tall Timbers Fire Ecology Conference*, **131**, 39–64.

Miller, G.R. & Watson, A. & Jenkins, D. (1970). Responses of red grouse populations to experimental improvement of their food. *Animal Populations in Relation to their Food Resource* (Ed. by A. Watson), pp. 323–335. Blackwell Scientific Publications, Oxford.

Miller, G.R., Miles, J. & Heal, O.W. (1984). *Moorland Management: a Study of Exmoor*. Institute of Terrestrial Ecology, Cambridge.

Muirburn Working Party (1977). *A Guide to Good Muirburn Practice*. HMSO, Edinburgh.

Mutch, W.E.S. (1984). Ecological principles in upland management. *Agriculture and Environment* (Ed. by D. Jenkins), pp. 135–139. Institute of Terrestrial Ecology, Cambridge.

NCC (1987). *Changes in the Cumbrian Countryside*. Research and Survey in Nature Conservation, No. 6, Nature Conservancy Council, Peterborough.

Ratcliffe, D.A. & Thompson, D.B.A. (1988). The British uplands: their ecological character and international significance. *This volume*.

Reed, T.M. (1985). Grouse moors and wading birds. *Game Conservancy Annual Report*, **16**, 57–60.

RSPB (1984). *Hill Farming and Birds: a Survival Plan*. Royal Society for the Protection of Birds, Sandy.

Staines, B.W. & Ratcliffe, P.R. (1987). Estimating the abundance of red deer (*Cervus elaphus* L.) and roe deer (*Capreolus capreolus* L.) and their current status in Great Britain. *Symposia of the Zoological Society*, **53**, 131–152.

Stewart, L.K. (1985). Red deer. *Vegetation Management in Northern Britain* (Ed. by R.B. Murray), pp. 45–50. BCPC Publications, Croydon.

Watson, J., Langslow, D.R. & Rae, S.R. (1987). *The Impact of Land-use Changes on Golden Eagles* Aquila chrysaetos *in the Scottish Highlands*. Chief Scientist Directorate Report 720, Nature Conservancy Council, Peterborough.

Welch, D. (1984). Studies in the grazing of heath moorland in north-east Scotland. II Responses of heather. *Journal of Applied Ecology*, **21**, 197–207.

Williams, J.C. (1974). *Mathematical analysis of red grouse populations*. B.Sc. thesis, University of York.

Conservation of leafy liverwort-rich *Calluna vulgaris* heath in Scotland

A. M. HOBBS

Nature Conservancy Council, 12 Hope Terrace, Edinburgh EH9 2AS

INTRODUCTION

Sub-montane heaths dominated by *Calluna vulgaris* cover about $1 \cdot 2 \times 10^6$ ha of Scotland (Tivy 1973). Most of these, however, are the large, dry grouse moors of central and eastern Scotland (Hobbs & Gimingham 1987). In the west, the cool and wet climate is more favourable for the development of damp heaths: *Calluna vulgaris* with an understorey of *Sphagnum capillifolium* and *Racomitrium lanuginosum*, rather than the *Hypnum*-type mosses which characterize the drier heaths. In particularly damp, shady places in the west there is a form of *C. vulgaris* heath with a ground flora of large leafy liverworts, first recognized by McVean & Ratcliffe (1962) and classified by them as Vaccineto-Callunetum, hepaticosum facies. Subsequently the National Vegetation Classification (J. Rodwell, pers. comm.) referred these heaths to its *Calluna vulgaris–Vaccinium myrtillus–Sphagnum* heath, *Pleurozia purpurea–Bazzania tricrenata* sub-community. The typical species, which include *Herbertus aduncus* (usually dominant), *Anastrepta orcadensis*, *Mastigophora woodsii*, *Bazzania tricrenata* and *Scapania nimbosa,* grow in large cushions, made conspicuous by the bright orange colour of *Herbertus aduncus.*

This hepatic heath is limited in its distribution by the rather narrow ecological tolerances of the liverworts. The species belong to the Northern Atlantic phytogeographical group (Ratcliffe 1968). They grow only in regions with an oceanic climate, which has high atmospheric humidity and equable temperatures throughout the year. Ratcliffe (1968) found that the liverwort community develops only in climatic regions where there are more than 220 wet days, days with more than 1 mm of rain, a year. Ratcliffe (1968) found a further marked bias towards the most shaded and humid places within this region; on steep, north-facing or east-facing slopes, cliffs and in boulder-fields. The heath is only found in the mountainous districts of western Scotland and western Ireland, making its occurrence in Britain internationally important (Ratcliffe & Thompson 1988).

Although formerly widespread, the heath has proved very susceptible to grazing and burning (McVean & Ratcliffe 1962; Ratcliffe 1968). This paper discusses and quantifies the distribution and decline of the hepatic heaths.

Nomenclature for vascular plants follows Clapham, Tutin & Warburg (1981) and for bryophytes either Jones (1958) or Watson (1978).

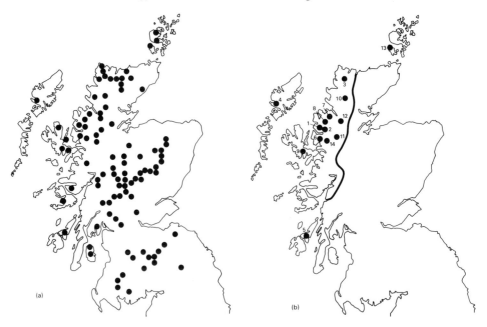

FIG. 1. The hill areas in Scotland surveyed by the Upland Vegetation Survey. (a) Location of all areas surveyed, and (b) areas where leafy liverwort-rich *Calluna vulgaris* heath was found. Numbers refer to areas listed in Table 1, and the line shows the eastern limit of 220 wet days year^{-1} (Ratcliffe 1968).

METHODS

The Nature Conservancy Council's Upland Vegetation Survey (UVS) has mapped the vegetation on 96 upland areas in Scotland (Fig. 1a), 81 of which are Sites of Special Scientific Interest. Examples of hepatic heaths which were found during this work were mapped on to aerial photographs at a scale of 1:25,000. These were then transferred to 1:25,000 scale Ordnance Survey maps. The surface areas of heath were measured using a Tamaya 'Planix' digital planimeter and converted from square centimetres to hectares.

RESULTS

Distribution of hepatic heath

Fig. 1b shows the distribution of hepatic heath on the hill areas mapped by the UVS. The heath is confined to the western parts of Scotland and has been found on only 14 of the 96 areas surveyed.

Table 1 shows that the heath is not evenly distributed on these 14 areas. The largest continuous patches are on the Torridon Hills; Liathach, Beinn Eighe and Beinn Alligin. On North Harris there are many small patches. The heaths on

TABLE 1. The number and surface area of patches of leafy liverwort-rich *Calluna vulgaris* heath on 14 upland areas. Site numbers are shown in Fig. 1

Site	Number of patches	Total area (ha)	Mean area (standard error)
1. Beinn Alligin	5	161	32·1 (10·1)
2. Beinn Eighe/Liathach	1	111	—
3. Foinaven	6	72	12·0 (5·5)
4. North Harris	14	65	4·5 (1·1)
5. Beinn an Oir	1	50	—
6. Beinn Bhàn	3	23	7·7 (2·5)
7. An Teallach	3	17	5·6 (1·3)
8. Letterewe Forest	3	17	5·6 (2·3)
9. Cuillin Hills	2	6	3·0 (1·1)
10. Ben More Assynt	1	5	—
11. Monar Forest	1	3	—
12. Fannich Hills	2	2·5	1·3 (0·4)
13. North Hoy	1	<1	—
14. Glas Cnoc	1	<1	—

North Hoy and Glas Cnoc cover too small an area to measure at the scale of these maps.

Table 2 shows that most of the hepatic heaths are on rocky slopes. They often grow on cliff ledges and in boulder-fields within the area of the slope. The altitudinal range is from almost sea-level (North Harris) to nearly 900 m (Liathach) but these are extremes and most lie between 300 m and 600 m above sea-level. The aspect in all cases was between north-west and east.

Loss of hepatic heaths

Altogether the UVS has mapped 44 patches of hepatic heath. Ratcliffe (1968) gave 55 records. McVean & Ratcliffe (1962) gave grid references for eleven heaths; the UVS team has visited 10 of these and has found that 4 no longer exist. Two more have been damaged by recent grazing and burning, with the loss of

TABLE 2. Occurrence of leafy liverwort-rich *Calluna vulgaris* heath according to aspect and habitat (there were no south-east, south, south-west or west occurrences)

	Habitat				
Aspect	Rocky slope	Boulder-field	Cliff	Corrie	Total
North-east	14	6	3	—	23
East	2	2	—	1	5
North-west	4	—	—	—	4
North	7	4	—	1	12
Total	27	12	3	2	44

most of the liverworts, and 4 remain. The damage must have been done in the last 25 years, between McVean and Ratcliffe's visit and our own.

DISCUSSION

The distribution and quantity of hepatic heath on each of the 14 massifs is probably a result of the topography and geographical position of each. For example, there is probably one large patch on Liathach because this mountain has a single long, steep, north-facing slope. North Harris is made up of many small hills, most of which provide a suitable but small site.

All 14 areas with the hepatic heath are within the climatic region where there are more than 220 wet days a year, and most of them are towards the west of this region. The more easterly localities, Monar Forest, the Fannich Hills and Ben More Assynt, do not have very much of the heath and it may be because these hills are close to the limit of the suitable climate.

All our examples were found on rocky slopes, cliffs or in boulder-fields facing between north-west and east, and always with tall *C. vulgaris*. In these places the effects of the oceanic climate are amplified. A northerly to easterly aspect is in shade for much of the day, and the rocks and tall shrubs further decrease the amount of light reaching the ground. Slopes with these aspects are sheltered from the prevailing winds, and the combination of still air and little direct sunlight makes evaporation into the atmosphere almost negligible (Ratcliffe 1968). Constant atmospheric humidity seems to be one of the most important factors in the ecology of the hepatic heath.

According to Ratcliffe (1968), the hepatic heath was once widespread from the Reay Forest and Ben Hope in Sutherland to Glen Coe and possibly to Ben Cruachan in Argyll. There is good evidence for this. Throughout the region there are fragments of heath on cliffs, in boulder-fields and on islands, existing as isolated patches in places which look suitable for more widespread development but which have obviously been burned, intensively grazed, or both. The effects of burning and grazing, although often obvious in the field, are difficult to quantify. The charred wood of *C. vulgaris* persists for several years after a fire. The liverworts are destroyed by burning and seem to be unable to recolonize without the shelter of tall *C. vulgaris*. Instead, they are typically replaced by *Racomitrium lanuginosum* and *Hypnum*-type mosses. For example, at NC 350495 on Foinaven there are signs of recent burning, with the *C. vulgaris* dead and the liverworts reduced to a few scattered plants (A. Brown, pers. comm.).

The long-term effect of grazing is to promote the conversion of *C. vulgaris* heath to grassland, a far less favourable habitat for liverworts (Ratcliffe 1968). Hepatic heaths on Ben More Assynt and Beinn Bhàn may have been reduced in extent by deer-grazing. In both cases hepatic heath is bordered by *Nardus stricta* grassland in which there are a few closely-grazed plants of *C. vulgaris* and *H. aduncus.*

D. A. Ratcliffe (pers. comm.) has recorded hepatic heath on the island of Loch an Eoin in the Beinn Damph Forest, Ross-shire, where there is neither grazing nor burning and the heath grows on flat ground. This suggests that the heaths were once more widespread on flat ground as well as on steep slopes and that their present distribution is an artefact of destruction in places which are more easily burnt and grazed. Boulder-fields, steep slopes and cliffs are difficult to burn because there is too little continuity of vegetation and in most cases shepherds try to discourage sheep from using these dangerous places.

The usual way of managing moorland in the west of Scotland is to burn large areas so as to improve the quality of the forage by encouraging the growth of young *C. vulgaris* and grasses. This, together with intensive grazing, is probably threatening the survival of the hepatic heath.

This heath is one of the most distinctive forms of upland vegetation, and in areas which can be safeguarded for nature conservation it could be located and protected from burning. Some grazing may be necessary to prevent colonization by trees. Britain is one of the richest places for Atlantic bryophytes in the world, and nowhere else do they form this type of vegetation. These liverworts are ideal candidates for the study of plant distribution in relation to climate (Ratcliffe 1968) and their scientific interest cannot be overestimated. Their survival probably depends entirely on the way that they are managed in future.

ACKNOWLEDGMENTS

I used the work of Alan Brown, David Horsfield, Lyndsey Kinnes, Robin Payne and others. Dr Desmond Thompson and Dr Christopher Sydes commented on the text. I am very grateful to Dr Derek Ratcliffe for his help and encouragement.

REFERENCES

Clapham, A.R., Tutin, T.G. & Warburg, E.F. (1981). *Excursion Flora of the British Isles,* 3rd edn. Cambridge University Press, Cambridge.

Hobbs, R.J. & Gimingham, C.H. (1987). Vegetation, fire and herbivore interactions in heathland. *Advances in Ecological Research,* **16,** 87–171.

Jones, E.W. (1958). An annotated list of British hepatics. *Transactions of the British Bryological Society,* 3, 353–374.

McVean, D.N. & Ratcliffe, D.A. (1962). *Plant Communities of the Scottish Highlands.* Monographs of the Nature Conservancy No.1, HMSO, London.

Ratcliffe, D.A. (1968). An ecological account of Atlantic bryophytes in the British Isles. *New Phytologist,* 67, 365–439.

Ratcliffe, D.A. & Thompson, D.B.A. (1988). The British uplands: their ecological character and international significance. *This volume.*

Tivy, J. (1973). *The Organic Resources of Scotland: Their Nature and Evaluation.* Oliver & Boyd, Edinburgh.

Watson, E.V. (1978). *Mosses and Liverworts of the British Isles,* 3rd edn. Cambridge University Press, Cambridge.

Comparison of agriculture and conservation in the uplands of New Zealand and Wales

J. B. DIXON*

*Department of Applied Biology, University of Wales Institute of Science and Technology,
Newbridge-on-Wye, Powys LD1 6NB. *Current address: Coed Cymru Experimental Farm Woodland
Project, The Plas, Machynlleth, Powys SY20 8EU*

SUMMARY

1 Nature conservation programmes in upland regions of New Zealand and Wales include protection of representative examples of indigenous ecosystems.

2 This paper compares the tussock grasslands in upland New Zealand with rough grazings in Wales. In both countries these habitats have been lost or modified because of farming practices, principally for sheep. Conservation of remaining representative areas is difficult.

3 Designation of Environmentally Sensitive Areas in Wales should enhance the conservation value of the vegetation whilst also favouring farming communities. In New Zealand, such designation is unlikely, but the recent formation of the Department of Conservation should be beneficial.

INTRODUCTION

Vegetation management is a sphere of study which has a particular relevance in the successful administration of nature conservation in marginal farming areas. In upland Wales as well as in the hill and high country of New Zealand, extensive agricultural management of indigenous vegetation resources and conversion of these to more intensively used pasture have, in the past, compromised their management for nature conservation. This paper reviews the management and current administrative framework for conservation of the major upland vegetation types of South Island, New Zealand, and Wales.

The islands of New Zealand lie on the rim of the Pacific basin 1600 km east of Australia. Covering an area of 26·9 million ha New Zealand is roughly similar in extent to the United Kingdom (24·1 million ha). Although it experiences a highly oceanic climate, the extensive Southern Alps on much of South Island intercept most of the precipitation from the prevailing westerlies, creating a strong rain shadow with relatively continental conditions over many inland areas. New Zealand extends from 34°S to 47°S, giving a wide latitudinal range from subtropical to cool temperate regions. Wide variation in altitude contributes further to strong climatic and biogeographic gradients which, although far more pronounced, are similar to those of the British Isles (Ratcliffe & Thompson 1988).

The high degree of endemism within the New Zealand biota is distinctly different from the British Isles. The large size of the islands compared with other oceanic islands, some 80 million years of isolation, and contemporary remoteness combined with pronounced biogeographic gradients all give New Zealand a rich and highly distinctive assemblage of natural ecosystems harbouring a high proportion of endemic species (Fleet 1986). The long historical progression of landscape modification throughout most of Britain (and much of Europe) with secondary, human influences is wholly different from the dichotomy in New Zealand between landscapes which are either intensively managed or largely natural.

Recently, there has been concern about the inadequacy of existing reserves and other protected natural areas in New Zealand (Dingwall 1982). There has been much protection of native forest and high mountains but little of other ecosystems, notably tussock grassland communities (Scott 1979; Dingwall 1982; Mark 1985). These latter are important for traditional high country pastoral farming, and are significant in New Zealand (Fig. 1). However, in many areas they have been replaced, largely by sown exotic grasslands and exotic conifers. The tussock grasslands include a wide range of indigenous communities,

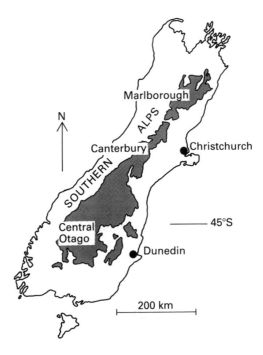

FIG. 1. Location of tussock grasslands (■) in the hill and mountain country, South Island, New Zealand.

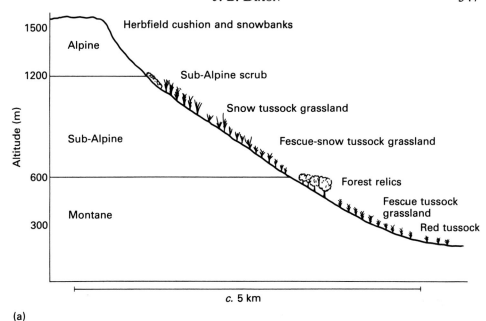

(a)

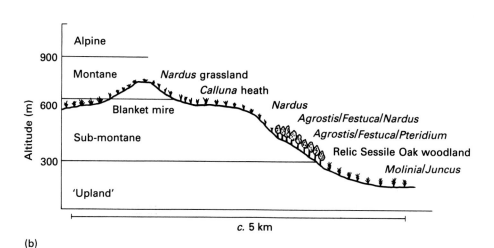

(b)

Fig. 2. Examples of changes in vegetation with increasing altitude. (a) Rock and Pillar Range, Otago, New Zealand. (b) Cambrian Mountains, Powys, Wales.

dominated by *Chionochloa* spp. (Poaceae), which vary in relation to a number of biogeographic gradients, notably altitude (Fig. 2a) but also latitude, geology, soils and climate (Connor 1964, 1965; Mark 1969; Scott 1979).

An opportunity to improve habitat representation in protected areas in New

Zealand has arisen recently through pilot schemes of the Protected Natural Areas Programme in the high country. Management prescriptions which lead to the retention of natural vegetation communities are clearly important in meeting the objectives of this programme, as indeed they are for protected areas in upland Wales (Ratcliffe 1977).

RESPONSE OF VEGETATION TO MANAGEMENT

New Zealand situation

Considerable autecological studies of *Chionochloa rigida* (Zotov.), dominant over much of the tussock grasslands, and specifically of its response to management practices, have been undertaken in New Zealand (Mark 1969; O'Connor 1980). Spring burning, a common high country practice, initiates several physiological changes in *C. rigida* which have been interpreted as adaptations to natural fires. These include an initially increased growth rate of surviving tillers and flowering. However, subsequent growth rates are lower after several years post-fire and grazing immediately after the fire kills the plant. (Some of these responses are similar to those found in British heather, *Calluna vulgaris* L.) Indeed, excessive grazing combined with too frequent burning (typically, 3–5 years) was the cause of much pastoral depletion and soil erosion associated with loss of the dominant tussocks in the high country. It was particularly prevalent in the late nineteenth and early twentieth centuries. Recent agricultural management has therefore attempted to maintain an open tussock cover, by periodic burning, to allow growth of inter-tussock species, including many exotics. Grazing value of the tussock grasses is low, but their role in soil stabilization necessitates their retention.

Mature dominant tussock grasses should be the climax vegetation cover over much of the New Zealand uplands (Fig. 1). Perhaps nature conservation objectives should therefore facilitate succession to this stage. However, the likelihood of accidental fires (which in mature *C. rigida* stands can burn at high temperatures and kill many plants) and the agricultural desirability of swards with better grazing value has produced a compromise management prescription of strictly controlled burning and grazing.

Upland Wales situation

In upland Wales *Calluna* and *Molinia caerulea* (L.) are both dominants on the sub-montane upland plateaux and constitute largely anthropogenic communities. Conservation objectives are generally the retention of examples of *Calluna* and grass heath within the uplands (Ratcliffe 1977; Ratcliffe & Thompson 1988). Physiological aging of heather (Gimingham 1972) is controlled partly by grazing and partly by rotational burning. This maintains *Calluna* in its most productive, 'building' phase and has been the traditional management of grouse moor and sheepwalk in Wales. Burning, or 'swaling', of *M. caerulea*, cutting for 'moorland' or 'rhos' hay and intensive grazing with shepherding have been practised. It is unclear to what extent *Calluna* has been replaced by *M. caerulea* on the Welsh hills (see statistics in Sydes & Miller 1988). Given a choice, *Calluna* will be preferentially grazed, whereas the deciduous habit and perennating buds enclosed within the tussock of *M. caerulea* allow it to survive fire. Acidic deposition, however, may also have favoured the spread of *M. caerulea* (Ratcliffe & Thompson 1988).

AGRICULTURE AND LAND USE HISTORY

Agriculturalists in both New Zealand and Britain have perceived severe mismanagement of extensive hill vegetation since the mid-1940s. Some have viewed reparation as a priority, although the methods used within agricultural policy in both countries have largely involved replacing indigenous vegetation by sown grasslands. In neither country has improved management of indigenous vegetation resources been considered an appropriate basis for supporting marginal farming, however.

The economic development of hill and high country in South Island has been underpinned by subdivision of indigenous grasslands combined with aerial oversowing (of pasture grasses and legumes) and top-dressing (with fertilizers), with an overall increase in stock carrying capacity. Land improvement in post-1945 upland Britain has primarily been by ploughing and reseeding moorland. This has been particularly widespread on the relatively fertile hill soils of the Welsh Marches and the Cambrian Mountains (Parry & Sinclair 1985; Dixon 1987). In Wales, especially in the 1970s and 1980s, financial incentives and technical innovation have led to the development of hill pastures with a marked increase in stocking rates, lamb production and farm incomes.

LAND TENURE AND ADMINISTRATION

In the administration of nature conservation, land tenure is one of two fundamental differences between upland Wales and New Zealand. In New Zealand, land is held by the Crown and leased, for 33 years, to pastoral farmers. The second difference is in the social and economic structure of hill farming communities. Table 1 shows New Zealand pastoral enterprises as considerably larger in area than those of Wales or even Scotland. Whilst stock carrying capacity on unimproved New Zealand hill and high country is lower than that on similar Welsh hill land, areas with a higher proportion of developed hill land probably make a better comparison. The hill farm areas in New Zealand support proportionately less labour than those in upland Wales (Table 2). The ancient history of settlement on the Welsh hills, the small size of farms and the intrinsic dependence on subsidy and other sources of income is different from the New Zealand situation. Economic support for social reasons is a consideration of much less importance in New Zealand than in Wales.

TABLE 1. Relative size of pastoral farms and stocking rates in New Zealand, Wales and Scotland. Sources: (1) O'Connor (1980); (2) Department of Lands and Survey (1986); (3) Parry & Sinclair (1985); (4) census returns (pers. comm.); (5) Welsh Office (1986); (6) HMSO (1987)

Country and area	No. of units	Area (ha)	Average farm size (ha)	% developed	Sheep ha⁻¹	Cattle ha⁻¹
New Zealand						
Hill farming, Canterbury (1978/79)[1]	450	625,100	1,389	26	3·3	0·2
High country, Canterbury (1978/79)[2]	90	1,033,400	11,482	7·5	0·6	0·03
Pastoral leases (1985)[2]	365	2,509,057	6,876	—	—	—
Wales						
Cambrian Mountains (1985) (Mainly hill farming)[3]	1,062	99,083	93	20	6·9	0·3
Radnor (district of Powys) (1981) (hill and upland farming)[4]	1,500	111,250	75	—	—	—
Wales LFA (1984) (cattle and sheep)[5]	9,459	1,200,000	127	—	—	—
Scotland (LFA cattle and sheep)[6]	9,309	4,370,721	469	—	—	—

TABLE 2. Agricultural workforce of upland New Zealand and Wales.
Sources: (1) New Zealand Meat and Wool Board (1987); (2) Parry &
Sinclair (1985); (3) Dixon (1987, unpublished)

Region	Labour units 10 km^{-2} (farmers and farmworkers)
New Zealand	
Hill farming (1986)[1]	1·17
High country (1986)[1]	0·29
Wales	
Cambrian Mountains (1981)[2]	17·68
Radnor district (1981)[3]	c. 18·00

NEW ZEALAND'S PROTECTED NATURAL AREAS PROGRAMME

Similar in principle to the Nature Conservation Review (Ratcliffe 1977) and Site of Special Scientific Interest (SSSI) network in Britain (Sydes & Miller 1988), the Protected Natural Areas (PNA) Programme has, until recently, been administered by the Department of Lands and Survey (with its legal basis in the Reserves Act 1977; Kelly & Park 1986). The 84 Ecological Regions (ERs) of New Zealand have been subdivided into 268 Ecological Districts (EDs) on the basis of landform, vegetation, biota and climate. EDs are the level at which the representativeness of major ecosystems is assessed for inclusion within a national network of protected natural areas. In this sense they are comparable, and preferable, to the administrative vice-counties used for SSSI selection in Britain as they follow ecological and not socio-political boundaries. Within a small number of EDs there have been rapid ecological surveys and preliminary selection of Priority Places for Protection (PPPs), using representativeness, rarity, diversity and other criteria similar to those of Ratcliffe (1977). Seven of the ten surveys completed by 1987 have been in South Island high country (Mackenzie ER, Umbrella ED, Kaikoura ER, Lindis–Pisa–Dunstan EDs, Old Man ED, Heron ED and Nokomai ED). For example, the Umbrella ED survey, reported in full by Dickinson (1986), covers 150,000 ha and includes 20 PPPs,14 as priority 1 and the remainder as alternative priority 2 areas. As yet, there is no framework for resolution of conflicts over these constraints on land use.

RECENT CHANGES AND THE FUTURE

British accession to the European Community and consequent implementation of the Common Agricultural Policy, as well as widespread surpluses in agricultural products, prompted the New Zealand Government to introduce support schemes for agriculture. These have been similar to European measures and, in the hill and high country, took the form of land development loans and grants, and guaranteed (or supplementary minimum) prices. Most recently, the escalating costs of this (to a primarily agricultural economy) and the election in 1984 of an administration committed to reducing public expenditure have caused the removal of almost all exchequer support for New Zealand agriculture (Kerr & Naylor 1986). Social justification for such support is much less significant in New Zealand than in upland Wales (Table 2).

The rate of land development in the New Zealand uplands has declined, at least temporarily, concurrently with falling farm incomes. In the Cambrian Mountains of Wales and elsewhere in Britain, a similar situation prompted the introduction of Environmentally Sensitive Area (ESA) status, with the provision of payments for maintaining traditional agricultural practices and thus conserving upland habitats and farming communities. This approach is likely to be inappropriate in New Zealand, whose agricultural exports are currently the mainstay of the economy.

A significant administrative change in New Zealand has been the institution of the Department of Conservation (DOC) within the bureaucracy (from April 1987). DOC also assumes responsibility for the remainder of the PNA Programme as well as for National Parks and other Crown lands of recognized conservation interest. Management of the majority of hill and high country pastoral leases has been transferred to corporations, although their ownership remains with the Crown. PPPs identified through PNA surveys would ideally be transferred to the control of the DOC. The pace of change within the New Zealand public service has been rapid; considerable debate has arisen over the division of land between DOC and the Land Corporation (which runs pastoral leases on economic grounds for profit). The New Zealand Government has pledged adequate financial support for the DOC and has even given it Cabinet weighting in Government.

ACKNOWLEDGMENTS

My study tour of New Zealand was made possible by the financial assistance of the Miss E. L. Hellaby Indigenous Grasslands Research Trust, the Stapledon Memorial Trust, the Garthgwynion Charities and UWIST. Prof. Alan Mark guided my introduction to tussock grasslands and I am indebted to him. I thank Dr Katharine Dickinson, Prof. Kevin O'Connor and others who gave so freely of their time during my stay in New Zealand. I acknowledge the constructive comments of an anonymous referee, the Department of Botany (University of Otago) and the Llysdinam Charitable Trust for facilities provided.

REFERENCES

Connor, H.E. (1964). Tussock grassland communities in the Mackenzie country, South Canterbury, New Zealand. *New Zealand Journal of Botany*, **2**, 325–351.

Connor, H.E. (1965). Tussock grasslands in the middle Rakaia valley, Canterbury, New Zealand. *New Zealand Journal of Botany*, **3**, 261–276.

Department of Lands and Survey (1986). *Annual Report of the Department of Lands and Survey to 31 March 1986.* Government Printer, Wellington.

Dickinson, K.J.M. (1986). *Umbrella Ecological District. Survey Report for the New Zealand PNA Programme.* Unpublished report, Botany Department, University of Otago, NZ.

Dingwall, P. (1982). New Zealand: saving some of everything. *Ambio*, **11**, 296–301.

Dixon, J.B. (1987). Ecology and management of improved, unimproved and reverted hill grassland in mid-Wales. *Agriculture and Nature Conservation in the Hills and Uplands* (Ed. by M. Bell & R.G.H. Bunce), pp. 33–37. Institute of Terrestrial Ecology, Merlewood.

Fleet, H. (1986). *The Concise Natural History of New Zealand.* Heinemann, Auckland.

Gimingham, C.H. (1972). *Ecology of Heathlands.* Chapman & Hall, London.

HMSO (1987). *Annual Review of Agriculture 1985/1986.* HMSO, London.

Kelly, G.C. & Park, G.N. (Eds) (1986). *The New Zealand Protected Natural Areas Programme: A Scientific Focus.* Department of Scientific and Industrial Research, Wellington, NZ.

Kerr, I.G.C. & Naylor, N.W. (1986). *The Impact of Recent Government Policies on the Economics of South Island Hill and High Country Farming.* Information Paper, Centre for Resource Management, Lincoln College and University of Canterbury, NZ.

Mark, A.F. (1969). Ecology of snow tussocks in the mountain grasslands of New Zealand. *Vegetatio*, **18**, 289–306.

Mark, A.F. (1985). The botanical component of conservation in New Zealand. *New Zealand Journal of Botany*, **23**, 789–810.

New Zealand Meat and Wool Board (1987). *Sheep and Beef Income and Expenditure Forecast to 1986/1987.* Paper 1958, New Zealand Meat and Wool Board, Wellington, NZ.

O'Connor, K.F. (1980). Land use in hill and high country. *Canterbury at the Crossroads* (Ed. by R. Bedford & A. Sturman), pp. 208–243. New Zealand Geographical Society, Christchurch, NZ.

Parry, M.L. & Sinclair, G. (1985). *Mid-Wales Uplands Study.* Countryside Commission, Cheltenham.

Ratcliffe, D.A. (Ed.) (1977). *A Nature Conservation Review, Vol. 1.* Cambridge University Press, Cambridge.

Ratcliffe, D.A. & Thompson, D.B.A. (1988). The British uplands: their ecological character and international significance. *This volume.*

Scott, D. (1979). Use and conservation of New Zealand native grasslands in 2079. *New Zealand Journal of Ecology*, **2**, 71–75.

Sydes, C. & Miller, G.R. (1988). Range management and nature conservation in the British uplands. *This volume.*

Welsh Office (1986). *Agriculture in Wales.* Welsh Office, Cardiff.

Acidic deposition and upland conservation: an overview and the way ahead

S. J. WOODIN

Nature Conservancy Council, Northminster House, Peterborough PE1 1UA

WHY ARE BRITISH UPLANDS SENSITIVE TO ACIDIC DEPOSITION?

The potential for acidification of an ecosystem is essentially dictated by the chemical buffering capacity of the system. Many British uplands have non-calcareous, almost unweatherable rock and thin acidic soil which is often on steep slopes, thus allowing little contact time between precipitation and soil. The water bodies have naturally low alkalinity and the buffering capacity of the whole ecosystem is very small. Large areas of British upland of varying geologies have a blanket peat cover which is naturally acidic and also has little buffering capacity (Fig. 1a).

Nutrient input to uplands is largely of atmospheric origin and therefore small. A change in the nature of the atmospheric supply may thus have greater consequences in the uplands than in other areas. In acidic deposition the hydrogen ions are accompanied by anions, predominantly sulphate and nitrate. Deposition of these ions can alter the nutrient status of a habitat, thus causing ecological changes (Lee, Tallis & Woodin 1988).

Land at high altitude or markedly elevated above the surrounding terrain is generally subject to high rates of precipitation and thus potentially subject to high rates of acidic deposition (Fig. 1b). Cloud droplets contain solutes at about ten times their concentrations in rain drops; cloud acidity can therefore be about ten times as great as that of rain. Where there is occluding cloud cover for a significant proportion of the time, rates of occult deposition (cloud moisture deposition) are high (Fig. 2a) and can be increased by turbulence over rough terrain. Thus in the British uplands occult deposition is an important component of acidic deposition (Hough 1986). Snow has a particular acidification potential. When it melts, all of the solutes contained within it are released in the first melt water, which can thus be extremely acidic, causing an 'acid shock' as it flushes through the ecosystem (Tranter *et al.* 1986).

Acidification in British uplands is greatly exacerbated by coniferous afforestation (Fig. 2b). The reasons for this are not completely understood; there may be many contributory factors. Ploughing and drainage cause rapid runoff of precipitation, reducing available buffering time. The trees are very efficient scavengers of pollutants from the atmosphere, particularly with regard to dry and occult deposition. Evaporation from the trees causes concentration of solutes and thus increased acidity in the water which remains. This water is further acidified

355

(a)

FIG. 1. Distribution of (a) blanket peat and soils of low buffering capacity and (b) land above 400 m.

(b)

by the needles, bark and litter, which are themselves acidic. Precipitation tends to be channelled down the tree trunk so that small areas of soil receive relatively high concentrations of acid. In base-poor soils this can lead to rapid depletion of their buffering capacity. Coniferous plantations may displace vegetation more able to withstand acidification, such as broadleaf woodland or grassland of types which form mull soils.

Regions of low land-surface buffering capacity and high elevation tend to coincide and occur predominantly in the north and west of Britain (Fig. 1). These factors combine with the rugged nature of the terrain, the wetness of the climate (Fig. 2a) and the high degree of afforestation (Fig. 2b) to render the north and west of the country more sensitive to acidic deposition than other areas.

EFFECTS OF ACIDIC DEPOSITION IN THE BRITISH UPLANDS

Ecological damage caused by acidic deposition may be directly due to the increased acidity of soil or water. Damage may also be caused by the associated increased mobility of metal ions and changes in buffering systems within ecosystems or by the increased availability of sulphur and nitrogen.

Many of the best-documented examples of acidification effects in Britain concern upland freshwater ecosystems. In many cases afforestation has contributed to freshwater acidification problems. Studies of diatoms and soot particles in sediments have enabled reconstruction of the acidification histories of lakes and have demonstrated the role of acidic deposition in the acidification of some lakes in south-west Scotland (Battarbee *et al.* 1985) and North and mid-Wales (R. W. Battarbee, pers. comm.). Increased concentrations of aluminium and decreased availability of calcium in waters in acidic catchments can be toxic to fish (Muniz & Leivestad 1980). There are examples of lakes in upland Britain that once supported healthy fish populations no longer doing so and in some cases populations of rare fish are now threatened by acidification (P. S. Maitland, pers. comm.). Fish with deformed tails have been caught in waters which were becoming more acid in south-west Scotland (Campbell, Maitland & Lyle 1986) and large fish kills have been reported in rivers in the Lake District as a result of acid 'flushes' caused by rain storms (Crawshaw 1986). Invertebrate communities in acid streams, particularly those draining afforested catchments, are greatly impoverished (Stoner, Gee & Wade 1984). The dipper (*Cinclus cinclus*), which feeds on upland stream invertebrates, is less successful on acidic streams (Ormerod *et al.* 1986) and preliminary work suggests that otter distribution may be limited by river acidity in the Welsh uplands (S. J. Woodin, pers. obs.; Mason & Macdonald 1987).

Less is known about the effects of acidic deposition in terrestrial environments. Nitrate and ammonium deposition in the southern Pennines have been shown to be supra-optimal for the growth of ombrotrophic *Sphagnum* species (Press, Woodin & Lee 1986) and it appears that the structure of *Sphagnum* communities in Wales may be being affected by increased rates of nitrogen deposition (R. Woods, pers. comm.). The Lobarion lichen community is considered to be sensitive to acidic deposition and a national survey of distribution and growth is identifying regions in which the community appears to be declining (J. H. Looney, pers. comm.). The debate about effects of acidic deposition on trees continues. However, observations are increasingly suggesting that there may be some effect in Britain and, if this is so, trees on poor soils in exposed sites receiving significant amounts of occult deposition may be worst affected.

ACIDIC DEPOSITION AND CONSERVATION STRATEGY

Whilst the Nature Conservancy Council can take action to protect National Nature Reserves and Sites of Special Scientific Interest from harmful land management practices or development, it cannot protect them from acidic deposition. There are, however, some measures which can be taken to ameliorate acidification in semi-natural habitats. The Scandinavians spend the equivalent of millions of pounds each year on the liming of water bodies. Although this reduces acidity it does not restore lakes to their original chemical composition and thus will not restore their natural microflora and microfauna. Lime application to water bodies has to be repeated with considerable frequency and an alternative method now increasingly employed is catchment liming, allowing slow release of lime into water systems. This is being experimented with in Britain but, although successful in that water quality can be improved sufficiently for fish to survive, the lime can kill *Sphagnum* in the catchment vegetation communities.

Measures can be taken by the forestry industry to prevent the amplification of acidification in afforested catchments. Obviously the ideal would be not to plant conifers in areas which are already susceptible to acidification. The planting of broadleaved trees which form mull humus — such as birch (*Betula*) — would be far more acceptable (Miles 1988). If conifers are planted, water acidification can be reduced by the creation of buffer zones around waterways through which no drains or plough furrows are cut and in which no trees are planted, allowing natural vegetation to remain. The use of wider tree spacing, respacing of young crops or heavier thinning would promote ground-layer vegetation and the breakdown of needle litter, thus reducing the acidifying potential.

Ameliorative measures cannot satisfactorily cure acidification. In some cases, if the continuation of a rare species is threatened by acidification, the only immediate answer to the problem may be to transplant individuals of the threatened population to less damaged sites.

The problem of acidification by acidifying pollutants can only be solved at its source through emission reduction. Economic considerations tend to govern decisions made about the extent to which emissions should be reduced. A different approach, which bases proposed reduction percentages entirely on ecological considerations, is the critical load concept. A critical load of sulphur,

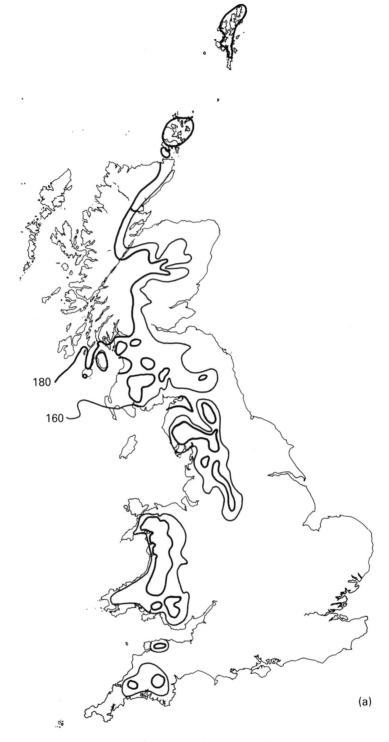

180

160

(a)

FIG. 2. (a) 'Wet days', the number of days per year when there is precipitation of at least 1 mm (from Nature Conservancy Council 1986). (b) Forestry in Britain, including Forestry Commission woodland and forest parks and all other grant-aided woodland (Forestry Commission, pers. comm.).

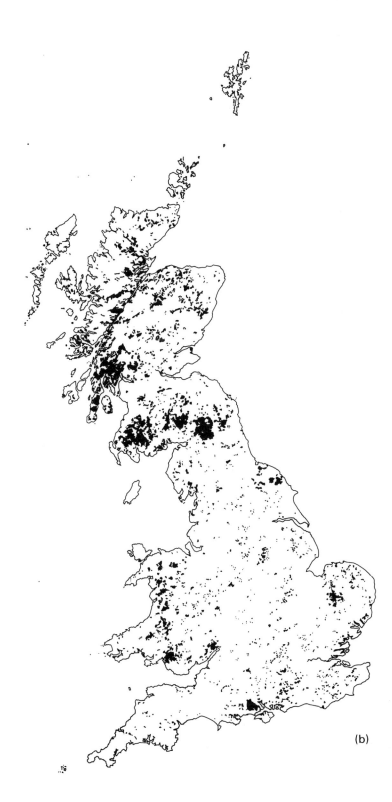

(b)

nitrogen or acidity is defined as 'the highest load that will not cause chemical changes leading to long-term harmful effects on the most sensitive ecological systems' (Nilsson 1986) and is calculated by ecological modelling. The difference between measured deposition rates and the critical load is used as a basis for emission reduction proposals. Current suggestions are for 80 per cent emission reductions. Observations have shown that sulphur emission reductions can lead to a decrease in surface water sulphate concentrations although it is not yet known whether the relationship between the two is linear. Quantitative effects of nitrogen emission reduction are even less certain, as emissions are still rising, but ecological effects could only be beneficial. Emission reductions should be planned on the basis of the critical load concept, initially assuming linearity of emission/deposition relationships. Once emissions are falling more rapidly, detailed monitoring of emissions and deposition rates should enable increased understanding of relationships between the two. In many habitats there will be a lag time, possibly of many years, between emission reduction and ecological restoration. However, biological monitoring should provide information on the recovery of ecosystems. Information gained from both emission/deposition monitoring and habitat monitoring should be used to redefine emission limits as necessary so as to enable complete recovery.

No Sites of Special Scientific Interest have been denotified because of recognized damage by acidification. Evidence of such damage, or of potential susceptibility to damage, would not preclude a site from notification. It may, however, be taken into consideration if a choice between sites as particular type examples was being made. In the countryside as a whole it is important that the conservation of semi-natural habitats continues, whether or not sites are acidified. It is to be hoped that those areas which have been, or are being, ecologically devalued by acidification will be given the eventual chance of recovery. In the mean time, we should ensure that such areas are protected from land-use practices which destroy or severely diminish their semi-natural character.

REFERENCES

Battarbee, R.W., Flower, R.F., Stevenson, A.C. & Rippey, B. (1985). Lake acidification in Galloway: a palaeoecological test of competing hypotheses. *Nature,* **314,** 350–352.

Campbell, R.N.B., Maitland, P.S.M. & Lyle, A.A. (1986). Brown trout deformities: an association with acidification? *Ambio,* **15,** 244–245.

Crawshaw, D.H. (1986). The effects of acidic runoff on streams in Cumbria. *Pollution in Cumbria* (Ed. by P. Ineson), pp. 25–32. Symposium 16, Institute of Terrestrial Ecology, Huntingdon.

Hough, A.M. (1986). *Acid Deposition at Elevated Sites, a Review.* United Kingdom Atomic Energy Authority, Harwell.

Lee, J.A., Tallis, J.H. & Woodin, S.J. (1988). Acidic deposition and British upland vegetation. *This volume.*

Mason, C.F. & Macdonald, S.M. (1987). Acidification and otter (*Lutra lutra*) distribution on a British river. *Mammalia,* **51,** 81–87.

Miles, J. (1988). Vegetation and soil change in the uplands. *This volume.*

Muniz, I.P. & Leivestad, H. (1980). Acidification — effects on fish. *Ecological Impact of Acid Precipitation, Proceedings of an International Conference* (Ed. by D. Drablos & A. Tollan), pp. 84–92. SNSF, Oslo – Ås.

Nature Conservancy Council (1986). *Nature Conservation and Afforestation in Britain.* Nature Conservancy Council, Peterborough.

Nilsson, J. (1986). *Critical Loads for Sulphur and Nitrogen.* Nordisk Ministerrad, Stockholm.

Ormerod, S.J., Allinson, N., Hudson, D. & Tyler, S.J. (1986). The distribution of breeding dippers (*Cinclus cinclus* (L.); Aves) in relation to stream acidity in upland Wales. *Freshwater Biology,* 16, 501–507.

Press, M.C., Woodin, S.J. & Lee, J.A. (1986). The potential importance of an increased atmospheric nitrogen supply to the growth of ombrotrophic *Sphagnum* species. *New Phytologist,* 103, 45–55.

Stoner, J.H., Gee, A.S. & Wade, K.R. (1984). The effects of acidification on the ecology of streams in the Upper Tywi catchment in West Wales. *Environmental Pollution (Series A),* 35, 125–157.

Tranter, M., Brimblecombe, P., Davies, T.D., Vincent, C.E., Abrahams, P.W. & Blackwood, I. (1986). The composition of snowfall, snowpack and meltwater in the Scottish highlands — evidence for preferential elution. *Atmospheric Environment,* 20, 517–525.

The structure and importance of invertebrate communities on peatlands and moorlands, and effects of environmental and management changes

J. C. COULSON

Department of Zoology, University of Durham, Science Laboratories, South Road, Durham DH1 3LE

SUMMARY

1 In a previous study, five major invertebrate communities have been identified on peatlands and moorlands. This paper examines these.

2 Similar numbers of invertebrate species were found in all five communities.

3 Standing crop showed marked differences between the communities. Worms (Lumbricidae and Enchytraeidae) formed the majority of the standing crop in four of the communities and 24 per cent of the total on the fifth. There is appreciable variation in standing crop between communities, attributable to Diptera and Lepidoptera, the latter declining and Diptera increasing with altitude. Hemiptera, Araneae and Coleoptera contribute little to the total standing crop.

4 There is much greater separation of species of herbivores and detritivores between communities than there is of predators.

5 Consideration is given to the important functions of the communities, i.e. nutrient cycling, decomposition and a food resource for vertebrates. All three are recognized as being of major ecological importance, whilst altitude (acting through temperature and rainfall) and drainage are important in modifying the structure of the invertebrate communities.

6 The paper considers the effects of drainage, sheep grazing, burning and forestry on the structure and diversity of the invertebrate communities.

INTRODUCTION

In a study of the invertebrate communities associated with peatlands and uplands, Coulson & Butterfield (1985) recognized five main invertebrate communities in northern England, four of which occurred on what is generally called moorland. These communities usually coincided with characteristic soils and plant communities.

Moorland is the term used here to describe large areas of Britain which have extensive areas of heather (*Calluna vulgaris*) usually, but not invariably, occurring on high ground and normally having peat or a peaty soil. Within 'moorland', there is a series of soil types and associated plant communities. On high rainfall areas, blanket bog occurs, with peat up to 2 m deep; but other areas of deep peat are also found, albeit more locally, in the form of raised or valley bogs at lower altitudes. Such areas of deep peat differ in many ways from the heath-like moorlands (called 'northern heaths' by Gimingham (1972)), which occur extensively at lower altitudes than blanket bog. The soils underlying the northern heaths are formed mainly over base-poor rock and there is often a high organic content in the top 50 mm. Each of these types has a distinctive and characteristic invertebrate community. Throughout the uplands, grasslands occur on brown earth soils with low organic content, overlying base-rich rock (often limestone in the north of England). Again, a characteristic invertebrate community occurs on these upland grasslands.

A further invertebrate community has been found on steep-gradient (over 15°) areas in regions of high rainfall. In our study, typical sites were restricted to the scarp slopes of the northern Pennines. The steep gradient prevents accumulations of deep peat and facilitates drainage. Such areas are dominated typically by *Juncus squarrosus*, *Festuca ovina* or *Agrostis tenuis*, and carry an invertebrate fauna with characteristic species found in both peat and mineral soils.

In this paper, the characteristics of the invertebrate communities are described and some of the environmental factors which determine the distribution of the component species are considered. Due emphasis is then given to the effects of upland communities and their likely impact on the whole ecosystem.

METHODS

Densities of invertebrates forming the communities on the five habitat types have been obtained from a minimum of eight quantitative soil samples taken from typical sites in each quarter of the year (spring and autumn only for Lumbricidae and Enchytraeidae), using appropriate extraction methods (high temperature gradient for Acari and Collembola; wet funnels for Enchytraeidae and small Diptera larvae; formalin treatment in the field for Lumbricidae; and Berlese funnels for the remaining groups). Comparisons have been made with other sites with the same communities by the use of pitfall trap captures and conversion ratios calculated from sites where soil samples had been taken. Detailed consideration has not been given to Collembola and mites (Acari), except to determine their standing crops. Data on standing crops are the average of measures taken in each of the four seasons of the year. Where a species or group shows a markedly synchronized seasonal emergence, the average standing crop appreciably underestimates the maximum standing crop, which in the case of insects (Insecta) is usually at the end of their larval stages. All values are given per square metre of ground surface. Species lists have been compiled from pitfall trap captures. Apart from the worms (Annelida), soil samples did not produce species which were absent from the pitfall traps. More details of the sites and techniques are given in Butterfield & Coulson (1983) and Coulson & Butterfield (1985, 1986).

THE INVERTEBRATE COMMUNITIES

The analysis of 42 sites in the north of England recognized five distinct invertebrate communities, characteristic of the following habitats:
(i)   lowland mires;
(ii)   northern heaths;
(iii) blanket bog;
(iv)   steep-gradient peaty soils;
(v)   upland limestone grasslands.

STRUCTURE OF THE COMMUNITIES

The communities of invertebrates considered in this paper differ in five ways, each now considered under a separate heading.

Densities of individuals from major groups

Table 1 gives estimates of the densities of invertebrate groups found on these areas. The densities of earthworms (Lumbricidae), Lepidoptera and Diptera, in particular, vary considerably between the communities. The differences in densities of Hemiptera are due mainly to the numbers of the very small psillid (*Strophingia ericae*).

TABLE 1. The mean densities of invertebrates (number m^{-2}) in each of five habitats which possesses different invertebrate communities. *Excluding winter densities which have not been measured

Invertebrate taxa	Blanket bog	Northern heath	Lowland mire	Steep-gradient peaty soils	Upland limestone grassland
Worms					
Lumbricidae	1	20	2	50	390
Enchytraeidae	80,000	20,000	25,000	70,000	80,000
Collembola	33,000	25,000	22,000	40,000	46,000
Mites	60,000	20,000	25,000	25,000	33,000
Insects					
Diptera					
Tipulidae	700	10	70	1,200	120
Other	320	30	100	350	130
Coleoptera					
Carabidae	1	3	3	5	3
Staphylinidae	21	25	c. 10	c. 20	45
Others	2	3	5	7	9
Lepidoptera*	4	30	44	15	1
Hemiptera	3,500	2,000	1,200	100	100
Araneae	130	30	34	64	40
Opiliones	3	1	1	1	1

Number of species

The number of lumbricid and enchytraeid worm species have not been considered in this section since all sites were not examined in the same detail, but the number of species involved was less than six, except on the upland limestone grasslands. Whereas the total numbers of invertebrate species found in each community did not differ markedly (Table 2), there were significantly fewer Diptera species at low altitudes (northern heaths and lowland mires) than in the other communities.

Standing crop

Pronounced differences between the communities are evident in the average weight of the standing crop (Table 3), these being much greater than the differences in the numbers of species. Worms are dominant in four of the five habitat types. The role of Diptera becomes important in the higher upland communities, averaging between 17 per cent and over 20 per cent of the total standing crop, but declining to only 4 per cent on northern heaths and 5 per cent on upland limestone grasslands. The Lepidoptera make a major contribution to the standing crop on the northern heaths (33 per cent) and lowland mires (38 per cent) and play a complementary role to the Diptera, in terms of both proportion and actual weight. Coleoptera and Hemiptera contribute little to the standing crop.

Table 4 contrasts the differences between the percentage of species and standing crop contributed by the insects, harvestmen (Opiliones) and spiders (Araneae) in these communities. The high proportion of the standing crop contributed by the Diptera in the uplands is almost entirely due to the change in abundance of craneflies (Tipulidae). Despite virtually no increase in numbers of species of Diptera in the higher upland communities, the abundance of many of the species increases dramatically at higher altitudes.

Seasonal abundance

The blanket bog communities produce an extremely sharply defined spring peak (late May and early June) in the emergence of insects after which time adult insects are scarce until late autumn (Fig. 1). This pattern is in contrast to the situation in the upland grasslands and northern heaths where a more sustained distribution of adult insects occurs throughout the year. The steep-slope peaty soil community has both peat and mineral soil species and accordingly is intermediate in the seasonal distribution of adult invertebrates.

TABLE 2. The mean number of invertebrate species (excluding mites and Collembola) recorded on five habitats. The numbers of species of Diptera, Coleoptera and Araneae & Opiliones are also presented and expressed as a percentage of the mean number of species recorded on that habitat

	Blanket bog	Northern heath	Lowland mire	Steep-gradient peaty soil	Upland limestone grassland
Number of sites examined	14	9	3	6	5
Mean number of					
species per site	137	137	134	154	155
Insects					
Diptera	40	30	36	45	45
Coleoptera	39	45	32	47	42
Others	9	11	8	10	12
Araneae & Opiliones	45	48	52	43	46
Per cent Diptera	29%	22%	27%	29%	29%
Per cent Coleoptera	28%	33%	24%	31%	27%
Per cent Araneae & Opiliones	33%	35%	39%	28%	30%

TABLE 3. Estimates of the mean standing crop (g dry weight m^{-2}) of invertebrates on five habitats

	Blanket bog	Northern heath	Lowland mire	Steep-gradient peaty soil	Upland limestone grassland
Within the soil					
A Lumbricidae	<0·01	1·0	<0·01	3·1	23·2
Enchytraeidae	2·21	0·40	0·40	2·20	4·11
B Mites and Collembola	0·60	0·22	0·26	0·40	0·50
Above ground					
C Diptera	0·71	0·11	0·22	1·23	1·41
Coleoptera	0·01	0·01	0·04	0·01	0·03
Hemiptera	0·02	0·02	0·01	0·01	0·01
Lepidoptera	0·04	0·91	0·63	0·14	<0·01
D Spiders and harvestmen	<0·01	<0·01	<0·01	<0·01	neg.
Total (including others)	3·62	2·74	1·66	7·14	29·46
Worms as % of total	61%	51%	24%	74%	92%
Lepidoptera as % of total	1%	33%	38%	2%	<1%
Diptera as % of total	20%	4%	13%	17%	5%
Lepidoptera as % of					
Lepidoptera & Diptera	5%	89%	74%	10%	1%

TABLE 4. A comparison of the proportion of species and standing crop (weight) contributed by major insects groups, spiders and harvestmen in the habitats

	Blanket bog	Northern heath	Lowland mire	Steep-gradient peaty soil	Upland limestone grassland
Diptera					
By species	29%	22%	27%	29%	29%
By weight	91%	10%	24%	88%	97%
Coleoptera					
By species	28%	34%	24%	31%	27%
By weight	1%	1%	4%	1%	2%
Lepidoptera					
By species	1%	5%	4%	2%	2%
By weight	5%	87%	70%	10%	<1%
Hemiptera					
By species	7%	6%	4%	6%	6%
By weight	3%	2%	1%	1%	1%
Araneae and Opiliones					
By species	33%	35%	27%	28%	30%
By weight	<1%	<1%	<1%	<1%	<1%

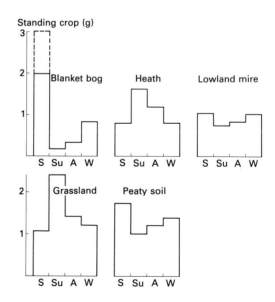

FIG. 1. The average standing crop (g dry weight m^{-1}) of invertebrates (excluding worms, mites and Collembola, which change very little seasonally) in each of the four seasons (S = spring, Su = summer, A = autumn, W = winter) on each of the five communities. The dotted line indicates the high-altitude blanket bog spring value (see Fig. 3).

Affinities

Blanket bog

The invertebrate species on blanket bog are typical of those found in sub-arctic regions of Scandinavia and have a northern European distribution. The typically short summer of most areas with blanket bog results in the late spring and early summer peak of activity and abundance above ground level (Coulson & Whittaker 1978). Truly arctic species are absent from this community, although several do occur on the high ground of the Cairngorm plateau, Scotland (Ratcliffe 1977; Thompson, Galbraith & Horsfield 1987).

Northern heaths

Although many of the species found on blanket bog occur locally and in low densities on the northern heaths, the fauna also shows affinities with that of the lowlands. The species of Lepidoptera are mainly determined by the presence of suitable food plants, but many moths do not penetrate into the blanket bog zone even when their food plants occur there.

Lowland mires

The fauna consists of a number of species commonly found on blanket bog, e.g. *Tipula subnodicornis* and *Tricyphona immaculata,* but also a few species not commonly found elsewhere on peatlands, e.g. *Agonum ericeti, Dolichopus atratus* and *Pirata uliginosus.* The common cranefly on blanket bog, *Molophilus ater,* is replaced by *M. occultus.*

Upland limestone grassland

Typically, there is an impoverished lowland grassland fauna with species occurring across central Europe. The fauna also includes several spiders and beetles which are typically found in woodland at low altitudes.

Steep-gradient peaty soil

This community has species typical of both upland grasslands and blanket peat, e.g. *Molophilus ater* and *Tipula varipennis.* The tendency to accumulate peat appears to be prevented by the steep gradient and the soil retains an appreciable mineral content. Areas with drainage from limestone have an exceedingly rich fauna which includes several rare and very restricted species, e.g. *Tipula gimmerthali.*

FACTORS INFLUENCING THE COMPOSITION OF THE COMMUNITIES

The separation of species between upland mineral and peat soils

Whilst lowland mires, northern heaths and blanket bog have many species in common, in some taxonomic groups few or even none of the species are found on both blanket bog and the adjacent grasslands. For example, there are no species of Tipulidae common to the two communities, although considerable overlap occurs in the Carabidae. The invertebrate species on four pairs of blanket peat and adjacent grassland sites have been compared for most taxonomic groups. Fig. 2 shows that the avoidance of peat or mineral soils is most pronounced in the herbivores and detritivores and, in contrast, the predatory species usually show little or no separation on the basis of the soil types. This conclusion was confirmed from a study of the Carabidae and spiders (Butterfield & Coulson 1983; Coulson & Butterfield 1986), where the architecture and physical environment of the site was found to be of major importance in determining invertebrate distribution, whilst the nature of the soil played little part.

Edge effects

Some animal species are restricted to peat or mineral soils for breeding but because they are mobile there is a tendency for them to move across the interface between two soil types and to 'invade' the neighbouring invertebrate community.

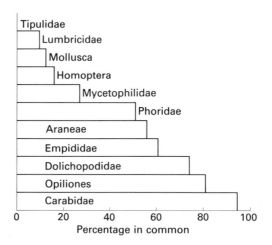

FIG. 2. The percentage of species found in blanket bog or adjacent upland grasslands which occurred regularly on both. The upper five taxa have mainly non-predatory species and the species in the lower five taxa are predatory. The Phoridae have a wide range of larval feeding habits, including parasitic and saprophytic feeding; they take an intermediate position between predators and herbivores in the figure.

This effect has become obvious in work on staphylinid and carabid beetles where species breeding only on limestone grassland have been frequently found more than 100 m beyond the interface between the grassland and blanket bog. L. Lloyd (pers. comm.) has shown that this effect appreciably increases the diversity of both carabid and staphylinid beetles for several metres on each side of the interface. The effect has also been noted in adults of *Tipula paludosa* and in earthworms, and is probably widespread. Accordingly, area mosaics of soil types are likely to have different species distribution and greater species richness compared with areas with a uniform soil type and the same plant community.

Precipitation

Blanket bog in Britain is distributed mainly in areas with an excess of 1500 mm rainfall per year and is closely associated with a high water-table throughout the year (Ratcliffe & Thompson 1988). In contrast, valley and basin mires are frequently in areas of lower rainfall although locally they have a high water-table (Ratcliffe 1977). Despite some similarities in water-table characteristics, the invertebrate fauna of blanket bog differs in many respects from that of oligotrophic mires. This may be due to differences in vegetation and nutrient input, although temperature may also play an important part.

The density of Tipulidae, even within the blanket bog community, increases with altitude and rainfall (see below). In this group, we know enough about the ecology of the group to state that rainfall, or rather the lack of drought spells, is important in determining high population densities. The exceptional drought of 1956 caused a marked decline in the distribution and density of *Tipula subnodicornis* in all but the wettest parts of blanket bog areas (Coulson 1962). Clearly, the greater the number of rain days in summer, the less the chance of drought affecting the survival of the eggs and larvae. Blanket bog invertebrate communities show evidence of marked differences in their structure in relation to rainfall. The difference does not show up in species composition to a major extent, but does in terms of the abundance of individual species. Numbers of specimens of Tipulidae occurring within the blanket bog community at different sites in the northern Pennines increase with mean rainfall in spring (Fig. 3). The increased abundance, evident only in the spring-emerging species, is almost entirely due to *Tipula subnodicornis, Molophilus ater* and *Tricyphona immaculata*. The first-larval instar of all these three species occurs in June and July and is particularly susceptible to desiccation. Obviously the risk of desiccation is greater in areas with low rainfall and it is for this reason that rainfall rather than temperature is considered here.

The effect of temperature and altitude

Throughout the year, the temperature is lower at higher altitudes (see Coulson *et al.* 1976; Ratcliffe & Thompson 1988). Though well known, the magnitude of

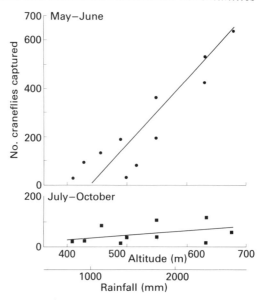

FIG. 3. The mean number of craneflies captured in ten pitfall traps (45 mm mouth diameter) per year on each of 11 blanket bog vegetation areas in relation to altitude (and annual rainfall (mm)). The upper graph shows the relationship for captures in May and June ($r_9 = +0.91$, $P < 0.01$) and the lower graph captures in the remaining part of the year ($r_9 = +0.26$, not significant). The regression slopes are significantly different from each other ($P < 0.01$).

the change is often not appreciated. In a study at Moor House it was shown that the average decrease in temperature was 0·5 °C per 100 m rise in altitude and that the effect was evident throughout the year, although slightly more pronounced in certain seasons. The difference in the average temperature (2·4 °C) between 370 m and 847 m, only 1 km apart, is equivalent to that at sea-level between Plymouth and Edinburgh, a distance of 1100 km (Coulson *et al.* 1976)! The effects of temperature are probably the most important factor limiting the distribution of invertebrates in the uplands. At lower temperatures species have to adapt in order to complete their life cycle. This can be achieved by (i) being active at low temperatures, e.g. *Nebria nivalis;* (ii) having a diapause which lasts for a shorter time at lower temperatures, e.g. several *Simulium* and *Tipula* species; (iii) lengthening the life cycle at lower temperatures, e.g. *Carabus problematicus* (Butterfield 1986), *Strophingia ericae* (Hodkinson 1973; Parkinson & Whittaker 1975), *Tipula rufina* (Coulson & Whittaker 1978); and (iv) possessing a low Q_{10} for growth, e.g. *Molophilus ater* (Butterfield 1976; Coulson *et al.* 1976). There is also a need to possess greater cold tolerance to cope with the lower winter temperatures in the uplands.

Lower temperatures do have some compensations, however. Levels of predation and parasitism are often lower. The activity of many parasitic Hymenoptera appears to be extremely temperature-limited and they do not

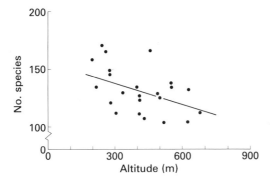

FIG. 4. The number of arthropod species recorded on 23 moorland (blanket bog and northern heath) sites in relation to altitude. The linear regression is $y = 155 \cdot 7 - 0 \cdot 06 \, (\pm 0 \cdot 025)x$, where y is the number of species and x is the altitude in metres.

extend to the upper altitude range of their hosts, e.g. parasitoids of *Coleophora alticolella* (Randall 1982).

An investigation into the number of arthropod species in relation to altitude shows only a modest decline in the numbers of species with increasing altitude, with considerable variation between sites (Fig. 4). When particular taxa are considered, a more variable pattern emerges (Table 5), with three predatory groups tending to be represented by significantly fewer species at higher altitudes, whereas the groups which feed on live and dead plants either show no significant trends or have an actual increase in the numbers of recorded species with increasing altitude (Lepidoptera and Acrididae provide exceptions). In effect, the

TABLE 5. The linear correlation coefficients (r) between altitude and the number of species ($\log_{10}$) recorded at 23 sites with peat or peaty soils, blanket bog and northern heaths. The altitudes ranged from 198 m to 675 m; the groups have been divided into predatory and non-predatory groups. $*P < 0 \cdot 05$, $**P < 0 \cdot 02$, $***P < 0 \cdot 01$

Non-predatory taxa	r	Predatory taxa	r
Tipulidae	+0·76***	Araneae	−0·60***
Trichoceridae	−0·26	Opiliones	−0·55***
Bibionidae	+0·32	Carabidae	−0·67***
Mycetophilidae	+0·47*	Staphylinidae	−0·13
Phoridae	+0·23	Empididae	−0·30
Homoptera	+0·19	Dolichopodidae	−0·33
Curculionidae	−0·27		
Elateridae	−0·17		
Apidae	−0·30		
Lepidoptera	−0·49**		
Acrididae	−0·60**		
Mollusca	−0·09		
Mean value	−0·02 ± 0·12		−0·43 ± 0·09

proportion of species which are predatory declines with altitude. However, the actual ratio of predatory to non-predatory individuals does not follow this trend and many even go in the opposite direction. Even within the same taxa, different groups may respond differently to altitude. In the spiders, the number of linyphiid species show no altitude trend, whereas the remaining families have markedly fewer species at higher altitudes (Coulson & Butterfield 1986). This difference is almost certainly related to their feeding habits, the former depending on soil fauna, such as mites, Collembola and perhaps enchytraeid worms, as food. Clearly, altitude has a substantial effect on invertebrate community composition, and in turn influences the structure of the upland food web.

IMPORTANT FUNCTIONS OF UPLAND INVERTEBRATES

The invertebrates forming each community utilize living plants, plant remains and other animals as food. The communities carry out important functions which have an impact on the rest of the ecosystem and I have selected three of these for further consideration.

Decomposition of primary production

Peat accumulates because of inefficient decomposition of the plant remains. Most of the decomposition must be carried out by organisms which can degrade cellulose, and most animals are unable to do this. Most of the decomposition is carried out by fungi and bacteria but the invertebrates play an important role in breaking up the plant material and increasing the surface area. Standen (1978) has shown that decomposition is greater in the presence of enchytraeid worms and tipulids, and the importance of the chemical quality of the plant material to decomposers has been considered by Coulson & Butterfield (1978).

Nutrient cycling

Nutrients are in short supply in the uplands, particularly in blanket bog, where the peat isolates the plants from the nutrients in the mineral soil and rock. Invertebrates concentrate nutrients by a factor of about 10 compared with those found in the vegetation (Coulson & Whittaker 1978). Although there is a much smaller biomass of invertebrates than plants, the high concentrations in the animals are of particular value to the plants because they are in an accessible form. These nutrients are returned to the soil on the death of the invertebrates and most are immediately available to the plants. Further, the exceptionally large spring emergence of insects on the blanket bog is followed almost immediately by a high mortality (most tipulids live only for about 3 days as adults) and therefore there is a sudden return of high concentrations of nutrients to the soil surface at the time of the spring surge in plant growth. The importance of this return requires detailed examination. Coulson & Whittaker (1978) estimated that the return averaged about 0.11 g nitrogen m^{-2} and 0.02 g phosphorus m^{-2} on blanket bog, but the values are higher in high rainfall areas. This cycling of nutrients through the blanket bog ecosystem may be important in maintaining high levels of cesium$_{137}$ in sheep grazing on moors following the Chernobyl disaster in 1986.

Food for invertebrate predators

The spring peak of insects produces a super-abundance of food for invertebrate predators on peatlands. The peak coincides with the hatching of grouse (*Lagopus lagopus scoticus*), many wading birds (Charadrii) and the meadow pipit (*Anthus pratensis*). The summer trough of insects is a problem for insect-feeding animals which remain on the moorland at that time. Meadow pipits use streams and grasslands as sources of food for their second broods, frogs (*Rana temporaria*), pygmy shrews (*Sorex minutus*) and some large carabid beetles 'migrate' from the blanket bog to limestone or stream-side grasslands, to utilize the summer emergence there. The change from spring peak to summer trough of insect abundance is dramatic and this may explain the virtual disappearance of most species of spiders, other than members of the Linyphiidae, on the high moors.

EFFECTS OF MANAGEMENT

Drainage

Of all human activities in the British uplands, drainage and heather burning are the most extensive (also discussed by Usher & Gardner 1988). Drainage removes rain-water from the site more rapidly. As a consequence, it simulates many of the effects of lower rainfall, since both tend to reduce the level of the water-table and determine the presence or absence of several of the plant species. Studies, to be reported in more detail elsewhere, show that drainage on a high rainfall site at Moor House did not produce significant changes in the plant community over a 30-year period, and that most invertebrates showed no changes as a result of drainage. This presumably reflects the high rainfall (2000 mm yr^{-1}) and its frequent input into this site, which overcomes the effect of drainage and maintains a high water-table to within a few centimetres of the ditches. Even so, small but significant changes occurred in the density of enchytraeid worms and in the distribution of one harvestman, *Nemastoma bimaculatum*, in relation to the position of the ditches. In contrast, the impact of open drainage ditches was marked at a site on the interface between blanket bog and northern heath and with only half the annual rainfall. In this case, the distribution of virtually every invertebrate species was changed, as was the cover of *Calluna*. Here, the effect of drainage was to move the composition of the invertebrate community from that found on blanket bog towards those on northern heath and grassland.

Sheep grazing

The immediate impact of sheep (and red deer, *Cervus elaphus*) grazing on moorland is twofold. First, they remove considerable quantities of the primary production, particularly from upland grasslands. Second, their dung is a microhabitat used by a number of coprophagous invertebrate species.

On blanket bog and lowland mires the existing primary production, even in the presence of sheep, is not totally consumed and so peat accumulates. Cessation of sheep grazing in this type of situation is unlikely to result in greater invertebrate populations since excess plant material as a potential food source already exists (see above). On the other habitats, grazing tends to lower the height of the vegetation and this is particularly so on upland limestone grasslands. The cessation of grazing is likely to increase the densities of the above-ground fauna, particularly spiders (responding to architecture of the habitat), Homoptera and other phytophagous species (responding to greater plant production).

Sheep dung supports a considerable fauna, particularly Diptera (Sepsidae, Sphaeroceridae, Scathophagidae) and certain beetles (*Aphodius* spp. and some Staphylinidae) which have no alternative food supply. In addition, several earthworms utilize dung; Svendsen (1957) demonstrated the strong dependence of the earthworms on dung in upland areas. Tipulidae larvae also concentrate in dung, although whether dung increases their overall densities is not known. The actual densities of moorland invertebrates that exploit dung are, however, very low (even on grasslands where the density of dung is some six times greater than on blanket bog). These animals make a negligible contribution to the total biomass, but contribute at least 15 species to each invertebrate community. The absence of sheep dung would not greatly influence the standing crop (or numbers of individuals present) of invertebrates. It would, however, decrease the species richness (and hence diversity) by some 10 per cent and would remove local high densities of invertebrates in and below the dung, a source of food which is important to some vertebrate predators.

Burning

Burning has the effect of removing most of the above-ground living plant material and has been considered in detail by Gimingham (1972) and Hobbs & Gimingham (1987). Burning is carried out more frequently on the northern heaths than on the blanket bog and therefore its impact is likely to be more wide-spread on the former. The main aim of burning is to produce more young, short *Calluna*; but obviously it has a destructive effect on the invertebrate fauna. The mobility of most invertebrates and the relatively small plots which are burnt at any one time raises no major problems for recolonization by invertebrates. However, it is difficult to separate the direct effects of burning from those associated with the loss of food for the invertebrates. Large and extensive accidental burning of moor, as occurred on the North York Moors in 1976, has had more pronounced effects on the whole ecosystem because of the much larger areas involved, the major effects of the hot fire on the vegetation and burning of the peat for many days.

Afforestation

There are two effects of afforestation on peatlands. First, ditching to assist tree growth increases runoff and lowers the water-table on the moors with lower rainfall. Second, the planting of young trees eventually changes the ground vegetation, with the loss of most of the invertebrate species in the heaths or blanket bogs. The introduction of small areas of woodland, whether conifer or broadleaved trees, increases the invertebrate species richness of an upland area because more species, particularly herbivores, are introduced. The wholesale conversion of moorland to commercial forestry clearly has a more marked effect, but whether this decreases the invertebrate species richness of the area remains to be demonstrated. We know that some birds disappear and others arrive following afforestation (Thompson, Stroud & Pienkowski 1988) but the extent to which changes in the invertebrate community composition contribute to this is not known.

ACKNOWLEDGMENTS

I wish to acknowledge the many people, who have worked on invertebrates on uplands and whose work is incorporated into this review, particularly Dr W. Block and Professor W. G. Hale. I am indebted to Professor J. B. Whittaker, with whom I wrote the synthesis of the Moor House studies and which resulted in the exchange of many thoughts and ideas concerning the fauna of the uplands. I am also particularly indebted to Drs J. Butterfield and V. Standen, who have been co-workers in much of the work on uplands.

The data used in this paper have been collected over many years and I am indebted to the Nature Conservancy Council and to Natural Environment Research Council for grants and contracts which allowed much of this work to be carried out.

REFERENCES

Butterfield, J.E.L. (1976). The response of development rate to temperature in the univoltine crane-fly *Tipula subnodicornis* Zetterstedt. *Oecologia*, **25**, 89–100.

Butterfield, J.E.L. (1986). Changes in life-cycle strategies of *Carabus problematicus* over a range of altitudes in northern England. *Ecological Entomology*, **11**, 17–26.

Butterfield, J. & Coulson, J.C. (1983). The carabid communities of peat and upland grasslands in northern England. *Holarctic Ecology*, **6**, 163–174.

Coulson, J.C. (1962). The biology of *Tipula paludosa* Meigen. *Journal of Animal Ecology*, **31**, 1–21.

Coulson, J.C. & Butterfield, J. (1978). An investigation of the biotic factors determining the rates of plant decomposition on blanket bog. *Journal of Ecology*, **66**, 631–650.

Coulson, J.C. & Butterfield, J.E.L. (1985). The invertebrate communities of peat and grasslands in the north of England and some conservation implications. *Biological Conservation*, **34**, 197–225.

Coulson, J.C. & Butterfield, J. (1986). The spider communities on peat and upland grasslands in northern England. *Holarctic Ecology*, 9, 229–239.

Coulson, J.C. & Whittaker, J.B. (1978). The ecology of moorland animals. *Production Ecology of British Moors and Montain Grasslands* (Ed. by O.W. Heal & D.F. Perkins), pp. 53–93. Springer-Verlag, Berlin.

Coulson, J.C., Horobin, J.C., Butterfield, J. & Smith, G.R.J. (1976). The maintenance of annual life-cycles in two species of Tipulidae (Diptera); a field study relating development, temperature and altitude. *Journal of Animal Ecology*, 45, 215–233.

Gimingham, C.H. (1972). *Ecology of Heathlands.* Chapman & Hall, London.

Hobbs, R.J. & Gimingham, C.H. (1987). Vegetation, fire and herbivore interactions in heathland. *Advances in Ecological Research*, 16, 87–173.

Hodkinson, I.D. (1973). The population dynamics and host plant interactions of *Strophingia ericae* (Curt) (Homoptera:Pyslloidea). *Journal of Animal Ecology*, 42, 565–583.

Parkinson, J.D. & Whittaker, J.B. (1975). A study of two physiological races of the heather psyllid *Strophingia ericae* (Curtis) (Homoptera:Psylloidea). *Biological Journal of the Linnaean Society*, 7, 72–81.

Randall, M.G. (1982). The ectoparasitization of *Coleophora alticolella* (Lepidoptera) in relation to its altitudinal distribution. *Ecological Entomology*, 7, 117–185.

Ratcliffe, D.A. (Ed.) (1977). *A Nature Conservation Review, Vol.1.* Cambridge University Press, Cambridge.

Ratcliffe, D.A. & Thompson, D.B.A. (1988). The British uplands: their ecological character and international significance. *This volume.*

Standen, V. (1978). The influence of soil fauna on decomposition by micro-organisms in blanket bog litter. *Journal of Animal Ecology*, 47, 25–38.

Svendsen, J.A. (1957). The behaviour of lumbricids under moorland conditions. *Journal of Animal Ecology*, 26, 423–439.

Thompson, D.B.A., Galbraith, H. & Horsfield, D. (1987). Ecology and resources of Britain's mountain plateaux: land-use issues and conflicts. *Agriculture and Conservation in the Hills and Uplands* (Ed. by M. Bell & R.G.H. Bunce), pp. 22–31. Institute of Terrestrial Ecology, Grange-over-Sands.

Thompson, D.B.A., Stroud, D.A. & Pienkowski, M.W. (1988). Afforestation and upland birds: consequences for population ecology. *This volume.*

Usher, M.B. & Gardner, S.M. (1988). Animal communities in the uplands: how is naturalness influenced by management? *This volume.*

Spatial variations, patterns and management options in upland bird communities

P. J. HUDSON

The Game Conservancy, Crubenmore Lodge, Newtonmore, Inverness-shire, Scotland PH20 1BE

SUMMARY

1 Two ecological approaches are explored in relation to management options for upland birds.

2 The first identifies correlates of distribution and abundance. There are no clear regional trends in bird species richness or associations with soil fertility. Species abundance is loosely associated with fertility, habitat structure and arthropod-rich patches.

3 Small areas of rough pasture and the construction of bog flushes may help to increase arthropod abundance and densities of upland birds. However, these could alter other factors and influence community stability.

4 The second approach proposes the application of empirical models. A host–parasite model is applied to the management of red grouse (*Lagopus lagopus scoticus*) to reduce the impact of the nematode parasite, *Trichostrongylus tenuis*.

5 Reducing either the life expectancy of free-living stages through environmental management or the size of the adult worm population through chemotherapy appears untenable. Indirect chemotherapy and an improvement in the hen's plane of nutrition through reduced competition with other herbivores are suggested as possible control methods.

6 Although both approaches have their disadvantages, they should be considered complementary and not mutually exclusive.

INTRODUCTION

For all animal populations there is at least one stable equilibrium, and that is extinction. Most, however, have multiple equilibria determined by their relationships with other organisms (stability and equilibria are defined by May 1973). This is an obvious but fundamental principle for the applied ecologist, who must estimate the parameters that determine the boundaries between continued existence and the decline to extinction. The task of the environmental manager is to manipulate these parameters and keep populations far enough away from critical boundaries so that environmental fluctuations, or man-induced perturbations, will not take the population to extinction. With limited resources, the manager must ensure that efforts are concentrated in the correct direction.

The manager of a natural animal resource has the option of two general approaches. The first is to conduct survey work, identify through a correlative approach the features associated with abundance and then from these correlations develop the practical techniques needed to manipulate a population. Such an approach may be productive, but might fail to identify the key problems. For example, shooting wood pigeons (*Columba palumbus*) during winter months together with high rates of predation on their eggs has little effect upon population stability since these mortalities occur before the key mortality (Murton, Issacson & Westwood 1974). The second approach is to develop or apply fundamental, dynamical models. A number of general models on predation, parasitism, herbivory, etc. have already been developed (reviewed in May 1981) which have identified the important parameters that should be manipulated to influence stability or the position of a stable equilibrium. Applying such models will incorporate a number of assumptions, some of which may be unrealistic within the system under consideration. Developing a dynamical model for a specific system is preferable (Peterman, Clark & Holling 1979) but requires the collection and analysis of large quantities of data. It should be noted that these two approaches are not mutually exclusive but simply different. In the first approach data are collected for intuitive reasons while in the second approach the models, which are based on *a priori* reasoning, indicate the data to be collected and the subsequent analysis.

This paper looks at these two approaches and how they can be used in the management of upland birds. General correlates of abundance and distribution for upland birds are examined first and it is proposed that a general rule of increasing habitat diversity could result in increases in both species composition and abundance within delimited areas. Second, a generalized dynamical model is applied to a specific management problem: how to reduce the impact of nematode parasites on red grouse populations. The general tenet of this paper is that abundance and distribution of natural animal populations can be altered to satisfy the objectives of conservation, but it is necessary to pin-point which ecological parameters should be manipulated to ensure that environmental managers select the most effective management techniques.

PATTERNS IN UPLAND BIRD COMMUNITIES

There is clearly a wide range of factors that influence the distribution and abundance of each upland bird species. As Pearsall (1950) and Ratcliffe (1977, in press) note, harsh environmental features coupled with poor soil fertility in many areas of the uplands are likely to play a predominant role. This section looks at the general patterns of distribution and abundance against these environmental features, before proposing a general rule which can be applied to upland bird management.

The oceanic climate of the British Isles imposes a heavy rainfall on the uplands, particularly in western areas, where rainfall can exceed 450 cm a^{-1}. In combination with this heavy rainfall most of the uplands are on hard acidic rocks that have produced base-poor soils that suffer leaching. On wet, flat areas the ground becomes waterlogged and dead plant material accumulates to form deep peat. On these blanket-bog areas most nutrients enter the ecosystem through rain, and few originate from the underlying substrata. Temperature decreases with altitude and latitude while precipitation and wind increase westwards to produce more oceanic conditions. Thus, within each area of the uplands there is a series of altitudinal life zones comprising particular vegetation and animal communities which tend to get lower towards the north and west (Ratcliffe & Thompson, this symposium). This is shown clearly in the arctic-alpine (or montane) areas, e.g. ptarmigan (*Lagopus mutus*) are present at altitudes above 750 m in the Cairngorms but breed at only 200 m near Cape Wrath, Sutherland.

Regional variations in species richness

An increase in oceanity and a fall in temperature are generally associated with a fall in bird species richness. For woodland bird communities this is quite clear, with more than 50 breeding species in south-eastern England and less than 25 in north-western Scotland (Fuller 1982); this trend is also present before the data are corrected for area of woodland. This same pattern is not obvious in the upland bird communities. There is little variation in species richness, with the greatest number (41) being found in the central Scottish Highlands (Fig. 1). The uplands in the context include blanket bog and heather and grass moorland and the bird species associated with these habitats (after Fuller 1982). Species that use the uplands as a secondary site, nest alongside water or are associated with the montane plateaux are not included in Fig. 1 (see full list in Thompson, Stroud & Pienkowski, this symposium).

There are not fewer species in the less fertile areas of western Scotland compared with the relatively rich limestone areas in North Yorkshire. However, other factors may contribute to species richness. The Scottish Highlands have more life-zones than northern England (Ratcliffe & Thompson 1988), although the effect of this has been removed from the analysis in Fig. 1 by incorporating only species that breed in comparable habitats. The Highlands also include a larger contiguous block of moorland than the Pennines and Southern Uplands so that the expected positive relationship with soil fertility may be lost through a species–area relationship (summarized by Begon, Harper & Townsend 1986). Furthermore, examination of each species' breeding distribution indicates that the additional ones found in Scotland are those at the edge of their range (e.g. golden eagle, *Aquila chrysaetos*; greenshank, *Tringa nebularia*), although regional differences in past land-use history, such as variations in grazing pressure and burning, have probably contributed to this.

Overall, there is no clear trend between soil fertility and bird species richness.

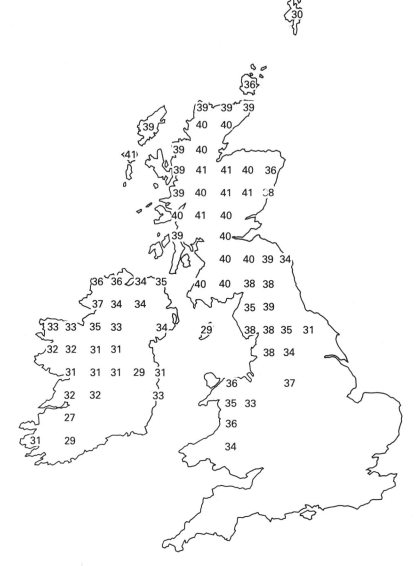

FIG. 1. Regional variation in species richness of upland birds expressed as number per 25 km²; total = 48 (after Fuller 1982). See text for species and habitat criteria. Data extracted from Sharrock (1976).

Such a relationship probably does not exist since many of the upland birds are insectivorous during the breeding season and arthropod availability is not simply associated with soil fertility. The total biomass (dry weight) of arthropods produced above ground is relatively low in upland habitats and is not greater on limestone grassland compared with the relatively infertile blanket bog (Coulson & Whittaker 1978; Coulson 1988).

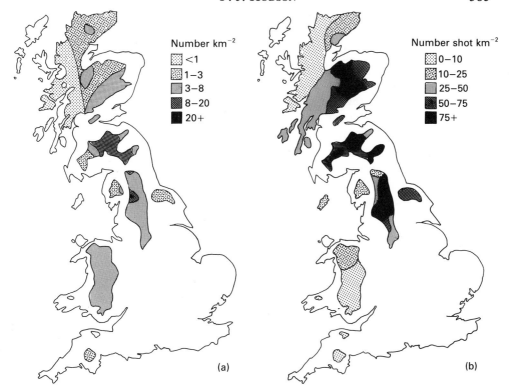

FIG. 2. (a) Breeding density (number km^{-2}) of golden plover (derived from Ratcliffe 1976). (b) Numbers of red grouse shot km^{-2} in the uplands (source: original data). Densities tend to be lower in the north and west and there is a weak positive association between the two species.

Regional variations in species abundance

Although there may be little variation in species composition between base-rich and base-poor areas there can be large variations in density. Ratcliffe (1976) recorded considerable variation in the density of nesting golden plover (*Pluvialis apricaria*), with 16 pairs km^{-2} on fertile, limestone grassland but less than 1 pair km^{-2} on the relatively barren moorlands of north-west Scotland (see Fig. 2a). Although the association between density and soil fertility is reasonable it fails to provide a total explanation; for example, in Scotland density on the relatively base-rich Breadalbane Hills is relatively low whereas the Moorfoots Hills carry a greater density than a comparable area in Dumfries-shire. Neither does the distribution necessarily reflect the abundance of arthropod food items. Ratcliffe (1976) found a lower density of golden plovers on blanket bog than on limestone grassland, yet Coulson & Whittaker (1978) and Coulson (1988) recorded an overall greater abundance of arthropods on blanket bog compared with limestone grassland. Nevertheless, earthworms (Lumbricidae) constitute a major part of the golden plover's diet and the availability of these within easy flight of nesting areas is likely to play an important role in influencing nesting density. In this respect,

Coulson & Whittaker (1978) recorded only 0·04 earthworms m^{-2} on blanket bog but 390 m^{-2} on nearby limestone grassland. Thompson (1984) compared invertebrate densities in blanket bog and improved pasture used by feeding golden plovers in northern Scotland and found significantly greater densities of coleopterans and lumbricids in the preferred pastures.

Although red grouse are herbivorous there is an association between densities of grouse and golden plover (Figs. 2a, b). Perhaps the base-rich areas producing nutritious heather shoots also have good food patches with earthworms. Historically, numbers of grouse may not have been associated with soil fertility since estates on the west coast shot as many grouse per unit area as estates on the east coast (Fig. 3). Although grouse numbers in all areas have fallen, there is a tendency for grouse numbers to have fallen at a greater rate in the west ($r = -0·50$,

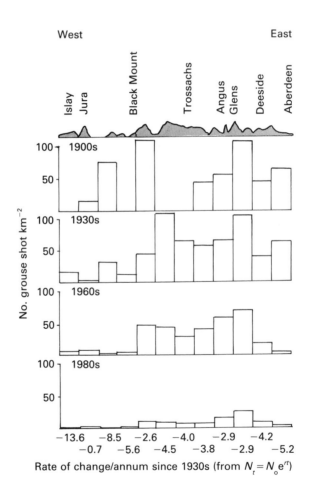

FIG. 3. Mean numbers of red grouse shot km^{-2} on an east–west transect across Scotland in different decades. While numbers of grouse have fallen in all areas there is a tendency for the rate of decline to increase towards the west. A blank space in the data set for the 1900s indicates absence of data.

$P < 0.10$, df $= 10$). Part of this decline can be attributed to the loss of heather habitat for grouse as a result of overgrazing by sheep (*Ovis aries*) and deer (*Cervus elaphus*) in combination with poor management practices. Even so, there has been an additional fall in density *per se* and areas of suitable habitat currently sustain a low density of grouse. Increased levels of predation from foxes (*Vulpes vulpes*) and crow (*Corvus corone*), coupled with prolonged leaching of nutrients from the soil leading to reduced food quality, are probably important contributory factors.

Habitat structure

A general pattern in many bird communities is an increase in species composition and density with increased heterogeneity in habitat structure (Fuller 1982). Upland habitats, such as blanket bog, heather moorland and sub-montane

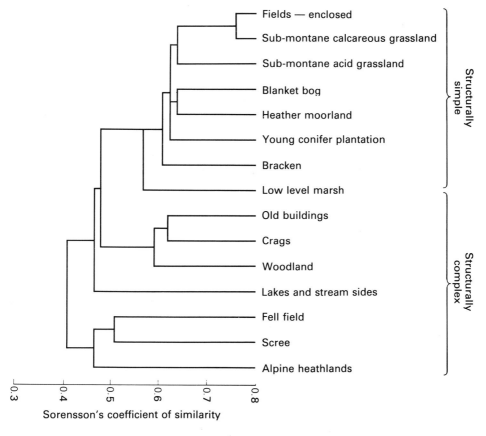

FIG. 4. Dendrogram indicating similarity between different habitat components in the upland bird community. The structurally complex habitats carry a different avifauna from the relatively simple habitats. The degrees of similarity are calculated using Sorensson's index. Data source: Ratcliffe (1977).

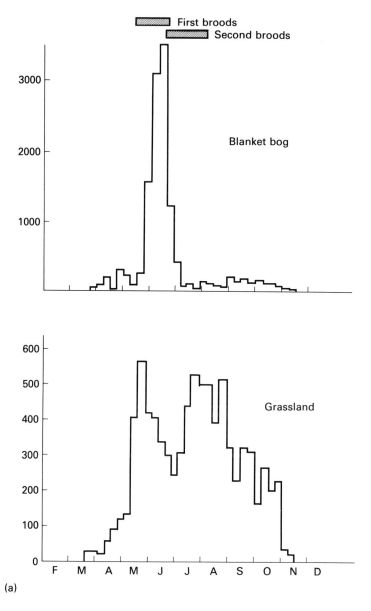

FIG. 5. (a) Changes in insect production from blanket bog and limestone grassland in relation to brood production (shaded). (b) Nesting dispersion of meadow pipits in relation to blanket bog (shaded) and mineral soils (unshaded). Nests are located at the interface of the two habitats and adults utilize insects from blanket bog for first broods and alluvial grassland for second broods. (Redrawn from Coulson & Whittaker 1978.)

grasslands, tend to be relatively homogeneous. An increase in habitat diversity naturally leads to an increase in species composition although the addition of structurally complex habitats such as crags, woodland, lakes and fellfield introduces a new range of species (Fig. 4). For example, crags on heather moorland can introduce peregrine falcon (*Falco peregrinus*) and stock dove (*Columba oenas*).

Increased habitat structure can also lead to higher densities. Red grouse tend to nest at greater densities on heather moorland when the ground is structurally more complex and where the heather has been burnt in small patches. Burning increases the abundance of the more nutritious young heather shoots and also the structural heterogeneity of the heather habitat, providing more shelter from predators and more 'edges' which cocks may use as reference points for territorial boundaries (Watson & Miller 1970).

A mosaic of habitats will increase the faunal composition of a moorland area and this can lead to an increase in the density of species. This was shown explicitly in a study of meadow pipits (*Anthus pratensis*) and their insect prey in the northern Pennines by Coulson (summarized in Coulson & Whittaker 1978). The breeding of meadow pipits on blanket bog is synchronized with the emergence of *Tipula subnodicornis*, a cranefly that constitutes 83% of the first broods' diet. However, emergence of *T. subnodicornis* lasts only a few weeks and has finished by the time the second broods are in the nest. Then the adults utilize arthropods found on alluvial grasslands (Fig. 5a). This switch in foraging sites is reflected in the distribution of the pipits' nests, which tend to be along the interface of the two habitats (Fig. 5b).

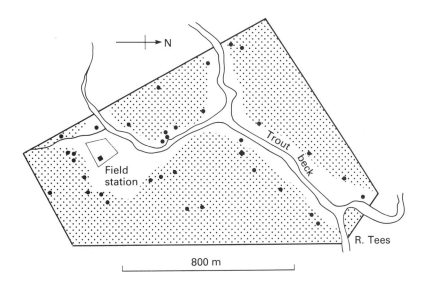

(b)

PATTERNS AND MANAGEMENT OPTIONS IN UPLAND BIRD COMMUNITIES

This brief summary of patterns within the upland bird community has concentrated on the interaction between general environmental features and both species composition and species density. There are, of course, other patterns within the upland bird community (e.g. in number of winter residents, preponderance of large species, altitudinal occurrence) but while these are ecologically interesting they generally cannot be altered by management. Compared with lowland communities the entire upland bird community appears as one of low species richness and low overall density (Ratcliffe & Thompson 1988). Often, areas of low species richness are evaluated as areas of low conservation value, although the inverse can be argued and generally species richness *per se* provides an insufficient measure of conservation value (Fuller & Langslow 1986). Similarly, population density is rarely used in the evaluation of ornithological sites. However, areas with low densities require large areas to be conserved and so tend to be threatened by fragmentation and local extinction (Usher 1986). Even so, most of the upland birds tend to have an aggregated distribution (Campbell 1983), and safeguarding an area of uplands for bird conservation requires the identification of areas where species density is both high and large enough to include a viable population (Thompson *et al.* 1988). The problem for the conservationists in this situation is how to maintain, or increase, density within an area and so retain a viable population, without adversely affecting other species of conservation value. In terms of management, the basic area resource is often predetermined and the object is to manipulate conditions to improve density within the resource (e.g. moorland surrounded by recently planted trees; Thompson *et al.* 1988).

Although soil fertility appears significant, it would seem that this alone is of less importance in determining species composition and density than is a mosaic of vegetation communities. For upland birds, the provision of invertebrate-rich patches or at least limited areas that provide an alternative feeding area is likely to benefit most of the waders, passerines and consequently raptors, within limits (Bibby 1988). Two ways in which invertebrate availability could be increased artificially involve (i) rough pasture and (ii) bog flushes.

Rough pasture

The availability of grassland areas close to blanket bog increases the period that invertebrates are available to upland birds and may also provide critical food types, such as earthworms (see above). Traditionally, areas within Scottish moorland have been improved by crofting tenants to produce agriculturally richer areas of grassland — nutrient-rich islands in a sea of nutrient-poor moorland. However, the area of croftland meadow has decreased as crofting interests have shifted from cattle to sheep, with a concomitant fall in hay and spread of bracken (*Pteridium aquilinum*). Duncan (1982, pers. comm.) reseeded moorland areas in Strathspey to produce artificial crofting fields of 1–2 ha close to blanket bog. There was an overall increase in spring density of golden plover over 5 years to a mean of 6·7 birds km^{-2} on improved areas, compared with only 1·5 km^{-2} on unimproved control areas.

Bog flushes

Bog flushes are arthropod-rich areas produced when spring water that has run over base-rich substrata emerges on to freely drained heather moorland and produces a suitable habitat for arthropods to lay eggs. Arthropod production from these areas is around 18 times greater than on nearby heather moorland (Hudson 1986a) and provides a rich source of food for upland birds. Bog flushes can also be produced when the rate of water flow along drainage channels (grips) is reduced with simple blockages. In areas without base-rich rock, artificial bog flushes can be constructed by excavating holes to which limestone chippings have been added, thus producing suitable nutrient-rich areas. Results in the first year after the construction of artificial bog flushes indicate production to be on average 26·0 arthropods per sweep net sample — significantly more than the 7·7 arthropods per sample from a control site.

STABILITY ANALYSIS, POPULATION DYNAMICS AND WILDLIFE MANAGEMENT

While the approach taken in the previous sections has looked for general ecological correlates with distribution and density and short-term management remedies, it has revealed little about the dynamic stability of the system. In some instances, slight environmental changes can result in large-scale changes in a population. For example, slight changes in the proportion of grasslands in merlin (*Falco columbarius*) territories can make large differences to merlin breeding success (Bibby 1988). In this respect, the development of models based on empirical findings has proved instructive in understanding the dynamical properties of populations. In some instances these have been applied to problems of wildlife management. One of the most enlightening approaches, pioneered in the ecological literature by May (1973), is first to examine the features of a dynamic model and then to determine the conditions that promote stability. Most population equations are non-linear and a locally stable equilibrium may not be globally stable; thus a significant perturbation of the system may result in equilibrium changes.

Applying this approach and some of the concepts developed from these models can be instructive to wildlife management since management often aims to change the position of a stable equilibrium or to move from one equilibrium to another. Such an approach has been used to produce predictive models for the management of the spruce budworm (*Choristoneura fumiferana*) and the Pacific salmon (*Oncorhynchus* spp.) (Peterman *et al.* 1979), as well as in the control of parasites (examples in Anderson 1982). This section presents an example of how a dynamical model of host–parasite interactions has been applied to the management of red grouse in the British uplands.

The grouse–Trichostrongylus tenuis *system*

The life cycle of the nematode parasite *Trichostrongylus tenuis* has been described by Hudson (1986a). It is direct and so consists of two basic populations: (i) the sexually mature worms living within the caeca of the red grouse and (ii) the free-living stages which pass out of the host in the caecal faeces (Fig. 6). The nematodes have been recorded in other gamebirds but, since these are relatively rare on heather moorland, the system effectively involves the red grouse as the only host. Parasite burdens do not increase within the host; infection occurs when the larval stages are ingested with food.

As with many parasite systems, the nematodes show an aggregated distribution within the host population, although in this system the degree of aggregation is relatively low (variance: mean ratio = 2·7) probably because of the high parasite burdens. Experimental manipulations of levels of infection have demonstrated that high parasite burdens reduce condition, survival (Hudson 1986a) and breeding production (Hudson 1986b) of grouse.

The dynamics of this system can be described using one of the models developed by May & Anderson (1978). The model consists of two differential equations that describe changes in the host and parasite populations respectively, and which are coupled through the effects of the parasite on the fecundity and mortality of the grouse. Details of this can be found in Hudson, Dobson & Newborn (1985), whilst Dobson & Hudson (in press) expand this first model to include a third coupled differential equation describing changes in the population of the free-living stages. Three important features of the system have been identified through stability analysis, namely

(i)   the life expectancy of the free-living stages,
(ii) the parasite's effects on grouse breeding, and
(iii) the parasite's effects on grouse survival.

When the life expectancy of the free-living stages is long and the sublethal effects on fecundity are greater than the effects on grouse survival, the population fluctuates in cycles. The period of these cycles is determined by the relative values of the free-living stages and intrinsic growth rate of the population, r. A fall in the survival rate of the free-living stages tends to reduce cycle length, but ultimately low values lead to the parasite being unable to sustain the adult parasite population. Stability analysis also provides an indication of how control techniques could be applied to reduce the impact of the parasites on the grouse population. The two parameters which could be reduced are, first, the life expectancy of the free-living stages (point (i) above) and, second, the size of the

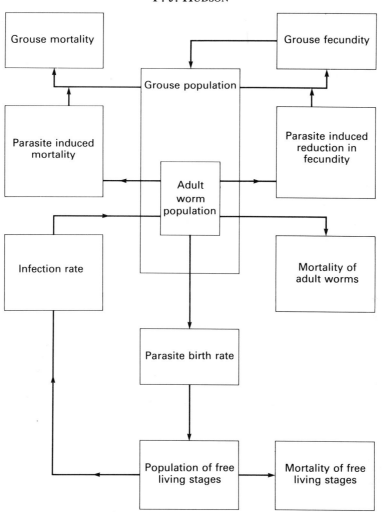

FIG. 6. Diagrammatic representation of the life cycle of *Trichostrongylus tenuis* in grouse. Further details are in the text and Hudson (1986a).

parasite population within the grouse population (the impact of (ii) and (iii) above). Both are now considered.

The life expectancy of the free-living stages

The free-living stages of nematodes are particularly susceptible to desiccation so that slight changes in humidity can influence the survival chances of the free-living stages. One possible way of reducing humidity would be to drain moorland areas, although in reality the efficacy of this technique is likely to be limited.

Moorland areas with rainfall that exceeds 1300 mm per annum have a blanket covering of peat which tends to act like a sponge, and drainage activities often fail to reduce the water-table (Coulson 1988). Drainage would also have other ecological consequences such as reducing the availability or abundance of arthropods, a factor which could reduce the survival of grouse chicks and abundance of other bird species (Coulson 1988).

Reducing the size of the adult worm population

Parasitic infections of man and his livestock are often reduced through the combined effects of chemotherapy and sanitation. In wild populations this approach is often impractical, although chemotherapy is attractive for macro-parasites since it is cheap and provides instant relief. On the other hand, the benefits are only short-term since treated animals are immediately susceptible to reinfection, the rate of which is dependent on the basic reproductive rate of the parasite, R. (Anderson 1981). In the grouse–*T. tenuis* system a technique for catching and treating grouse has been developed and it is feasible to catch large numbers for treatment. However, the value of R in the system has been estimated as lying between 5 and 10 (Dobson & Hudson in press) and if the objective of chemotherapy is to eradicate the parasite then every individual will need to be treated. This is clearly impractical. A more realistic approach would be to reduce average parasite burdens to below the point of significant impact (about 1000–2000 worms per bird) although the proportion that would need treating would depend on the initial parasite burdens (Fig. 7). Since the parasites are aggregated within the host population, selectively treating the heavily infected individuals would improve the efficacy of chemotherapy; from the distribution of parasites within the grouse population efficacy could be improved fourfold.

Indirect chemotherapy is possible in some systems and some success has been achieved by incorporating the anthelmintic with feed. Grouse rarely take any form of artificial feed although they do take grit as an aid to digestion. Current research has managed to incorporate an anthelmintic with a tallow fat and this mixture is used to coat quartz grit. Sufficient anthelmintic could be taken to reduce worm burdens, although lower levels may be sufficient to reduce survival of infective larvae soon after ingestion, or to damage adult worms and so reduce the worms' egg production. The combined and continuing effects of this technique may well provide a suitable control method.

The action of parasites is often synergistic with the effects of nutrition, so that when the two act together the total effect is greater than it would be if the two acted independently. In management terms, slight changes in the plane of nutrition can have dramatic effects on the equilibrium of the host–parasite relationship (Anderson 1981). In the grouse–*T. tenuis* system there is an association between the hen's plane of nutrition and breeding success. Competition with other moorland herbivores could be reduced to increase the availability

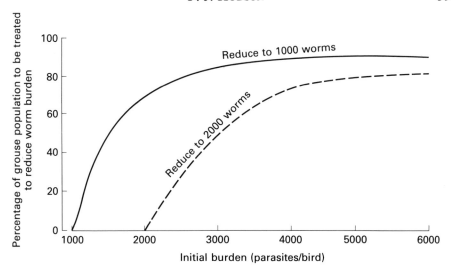

FIG. 7. The proportion of a grouse population that would need to be chemotherapeutically treated to reduce mean worm burdens to a level of 2000 and 1000 worms per bird (the level at which the parasite's impact would be insignificant).

of the nutrient-rich flowers of *Eriophorum vaginatum* in order to reduce relative pathogenicity of the parasite. This would increase the relative position of the stable equilibrium.

DISCUSSION

The object of this paper was to examine two different approaches to the management of upland birds. Both have certain advantages and disadvantages, and a combination of the two probably represents the most efficient approach within financial constraints. The major problem with correlative studies is that they may fail to identify the important features and instead lead workers to concentrate on the obvious. In particular, slight manipulative changes could alter the stability of a system and ultimately lead to local extinction. For example, increasing the abundance of meadow pipits by improving areas of grassland should lead to an increase in food availability for merlins. However, other features will also be influenced and these may ultimately result in a fall in merlin density (Bibby 1988).

The second approach, that of applying fundamental ecological theory to management problems, can help to identify the key parameters that should be measured during survey work and also the most effective approaches that should be taken in management. The example of the grouse–*T. tenuis* system suggested that management options consisted of reducing the survival of the parasite's free-living stages or improving the host's plane of nutrition. Moreover, all aspects of environmental management are concerned with changing the position of a population relative to equilibria, so identifying the boundaries between stable and unstable equilibria should make management techniques more efficient.

One of the principal problems associated with applying dynamical models is they often require detailed data obtained from a range of densities. In animal populations that show large fluctuations in numbers or have economic import-ance, such data can be readily available. However, for some of the specialized and rare species such data are often poor or lacking. Nevertheless, workers like Lack (1954) have stimulated long-term population studies of many species of birds, and these detailed data, coupled with extensive survey data, should allow the fundamental models to be applied to management problems. Moreover, the results of management procedures should be monitored carefully, since they provide additional experimental information which can lead to more robust models. The principle of 'adaptive management procedures' is not new to fisheries (Walters & Hilborn 1978), and some workers go so far as to propose over-harvesting strategies in order that additional data can be obtained to make the models more robust. Further problems and the application of dynamical models are discussed by Peterman *et al.* (1979).

In conclusion, I propose that the traditional techniques of ecological surveys, correlations and application to conservation problems have a role but may fail to provide the most effective solutions. In contrast, the application of empirical models provides a constructive approach which can pin-point the specific parameters that management should aim to alter. Such an approach should help conservationists to manipulate conditions and keep populations away from the critical boundaries that could lead to extinction.

ACKNOWLEDGMENTS

I would like to thank Dick Potts, Kenny Taylor, Stephen Redpath, David Hill and Des Thompson for discussion and comments on this paper. I would also like to thank the anonymous referees for making constructive comments on the first draft.

REFERENCES

Anderson, R.M. (1981). Population ecology of infectious disease agents. *Theoretical Ecology, Principles and Applications* (Ed. by R.M. May), pp. 318–355. Blackwell Scientific Publications, Oxford.
Anderson, R.M. (1982). *Population Dynamics of Infectious Diseases. Theory and Applications.* Chapman & Hall, London.
Begon, M., Harper, J.L. & Townsend, C.R. (1986). *Ecology. Individuals, Populations and Commu-nities.* Blackwell Scientific Publications, Oxford.
Bibby, C.J. (1988). Impacts of agriculture on upland birds. *This volume.*

Campbell, L. (1983). Characteristics of the moorland breeding bird community with particular implications for nature conservation. *Recreation Ecology Research Group Report*, **8**, 55–65.

Coulson, J.C. (1988). The structure and importance of invertebrate communities on peatlands and moorlands, and effects of environmental and management changes. *This volume.*

Coulson, J.C. & Whittaker, J.B. (1978). The ecology of moorland animals. *Production Ecology of British Moors and Mountain Grasslands* (Ed. by O.W. Heal & D.F. Perkins), pp. 52–93. Springer-Verlag, Berlin.

Dobson, A.P. & Hudson, P.J. (in press). The population dynamics and control of the parasitic nematode *Trichostrongylus tenuis* in red grouse in the north of England. *Applied Population Biology.*

Duncan, J. (1982). Sheep tick project. *Game Conservancy Annual Review*, **13**, 107–110.

Fuller, R.J. (1982). *Bird Habitats in Britain.* Poyser, Calton.

Fuller, R.J. & Langslow, D.R. (1986). Ornithological evaluation for wildlife conservation. *Wildlife Conservation Evaluation* (Ed. by M.B. Usher), pp. 247–270. Chapman & Hall, London.

Hudson, P.J. (1986a). *The Red Grouse. The Biology and Management of a Wild Gamebird.* Game Conservancy Trust, Fordingbridge.

Hudson, P.J. (1986b). The effects of a parasitic nematode on the breeding production of red grouse. *Journal of Animal Ecology*, **55**, 85–92.

Hudson, P.J., Dobson, A.P. & Newborn, D. (1985). Cyclic and non-cyclic populations of red grouse: a role for parasitism? *Ecology and Genetics of Host–Parasite Interactions* (Ed. by D. Rollinson and R.M. Anderson), pp. 77–90. Academic Press, London.

Lack, D. (1954). *The Natural Regulation of Animal Numbers.* Oxford University Press, Oxford.

May, R.M. (1973). *Stability and Complexity in Model Ecosystems.* Princeton University Press, Princeton.

May, R.M. (Ed.) (1981). *Theoretical Ecology. Principles and Applications.* Blackwell Scientific Publications, Oxford.

May, R.M. & Anderson, R.M. (1978). Regulation and stability of host–parasite population interactions. Part II. Destabilising processes. *Journal of Animal Ecology*, **47**, 249–268.

Murton, R.K., Issacson, A.J. & Westwood, N.J. (1974). A study of woodpigeon shooting: the exploitation of a natural animal population. *Journal of Applied Ecology*, **3**, 55–96.

Pearsall, W.H. (1950). *Mountains and Moorland.* Collins, London.

Peterman, R.M., W.C. Clark & C.S. Holling (1979). The dynamics of resilience: shifting stability domains in fish and insect systems. *Population Dynamics* (Ed. by R.M. Anderson, B.D. Turner & L.R. Taylor), pp. 321–342. Blackwell Scientific Publications, Oxford.

Ratcliffe, D.A. (1976). Observations on the breeding of the Golden Plover in Great Britain. *Bird Study*, **7**, 81–93.

Ratcliffe, D.A. (1977). Uplands and birds — an outline. *Bird Study*, **24**, 140–158.

Ratcliffe, D.A. (in press). *Upland Birds.* Cambridge University Press, Cambridge.

Ratcliffe, D.A. & Thompson, D.B.A. (1988). The British uplands: their ecological character and international significance. *This volume.*

Sharrock, J.T.R. (Comp.) (1976). *The Atlas of Breeding Birds in Britain and Ireland.* Poyser, Berkhamsted.

Thompson, D.B.A. (1984). *Foraging economics in flocks of lapwings* (Vanellus vanellus), *golden plover* (Pluvialis apricaria) *and black-headed gulls* (Larus ridibundus). Ph.D. thesis, University of Nottingham.

Thompson, D.B.A., Stroud, D.A. & Pienkowski, M.W. (1988). Afforestation and upland birds: consequences for population ecology. *This volume.*

Usher, M.B. (1986). Wildlife conservation evaluation: attributes, criteria and values. *Wildlife Conservation Evaluation* (Ed. by M.B. Usher), pp. 3–44. Chapman & Hall, London.

Walters, C.J. & Hilborn, R. (1978). Ecological optimization and adaptive management. *Annual Review of Ecology and Systematics*, **9**, 157–189.

Watson, A. & Miller, G.R. (1970). *Grouse Management.* Game Conservancy Trust, Fordingbridge.

Assessing and managing the effects of recreational use on British hills

N. G. BAYFIELD, A. WATSON AND G. R. MILLER

NERC Institute of Terrestrial Ecology, Banchory, Kincardineshire AB3 4BY

SUMMARY

1 Recreational activities have had substantial impacts on heavily used hills. The most widespread impact is probably the proliferation of footpaths, many of which have greatly deteriorated in condition.

2 In Scotland, there has been considerable penetration of previously remote ground by bulldozed vehicular tracks, and the number and size of ski areas has increased steadily. These developments have had substantial adverse visual impacts as well as disturbing soils, plants and animal life.

3 The management of impacts may require some manipulation of use and the development of techniques to reinstate damaged upland plant communities.

4 To assess trends and provide a rationale for management responses, it is important to develop programmes for monitoring change in resource and use characteristics. Examples of approaches to monitoring are discussed, together with the need for monitoring frameworks that provide a direct input into management.

INTRODUCTION

This paper describes some of the impacts of heavy recreational use on British hills and identifies a need for more monitoring to establish trends of change and for monitoring to be more closely related to possible management responses.

During the last two decades, the impacts of recreational activities on upland landscapes have become increasingly noticeable. Car-parks, ski centres, long-distance footpaths and other facilities have all tended to concentrate use. In many cases, heavy use has resulted in conspicuous wear, for example along footpaths, and around the pistes and tows of ski areas. Until now, resources to manage upland sites to minimize or alleviate these impacts have tended to lag behind the provision of access. One reason is that the data available to planners tend to be concerned mainly with the use of roads and buildings (Fishwick 1985). Few local authorities have comparable data on numbers of walkers or other countryside visitors. Data on changes in site condition are also uncommon, and it is difficult to justify expenditure on management if the extent and pattern of resource deterioration are largely unknown.

In the few cases where monitoring has been undertaken, the data collected have often been of little value because of failure to clarify objectives, selection of inappropriate methods, or a lack of any formal mechanism for the results to influence management. The last few years, however, have seen instances of monitoring schemes that include broad consultation to ensure relevance and that provide a direct input to management. Such schemes provide new opportunities for applying ecological information to land use planning and management.

SOME EXAMPLES OF RECREATIONAL IMPACT

Footpaths

The most widespread impact is probably on footpaths. A number of workers in Britain and overseas have recorded wear along footpaths, and related damage to soil, terrain, vegetation and use (Cole & Schreiner 1981). It is clear that footpaths deteriorate with heavy use, and research in several countries shows that certain types of vegetation and soils are more vulnerable to disturbance than others (Pryor 1985; Emmanuelsson 1984; Grabherr 1985; Hammitt & Cole 1987).

When deciding on management action it is, however, important to know the rate as well as the cause of deterioration. For instance, is the level of use continuing to increase, and is damage continuing to increase or has it peaked? Only a few studies have attempted this type of analysis. One retrospective approach is to analyse footpaths using aerial photographs or maps. Using a combination of this technique and ground observations, Watson (1984a) demonstrated a large increase in the total length of footpaths in the Cairngorms massif, Scotland, since 1946, with the greatest increases occurring in heavily visited areas (Fig. 1). Bayfield (1986) recorded changes in the width of some of the paths noted by Watson, making ground measurements between 1969 and 1985. He found a continuing increase in path widths on routes that were relatively recent in origin (mainly since development of the Cairn Gorm area for skiing in the early 1960s). However, he found little change on older, traditional paths over a period of ten years. These differences between recent and traditional paths have quite important management implications, since a wide but stable path will be less likely to need management than one of the same width that is growing wider each year. Unfortunately, with the exception of some work by the Nature Conservancy Council (NCC) there has been no continued monitoring of use of these Cairngorms footpaths, so it remains uncertain whether the continuing deterioration is a response to increasing use, or to constant or even falling use.

Concern has been expressed about the condition of footpaths in several other upland areas of Britain. Deterioration in the Three Peaks, Yorkshire Dales, has been particularly serious, with mean trampled widths of 11·4 m and bare widths of 2·7 m for the 20 km path network (Bayfield & McGowan 1986). These widths are two to three times greater than for footpaths in the Cairngorms, at Stac Polly in Sutherland, and on the Pennine Way, northern England, recorded in 1983 (Bayfield 1985).

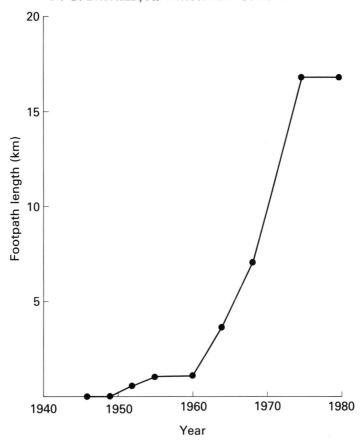

FIG. 1. Increases in footpath length on the Cairn Gorm–Ben Macdui plateau from 1946 to 1980 (after Watson 1984a).

Bulldozed hill tracks

Bulldozed hill tracks giving easy access to remote ground have proliferated since about 1960. A few tracks have been well made, but most are visually intrusive ribbons of bare ground and piles of spoil.

Although causing dismay to many hill walkers and conservationists, bull-dozed tracks are seen by landowners as essential for the economic management of hill areas. In assessing the impact of such tracks and considering management options, it is important to know the extent of the problem at different dates and the apparent trends for the future. Watson (1984b) noted a large expansion in the total length of tracks between 1960 and 1973, but a slower increase since then. If the main period of construction is over, then the main management concern should be to minimize the impact of existing routes and consider closing off and

rehabilitating those in the most outstanding areas of scenery or nature conservation. Alternatively, if, as seems to be the case, new routes continue to be built,
there is reason for greater guidance and control over their construction and siting
(Thompson, Galbraith & Horsfield 1987), and over the reinstatement of
damaged ground (Countryside Commission for Scotland (CCS) 1986). Complete
reinstatement is often assumed to be out of the question, but we have seen
examples of reinstatement of ski roads on slopes of up to 45° in the Western USA;
it ought to be technically feasible in Britain. All-terrain vehicles also damage
some of the more fragile plant communities, particular those associated with
peat, and their increasing use may pose localized problems in the future.

Ski resorts

Evidence of damage

Commercial provision of downhill skiing in Scotland dates from 1961 at Cairn
Gorm and has been expanding ever since. There are great fluctuations in demand
from year to year, but a recent estimate is that growth in demand at Cairn Gorm
is about 13 per cent (average of 27·6, 1·1 and 9·2 per cent) per annum over a
period of three years (ASH 1986). In the early years, the most severe damage was
caused by construction activities and the bulldozing of pistes (Watson, Bayfield
& Moyes 1970). Bare ground was exposed, and some of it eroded, depositing
sediment on to vegetation downslope. The visual impact of the damage was
substantial, and most buried species were unable to grow through more than 5 cm
of sediment (Bayfield 1974). Since the late 1960s there have been major changes
in the design and management of ski areas. New tows and other installations are
now placed by helicopter, and care is taken to avoid creating bare ground.

Some of the changes in the way that resorts are managed have, however,
resulted in new forms of impact. Hollows are no longer bulldozed to retain snow,
and pistes are created by lines of snow fences to catch drifted snow, and by the use
of piste machines to 'groom' and redistribute snow. Piste machines should not be
driven across snow-free vegetation, but this practice is difficult to control, and it
is quite common for machines to cross exposed vegetation when snow cover is
patchy. Piste machines have broad rubber tracks with alloy grips, and these can
tear up vegetation and cut grooves in soil. Ski edges can also slice off vegetation
when snow cover is patchy, but they have generally had much less impact than
piste machines. Damage in recent years has been particularly severe at the Lecht
ski area, where the extensive deep peat and heathery vegetation are vulnerable to
disturbance (Lecht Ski Company 1986).

Snow fencing has two types of impact. First, there is often local damage to
vegetation during construction, even when materials are delivered by helicopter.
Second, more extensive impacts on vegetation can result from changes in the persistence of snow cover and from the very dense snow created along pistes. These
changes alter the water regime of affected surfaces, and create conditions more
favourable for development of snow mould on heather. Watson (1985a) has

recorded up to 32 per cent dieback of heather (*Calluna vulgaris*) due to snow mould (*Fusarium* spp.) on pistes at the Lecht. Snow mould is common at high altitudes, and figures of around 10 per cent are quite typical of high-altitude ground unaffected by ski developments, but 32 per cent is so high that the heather there is unlikely to be able to make good the damage by new growth. After a few years, the heather tends to be replaced by more resilient grasses and blaeberry (*Vaccinium myrtillus*).

Impacts of snow fencing appear to be relatively unimportant on the continent and in the USA, where ski resorts are mainly in the forest zone and comparatively little use is made of snow fencing.

Impacts on birds have been studied by Watson (1979), who found that the population density and breeding success of ptarmigan (*Lagopus mutus*) and red grouse (*Lagopus lagopus scoticus*) were as high on heavily used ski grounds as on nearby areas that were seldom visited by people. However, both species bred unusually poorly at Cairn Gorm, where crows (*Corvus corone*), attracted to the high ground by tourists' food scraps, have robbed many nests. The number of breeding ptarmigan has been virtually nil in recent years on the highly developed ski grounds at Coire Cas, due to birds flying into the many overhead wires (Watson 1982) and possibly to less recruitment. Other birds do not seem to have been adversely affected. The number of the rare snow buntings (*Plectrophenax nivalis*) summering on the plateau has been no less (Milsom & Watson 1984) than in years before the ski developments (Nethersole-Thompson 1966). Naturalists have worried about whether the dotterel (*Charadrius morinellus*) would be affected by more people on the hills. However, the evidence from a 20-year study shows no effect of increased human impact on dotterel numbers or breeding success so far, probably because of the bird's exceptionally confiding behaviour when people are near (Watson, in press). R. C. Welch (pers. comm.) found very low numbers of craneflies (*Tipula* spp.) and other invertebrates on heavily trampled ground at Cairn Gorm in comparison with seldom visited ground (Watson 1976). Perhaps because of such reductions in invertebrates, there is some evidence that dotterel tend to avoid severely damaged patches (Watson, in press).

Indirect impacts

At Cairn Gorm, easy access to high ground by chairlift has greatly increased recreational use of the Cairn Gorm–Ben Macdui plateau and other high ground lying beyond the ski grounds. Increases in path length and wear have already been mentioned. Watson (1985b) has also found evidence of more bare ground, soil erosion and damage to vegetation in heavily visited parts of the plateau; and Bayfield, Urquhart & Cooper (1981) detected slight but statistically significant damage to lichens up to 50 m from footpaths. Because of such impacts and the high nature conservation status of the plateau area, the Nature Conservancy Council (NCC) initiated a mountain plateau research and monitoring project in 1987, which includes the recording of both impact and use (Thompson *et al.* 1987).

Possible beneficial impacts

Sometimes a by-product of the development of an area of recreation is an increase in its ecological interest. Large flocks of snow buntings are attracted to ski areas in winter, where scraps of food left by visitors attract the birds. Food scraps at car-parks and lay-bys also attract breeding pied wagtails (*Motacilla alba*) in summer (Watson 1979). These are perhaps marginal benefits. Scraps also attract crows and gulls (*Larus* spp.), which can prey on the eggs of hill birds such as red grouse and ptarmigan.

A better example is the regeneration of montane scrub, up to about 750 m, near the Cairn Gorm ski grounds. Montane scrub on open hill land is very rare in Scotland because of widespread grazing and burning (Miller & Cummins 1982; Sydes & Miller 1988). Ground below about 450 m on Cairn Gorm was fenced for afforestation, so preventing its use as a sheltered winter feeding area by hill red deer (*Cervus elaphus*); deer from surrounding areas rarely visit the ground above the fence-line, probably because of the high level of disturbance by visitors. The low browsing pressure and an absence of burning since the early 1950s have permitted the partial development of montane scrub. The principal colonizing species is Scots pine, *Pinus sylvestris*, with smaller numbers of birch (*Betula* spp.) and other deciduous species (Miller 1986). As yet, the scrub development is vestigial, with most trees less than 20 years old and at comparatively low density, but there are many thousands of trees in total, and the extent of colonization (several square kilometres) is of great scientific and nature conservation interest.

MANAGING RECREATION-RELATED IMPACTS

Regulation of use

In North America, where there is extensive public ownership of recreational areas such as National Parks, regulation of use by charging admission or by limiting the numbers of visitors is a common management tool (Hammitt & Cole 1987). It is often assumed that the restriction of use is not an option in most upland parts of Britain because of the extensive network of traditional rights of way and a strong lobby for public access. In fact, restriction by limiting parking, by not advertising routes, and by other covert techniques is quite commonly used, and seems likely to increase in future. At the proposed ski resort at Aonach Mor (near Fort William, west Scotland), for example, increased visitor use of the summit ridge would be likely to damage the vegetation and disturb dotterel. If the resort goes ahead, the most likely policy for access is to discourage use by (i) erecting signs discouraging use, (ii) interpretive displays explaining the sensitivity of the mountain vegetation and birds, and (iii) the construction of waymarked walks to two lower spurs with good views (Sgurr Finnisg-aig and Meall Beag).

Footpaths are increasingly being surfaced with gravel, stone or wood, enabling them to carry heavy traffic without much maintenance. Such reinstatement programmes are under way in Snowdonia (Wales), in the Peak District, the Lake

District and the Yorkshire Dales (England) and in several Scottish mountain areas, using a variety of techniques (Archer 1985; Aitken 1987; Allinson 1987; Speakman 1987). Good walking surfaces help to concentrate use and may then permit the closure of superfluous paths (Bayfield & Bathe 1982). It is, however, worth noting Cole's (1985) observation on wear at trampled sites in the Western USA. He concluded that even very light traffic can cause damage, and the restriction of use may be effective only if use can be reduced to very low levels or completely eliminated.

Interestingly, on two Cairngorms paths where use has substantially fallen, Bayfield (1986) noted marked declines in footpath widths (Fig. 2). It seems that

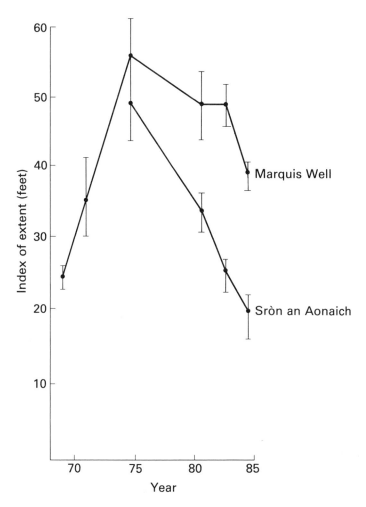

FIG. 2. Changes in width ('index of extent' = total width + bare width) on the Marquis Well path (heavily used to 1975, and lighter use since then) and Sron an Aonaich path (falling use since 1960, now little used) in 1969–85. Bars show S.E. (after Bayfield 1986).

progressive colonization of bare ground can occur even up to 1200 m, provided that disturbance is at a low level.

Reinstatement of damaged vegetation

If further disturbance can be prevented or minimized, then lightly damaged vegetation may recover completely within a single season, and no intervention may be required. However, when there is little or no intact vegetation, recovery must be much more prolonged. Factors that can militate against rapid spontaneous colonization include infertile soils, inclement climate, damage to or removal of organic matter, surface conditions being stony, gravelly or crusted due to rain splash, and a shortage of viable seed. Nevertheless, Bayfield, Urquhart & Rothery (1984) found progressive colonization of the verges of bulldozed hill tracks in the Cairgorms even about 1000 m, and concluded that at most sites the vegetation cover should be nearly complete after two or three decades, except on very steep or unstable surfaces. Similarly, gradual colonization has been noted on bared ground at arctic oil-well drilling sites (Hernandez 1973), on high-altitude road verges in the Rocky Mountains (Greller 1974) and on old paths in Lapland (Emmanuelsson 1984).

It is often desirable to speed up colonization so as to minimize visual impacts and reduce soil erosion. At Cairn Gorm, agricultural grasses have been sown successfully up to nearly 1200 m (Bayfield 1980, 1986), but the species are not those that occur naturally on the hill. Although invasion of sown swards by native species is fairly rapid below 750 m, with almost complete replacement of sown species after 8 years, it is much slower on higher ground. It is clearly preferable to try reinstating native species where feasible, for nature conservation reasons, for the appearance of the site, and because local species should be better adapted to local conditions. *Deschampsia flexuosa* is a useful species for sowing on peat (Gore & Godfrey 1981; Tallis & Yalden 1983) and *Festuca ovina, Agrostis tenuis* and *Poa annua* have also been used successfully (Gemmell & Crombie 1978; Phillips, Yalden & Tallis 1981). In recent trials on damaged footpath margins in the Yorkshire Dales, Bayfield & Miller (1986) found good initial establishment of mixtures of *Festuca ovina, Deschampsia flexuosa, Nardus stricta, Agrostis tenuis* and *Anthoxanthum odoratum*. These were all species found in adjacent, undamaged ground, and the cover produced was similar to that provided by an agricultural mixture like that used on the Cairn Gorm ski slopes (Table 1). Seeding may thus be capable of producing a sward not too dissimilar in composition to that on the surrounding ground. It is, however, likely to be a uniform sward, whereas pronounced patterning and heterogeneity are typical in much moorland vegetation. In the Yorshire Dales and Scotland, we are examining the suitability of common moorland plants for transplanting as a means of rapidly creating pattern in reinstated ground. Transplanting can also be used to redistribute patches of residual surviving vegetation in damaged ground. In preliminary trials, *Nardus stricta, Agrostis tenuis* and *Juncus effusus* have

TABLE 1. Comparison of cover produced by two 'native' seeds mixtures (I, N) and one mixture of 'agricultural' species (G), in fenced enclosures on Simon Fell, Ingleborough. Seeding was carried out in July 1986, and recorded in September 1986 (after 2·5 months) and May 1987 (after 10·0 months). Data on point quadrat estimates of cover (back transformed from angular transformation, based on 100 drops of a 3 mm point) with four replicates of each mixture. 95 per cent confidence limits are given in parentheses

Seed mixture	Composition (by weight)	% Cover Autumn 1986	% Cover Spring 1987
I	*Deschampsia flexuosa* 27%, *Festuca ovina* 27%, *Agrostis tenuis* 10%, *Anthoxanthum odoratum* 34%, *Potentilla erecta* 1%, *Campanula rotundifolia* 1%	91 (82–97)	85 (75–93)
N	*Deschampsia flexuosa* 70%, *Nardus stricta* 14%, *Agrostis tenuis* 14%, *Potentilla erecta* 1%, *Campanula rotundifolia* 1%	90 (80–96)	80 (69–90)
G	*Festuca rubra* 48%, *Poa pratensis* 18%, *Agrostis tenuis* 15%, *Cynosurus cristatus* 18%	90 (81–96)	84 (74–93)
Control		31 (19–44)	12 (5–21)

been found to transplant well, and benefit from dressings of fertilizer. Others, including *Juncus squarrosus*, did not transplant so satisfactorily; *Eriophorum angustifolium* was a failure.

A third approach to reinstatement is to make use of any buried seeds present in the soil. In the Yorkshire Dales, two disturbed peat sites used for reinstatement trials were found to have buried-seed reservoirs of about 20,000 seeds m^{-2} (mainly *Juncus effusus*), and germination of these seeds contributed up to 30 per cent cover on some plots. We hope to explore ways in which germination and establishment might be enhanced, including the use of nitrogenous fertilizers and the retention of soil moisture at the surface by using alginates and polymers. The size and distribution of any buried-seed reservoir is of course crucial to success or failure. We are currently surveying the extent of buried-seed reservoirs in different plant communities with various levels of disturbance (for the Yorkshire Dales National Park Committee).

A further technique for speeding reinstatement is to use fertilizer dressings to increase the growth of intact vegetation. This is clearly applicable only where there is some residual vegetation, but it may be a means of increasing the resilience of vegetation where disturbance is continuing. In the Yorkshire Dales we have found mean reductions of up to 62 per cent in bare ground on paths treated with a slow-release fertilizer (Table 2). The fertilizer tended to boost the growth of grasses such as *Festuca ovina* and *Agrostis* spp., with little response by sedges, *Nardus stricta* and mosses (Musci). Similar patterns of response to fertilizer by mountain vegetation in Norway have been reported by Garmo (1986).

TABLE 2. Effects of fertilizer (Enmag: 4·5:20:10, NPK) on the percentage cover of bare ground on sections of footpath in continuing use on Simon Fell, Ingleborough. Fertilizer was applied in July 1986, and cover analysed in May 1987. The effect of fertilizer was highly significant (Friedman Test, $P = 0·0008$)

Path section	Fertilizer $(g\ m^{-2})$			Principal species on unfertilized plots
	0	35	105	
1	28	5	4	*Festuca ovina, Eriophorum* spp., mosses
2	29	20	12	Mosses, *Festuca ovina, Eriophorum* spp.
3	27	11	10	*Festuca ovina, Carex nigra, Eriophorum* spp.
4	30	27	15	*Eriophorum* spp., *Festuca ovina,* mosses
5	32	20	14	*Nardus stricta, Festuca ovina, Juncus squarrosus*
Mean	29	17	11	

THE NEED FOR IMPROVED MONITORING FRAMEWORKS

Many of the uplands are used for purposes besides recreation, and conflicts of interest are common (Thompson *et al.* 1987). At Cairn Gorm, for example, which is the largest ski resort in Scotland, part of the mountain range is owned by the Royal Society for the Protection of Birds (RSPB) and part by the Highlands and Islands Development Board (HIDB) while part is in private ownership. The NCC, Forestry Commission (FC), CCS, district and regional councils, a farmer and a reindeer herder all have an interest in different aspects of the management of the ground. Then there are the various user groups, such as climbers, skiers, walkers, bird watchers, geologists, educationists and others. This is a particularly complex example, but by no means unique. Monitoring is useful to all groups, by helping to evaluate the extent and change in (i) the nature of site resources, (ii) impacts on these resources, (iii) the pattern of use on the site, and (iv) the effectiveness of management. In practice, it is usually feasible to monitor only a few key characteristics, but these should ideally take account of the broader context and tackle issues of general interest if they are to be of value to management. Schemes should contribute to decisions by managers and also indicate the effectiveness of management actions. We can thus draw up a checklist of desirable features:

(i) Monitoring should tackle issues identified by a consensus of interested parties, and not just by individuals or single agencies.
(ii) The methods and frequency of recording should be agreed by consensus.
(iii) There should be a regular review of results, to decide on necessary management responses and to evaluate the continuing relevance of the methods in use. Issues change, so flexibility of response is essential.
(iv) Where feasible, monitoring should be linked to management action so that detrimental changes trigger appropriate management responses.

Some recent examples of the application of these principles are to be found at Scottish ski resorts, where there has been a substantial change in attitudes to monitoring. Prior to 1981, some recording of characteristics such as path widths and bird numbers was undertaken by ecologists, but this was independent of the management of the ski resort. Since then, the value of monitoring has been increasingly recognized, and there are now formal monitoring schemes under way or proposed for most resorts. The following sections describe two examples.

Lecht monitoring

At the Lecht ski centre, a monitoring scheme has been in operation since 1984 as one of the conditions of planning approval. The scheme incorporates both broad consultation and a regular review of results. The monitoring is aimed principally at assessing trends in the extent of bare ground, major impacts on vegetation and soils, and the effectiveness of remedial work. There is a consultative working group consisting of representative of the Lecht Ski Company, CCS, NCC, Grampian regional council and Gordon and Moray district councils, and Grampian police. This group deals with general problems at the Lecht, and not just with monitoring. The monitoring procedure is as follows:

(i) In spring when the snow cover has gone, the ground is inspected with the ski company manager, problems discussed, remedial work agreed, and a short report made available immediately;
(ii) In the autumn, detailed recording of damage and other features is undertaken, the effectiveness of the summer's remedial work assessed, and the overall condition of the hill reviewed in a detailed report; and
(iii) Both reports are circulated to all those on the working group other than the police. Members of the group may discuss results, consider the implications for further proposals for snow-fence or tow installation, and suggest adjustment to the methods or frequency of recording. The ski company pays for the costs of the work.

One of us (AW) acts as an independent ecological consultant, as a member of staff of ITE, undertaking the spring inspection, supervising the detailed recording in autumn, and compiling both reports. So far this scheme has worked well, and a similar arrangement started at Glenshee in autumn 1986.

Aonach Mor monitoring

Aonach Mor is a proposed new ski area for which planning approval has been obtained but where construction has not yet started. Planning conditions have been stringent, and management, development and monitoring plans are a requirement of approval. The monitoring scheme, agreed in principle, is a little more complex than at the Lecht, and aims to record changes in use, visual appearance, snow lie and other characteristics, as well as impacts on vegetation and soils.

Possible changes in the hydrological characteristics of the hill are an additional concern, arising from the abstraction of water from the catchment for the generation of hydroelectric power at the British Alcan factory in Fort William.

As at the Lecht, there is a provisional working group of interested parties. The scheme is based on proposals in the CCS's (1986) management handbook for ski areas, which calls for broad consultation to identify issues to be tackled by monitoring and to agree on the methods and frequency of recording. It also suggests the setting of agreed quality standards for key parameters being monitored, based on the 'limits of acceptable change' (LAC) concept which was developed by the USDA Forest Service for wilderness management (Stankey *et al.* 1985). The purpose of these standards is to identify threshold levels at which management action should be triggered. These trigger points (the LAC values) represent a consensus view, not of what is ideal, but of what is acceptable in the light of current circumstances. When exceeded, they should automatically trigger a management response which is agreed in advance by the working group. LAC values and responses need to be flexible so as to match changing circumstances, and should therefore be reviewed at regular intervals. Figure 3 illustrates the stages in the Aonach Mor scheme.

The monitoring schemes at the Lecht and Aonach Mor actually have much in common, from the initial review of site information through to the monitoring and review stages. The main difference is the range of characteristics being monitored, which to some extent varies according to the site, and the absence of the LAC stage at the Lecht. However, in practice there is agreement at the Lecht about which areas should be rehabilitated, even though they amount to only a few square metres in some cases.

DISCUSSION

Although the principal impacts of recreation have probably been identified, and many have been the subject of scientific description, there remain substantial gaps in our understanding. For instance, there have been a number of studies of plant susceptibility to trampling (Cole & Schreiner 1981), but few are comparable and the range of communities and site conditions that have been examined is quite limited. There is scope for much more extensive comparison using standardized methods.

Certain forms of impact, such as those on soil and on invertebrate populations, involve technical or sampling difficulties, and have been little studied to date. Many people have tended to assume that direct human disturbance and indirect effects on bird species in the uplands are bound to reduce numbers, and some have even published statements to this effect without any data (references in Watson 1979). A reliable assessment of impacts on bird populations usually requires careful, quantitative study over several seasons and objective sampling. This is necessary if we are to fully appraise the impacts of recreation, and not just be concerned with the more obvious changes to landscape and vegetation. In

Year 1 (pre-development)

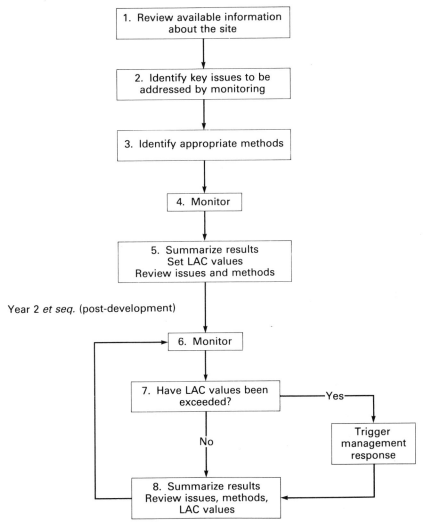

Year 2 *et seq.* (post-development)

FIG. 3. A framework for monitoring at Aonach Mor.

addition, it is important not only to identify and characterize impacts, but to devise practical monitoring techniques that will permit managers to identify where changes are actually occurring.

The reinstatement of damaged ground with appropriate native species poses a special challenge to ecologists. Although there are some promising trials under way, the techniques are very much in their infancy. The problems of reinstating high ground above the heather zone are particularly formidable, and almost nothing is known about re-creating the strongly patterned and often cryptogam-rich communities of the high tops.

We are now seeing the development of some interesting new frameworks which integrate monitoring and management, and which also encourage broad consultation with interested parties. This is a new opportunity, offering considerable scope to apply ecological principles to planning and management. One of the most useful spin-offs of the integrated approach is the greatly increased dialogue between different interests, and the consideration of contrasting views. An effective scheme of monitoring linked to management creates a greater awareness of both problems and solutions by all interested parties as well as creating an effective dialogue. Certainly at Scottish resorts we are experiencing a two-way growth in knowledge, expertise and respect, as ecologists and company staff tackle issues together.

The integrated approach to monitoring and management is not, of course confined to ski resorts. None of these approaches will necessarily prevent site deterioration, but they could result in quicker recognition of problems, and in more rapid implementation of solutions.

REFERENCES

Aitken, R. (1987). Taking up the challenge. *Great Outdoors*, **July**, 20–22.

Allinson, N. (1987). Freedom is the hills. *Great Outdoors*, **July**, 25–28.

Archer, D. (1985). Managing public pressures on Snowdon. *The Ecological Impacts of Outdoor Recreation on Mountain Areas in Europe and North America* (Ed. by N.G. Bayfield & G.C. Barrow), pp. 155–161. Recreation Ecology Research Group Report 9, RERG, Wye.

ASH (1986). *Proposed Ski Development. The Snowgoose: Aonach Mor Development Plan.* A report to Lochaber District Council by Anderson Semens Houston Environmental Design Partnership, Glasgow.

Bayfield, N.G. (1974). Burial of vegetation by erosion debris near chairlifts on Cairngorm. *Biological Conservation*, **6**, 246–251.

Bayfield, N.G. (1980). Replacement of vegetation on disturbed ground near ski lifts in the Cairngorm Mountains, Scotland. *Journal of Biogeography*, **7**, 249-260.

Bayfield, N.G. (1985). The effects of extended use on mountain footpaths in Britain. *The Ecological Impacts of Outdoor Recreation on Mountain Areas in Europe and North America* (Ed. by N.G. Bayfield & G.C. Barrow), pp. 100–111. Recreation Ecology Research Group Report 9, RERG, Wye.

Bayfield, N.G. (1986). Penetration of the Cairngorms Mountains, Scotland, by vehicle tracks and footpaths; impacts and recovery. *Proceedings of the National Wilderness Research Conference: Current Research* (Ed. by R.C. Lucas), pp. 121–128. Forest Service General Technical Report INT-212, USDA.

Bayfield, N.G. & Bathe, G.M. (1982). Experimental closure of footpaths in a woodland National Nature Reserve in Scotland. *Biological Conservation*, **22**, 229–237.

Bayfield, N.G. & McGowan, G.M. (1986). *Footpath Survey 1986. The Three Peaks Project.* ITE Report No. 1 to Yorkshire Dales National Park, YDNP, Grassington.

Bayfield, N.G. & Miller, G.R. (1986). *Reinstatement Trials 1986.* ITE Report No. 2 to Yorkshire Dales National Park, YDNP, Grassington.

Bayfield, N.G., Urquhart, U.H. & Cooper, S.M. (1981). Susceptibility of four species of *Cladonia* to disturbance by trampling in the Cairngorm Mountains, Scotland. *Journal of Applied Ecology*, **18**, 303–310.

Bayfield, N.G., Urquhart, U.H. & Rothery, P. (1984). Colonization of bulldozed tracks in the Cairngorm Mountains, Scotland. *Journal of Applied Ecology*, 21, 343–354.

Cole, D.N. (1985). *Recreational Trampling Effects on Six Habitat Types in Western Montana*. Forest Service Research Paper, INT-350, USDA.

Cole, D.N. & Schreiner, E.G.S. (1981). *Impacts of Back Country Recreation: Site Management and Rehabilitation — an Annotated Bibliography*. Forest Service General Technical Report INT-121, USDA.

Countryside Commission for Scotland (1986). *Environmental Design and Management of Ski Areas in Scotland: a Practical Handbook*. CCS, Perth.

Emmanuelsson, U. (1984). *Ecological effects of grazing and trampling on mountain vegetation in Northern Sweden*. Ph.D. thesis, University of Lund, Sweden.

Fishwick, A. (1985). Planning and management responses to recreational pressures in the Lake District National Park. *The Ecological Impacts of Outdoor Recreation on Mountain Areas in Europe and North America* (Ed. by N.G. Bayfield and G.C. Barrow), pp. 18–33. Recreation Ecology Research Group Report 9, RERG, Wye.

Garmo, T.H. (1986). Mountain pasture improvement by shrub clearing and fertilization. *Grazing Research in Northern Latitudes* (Ed. by O. Gudmundsson), pp. 87–96. Nato ASI Series, Series A: Life Sciences 108, Plenum Press, New York.

Gemmell, R.P. & Crombie, S.A. (1978). Beacon Fell's recovering from the patter of many feet. *Surveyor*, August, 17–18.

Gore, A.J.P. & Godfrey, M. (1981). Reclamation of eroded peat in the Pennies. *Journal of Ecology*, 69, 85–96.

Grabherr, G. (1985). Damage to vegetation by recreation in the Austrian and German Alps. *The Ecologocial Impacts of Outdoor Recreation on Mountain Areas in Europe and North America* (Ed. by N.G. Bayfield & G.C. Barrow), pp. 74–91. Recreation Ecology Research Group Report 9, RERG, Wye.

Greller, A.M. (1974). Vegetation of road cut slopes in the tundra of Rocky Mountain National Park, Colorado. *Biological Conservation*, 6, 84–93.

Hammitt, W.E. & Cole, D.N. (1987). *Wildland Recreation: Ecology and Management*. John Wiley & Sons, New York.

Hernandez, H. (1973). Natural plant recolonization of surficial disturbances, Taktoyaktuk Peninsula Region, Northwest Territories. *Canadian Journal of Botany*, 51, 2177–2196.

Lecht Ski Company (1986). *Lecht Ski Company Development and Management Plan 1986–91*. Lecht Ski Company, Aberdeen.

Miller, G.R. (1986). *Development of Subalpine Scrub at Northern Corries, Cairngorm SSSI*. Institute of Terrestrial Ecology Internal Report, Banchory.

Miller, G.R. & Cummins, R.P. (1982). Regeneration of Scots pine *Pinus sylvestris* at a natural tree-line in the Cairngorm Mountains, Scotland. *Holarctic Ecology*, 5, 27–34.

Milsom, T.P. & Watson, A. (1984). Numbers and spacing of snow buntings and snow cover in the Cairngorms. *Scottish Birds*, 13, 19–23.

Nethersole-Thompson, D. (1966). *The Snow Bunting*. Oliver & Boyd, Edinburgh.

Phillips, J., Yalden, D. & Tallis, J. (Eds) (1981). *Moorland Erosion Study. Phase 1 Report*. Peak Park Joint Planning Board, Bakewell.

Pryor, P. (1985). The effects of disturbance on open *Juncus trifidus* heath in the Cairngorm mountains, Scotland. *The Ecological Impacts of Outdoor Recreation on Mountain Areas in Europe and North America* (Ed. by N.G. Bayfield & G.C. Barrow), pp. 53–62. Recreation Ecology Research Group Report 9, RERG, Wye.

Speakman, C. (1987). Repairing the Three Peaks. *Great Outdoors*, August, 45–46.

Stankey, G.H., Cole, D.N., Lucas, R.C., Peterson, M.E. & Frissell, S.S. (1985). *The Limits of Acceptable Change (LAC) of Wilderness Planning*. Forest Service General Technical Report, INT–176, USDA.

Sydes, C. & Miller, G.R. (1988). Range management and nature conservation in the British uplands. *This volume*.

Tallis, J.H. & Yalden, D.W. (1983). *Peak District Moorland Restoration Project. Phase 2 Report: Re-vegetation Trials*. Peak Park Joint Planning Board, Bakewell.

Thompson, D.B.A., Galbraith, H. & Horsfield, D. (1987). Ecology and resources of Britain's mountain plateaux: land use conflicts and impacts. *Agriculture and Conservation in the Uplands* (Ed. by M. Bell & R.G.H. Bunce), pp. 22–31. Institute of Terrestrial Ecology, Grange-over-Sands.

Watson, A. (1976). Human impact on animal populations in the Cairngorms. *Landscape Research News 1*, 12, 14–15.

Watson, A. (1979). Bird and mammal numbers in relations to human impacts at ski lifts on Scottish hills. *Journal of Applied Ecology*, 16, 753–764.

Watson, A. (1982). Effects of human impact on ptarmigan and red grouse near ski lifts in Scotland. *Annual Report of Institute of Terrestrial Ecology 1981*, 48–50.

Watson, A. (1984a). Paths and people in the Cairngorms. *Scottish Geographical Magazine*, 100, 151–160.

Watson, A. (1984b). A survey of vehicular hill tracks in North-east Scotland for land use planning. *Journal of Environmental Management*, 18, 345–353.

Watson, A. (1985a). *Monitoring at the Lecht 1985.* Report to Gordon District Council, Institute of Terrestrial Ecology, Banchory.

Watson, A. (1985b). Soil erosion and vegetation damage near ski lifts at Cairn Gorm, Scotland. *Biological Conservation*, 33, 363–381.

Watson, A. (in press). Dotterel *Charadrius morinellus* numbers in relation to human impact in Scotland. *Biological Conservation.*

Watson, A., Bayfield, N.G. & Moyes, S.M. (1970). Resarch on human pressures on Scottish mountain tundra, soils and animals. *Productivity and Conservation in Northern Circumpolar Lands* (Ed. by W.A. Fuller & P.G. Kegan), pp. 256–266. International Union of Conservation of Nature, Morges.

The extent of land under different management regimes in the uplands and the potential for change

R. G. H. BUNCE AND C. J. BARR

Institute of Terrestrial Ecology, Merlewood Research Station, Grange-over-Sands, Cumbria LA11 6JU

SUMMARY

1 The determination of the extent and composition of upland Britain has proved difficult because of the problems of defining upland character and the requirement for national coverage. A two-stage approach is described which uses broad environmental classes to define the areas involved, and then ecological criteria to determine the local character and extent of upland vegetation.
2 Using this approach, the area of the uplands is estimated at 7.7×10^6 ha (39 per cent of Britain as a whole), comprising 1.4×10^6 ha of upland grassland, 1.7×10^6 ha of moorland and 1.5×10^6 ha of bog, the residual being mainly agricultural grassland and conifer forest.
3 An alternative breakdown is 3.8×10^6 ha for land mainly in agricultural use, 1.3×10^6 ha under mainly deer forest and 0.5×10^6 ha for grouse moor, the residual being largely conifer forest.
4 Both approaches show agriculture to be the most extensive land use, so changes in the uplands are most likely to arise from changes in agricultural policy, with forest expansion secondary.
5 Changes in management may well be buffered by institutional constraints and the individual views of landowners.

INTRODUCTION

Ecologists have traditionally concentrated upon fundamental research on ecosystems and their constituent species, as the majority of papers in this symposium demonstrates. The studies are influenced by the land use of the areas concerned, but this is usually incidental to the main study objectives. However, when it comes to determining the wider application of detailed studies, the extent of the resource to which they refer becomes a critical factor. It is the prime purpose of this paper to present basic statistics for upland land use, vegetation composition and management.

There are two main reasons for the lack of an adequate estimate of the extent of the uplands — firstly, the problem of determining exactly what is implied in the term 'upland' and, secondly, the difficulty of obtaining adequate consistent national data over the diverse landscapes concerned. As with many widely used ecological terms such as 'bog' or 'marsh', there is no generally accepted definition, although most ecologists have a general view of the likely conditions involved. Useful summaries of the situation are given by MacEwan & Sinclair (1983) and Sinclair (1983). Bunce (1987) has summarized views from a variety of sources, including the environmental methods mentioned below. Most ecologists use the species present on a given area of land in conjunction with environmental characteristics in order to determine upland character. Thus the presence of species such as *Potentilla erecta*, *Eriophorum vaginatum* and *Nardus stricta* on the one hand, in conjunction with more exacting species such as *Carex bigelowii* and *Empetrum nigrum* on the other, would indicate upland conditions. However, many of the former are found in the lowland heaths of the south, whereas the latter are found on the shore in Shetland. This confusion is due to the wide latitudinal range in Britain. Alternatively, agriculturally improved grasslands, containing species such as *Lolium perenne* or *Trifolium repens*, would generally be excluded even if these were in upland situations. The term 'uplands' does therefore have an element of circularity involved because of the frequent use of species to infer an environmental criterion.

The alternative approach is to use environmental parameters directly, usually in various combinations of altitude, climate and soil type. These approaches have been used to provide various regional estimates, for example of England and Wales or Scotland, but have not been applied consistently over Great Britain. The best figure currently available, derived from the Less Favoured Areas (LFA) directive of the European Economic Community in 1983, is due to the indirect correlation between the definition used (that of declining population) and upland vegetation. The map of LFAs corresponds closely with the generally accepted distribution of upland areas, as shown by Ball, Radford & Williams (1983).

METHODS

The problem of obtaining national estimates of rural parameters is not confined to the uplands, and the approach of Bunce & Heal (1984), using the Merlewood Land Classification System to obtain an integrated picture of land-use composition and environmental character, is advocated. The methodology and its objectives have been widely reported elsewhere, e.g. Bunce *et al.* (1984) and Bunce & Heal (1984), so only a brief summary is presented here. Multivariate analysis of environmental data derived from a 1 km^2 sampling framework (e.g. of climate and topography as recorded from maps) is used to produce strata (termed land classes) which are then sampled in the field for ecological parameters.

The classification of GB used to produce the classes was derived from 282 environmental attributes recorded from 1228 1 km^2 situated at the intersections of a 15 × 15 km grid. The classes were produced by Indicator Species Analysis (Hill, Bunce & Shaw 1975), which at the time was the best method available (see below) for classifying large sets of data which would produce an indicator key that could be used to identify subsequent squares. The analysis defined 32 land classes and identified 76 indicator attributes which have been used to classify a further 4800 sampling squares. The field data were collected in 1977/78, based on 8 random 1 km² drawn from the 32 land classes. In some regions, e.g. the Highland region in Scotland, the sampling squares were extended to cover the whole region.

Subsequently, Indicator Species Analysis was generally replaced by TWINSPAN (Hill 1979); although the classification algorithm is basically the same, it is held within a wider data-manipulation package. This method uses a divisive polythetic strategy, identifying indicators to define relatively homogeneous groups of sample squares (or quadrats if the data are in the form of species) at each progressive stage of the division. An alternative strategy would have been to use an agglomerative method, but in 1976 a suitable program that could analyse such a large data set was not available. Moreover, experience at that time with the packages available suggested that divisive methods were more appropriate for dealing with such continuous environmental data. The weighted pair-group method using arithmetic averages (WPGMA) and unweighted pair-group method using arithmetic averages (UPGMA) would also be appropriate. These methods have minimum variance properties and are space-conserving, i.e. they do not have much tendency to impose a structure on the data. Another hierarchical method which is already in use for vegetation data is Ward's error sum of square method, which also has minimum variance properties. An alternative strategy is to use a non-hierarchical iterative relocation procedure, which has been applied to comparable environmental data by Moss (1985).

An optimum strategy would be to carry out a series of classifications using the above procedures, together with various combinations of environmental data structures. An independent set of field data could then be used to assess the efficiency of the resulting classifications by correlation with the groupings produced, and the structure that explains the most variation could then be used in subsequent work. However, the majority of applied studies do not have the time or resources to carry out such a technically satisfying procedure and, as in the present case, are likely only to be able to carry out a single classification. The method of classification is objective and, although improved techniques have since been developed, this is reflected in higher standard errors for the survey parameters rather than incorrect results.

Sufficiently high correlations have been shown between the 32 land classes and the field data to enable the small field sample to be used as a representative series for deriving national statistics. A variety of tests have been carried out on the predictions, e.g. statistical estimates of standard errors, comparisons with other official figures and orthogonal regression (Bunce & Heal 1984). These studies have demonstrated its validity as a national sampling framework, leading to its adoption in a variety of applied studies, e.g. Mitchell *et al.* (1983).

In view of the intuitive nature of the system used by ecologists to define areas as upland, the land classes have been used as an objective method, first defining the types that are upland in environmental terms and then separating the land where the vegetation is upland in ecological character. The areas of the major management regimes of the uplands have then been estimated using the same framework.

THE DEFINITION OF UPLAND BRITAIN AND ITS VEGETATION COMPOSITION

The figures presented in Table 1 are derived from:

(i) Combining the areas of the land classes which have high rainfall, low insolation, low evapotranspiration and generally poor soils. Full numerical definitions are provided by Bunce, Barr & Whittaker (1981) and summaries of the principal characteristics of the 13 classes (out of the 32 described for the whole of Britain) are given in Table 2. The figure is the same as that given by Bunce (in press a).

(ii) Examining the composition of the vegetation; upland vegetation was considered as being comprised of species such as *Agrostis tenuis*, *Nardus stricta*, *Molinia caerulea* and *Eriophorum vaginatum* as distinct from the predominantly agriculturally managed area (either grass or crops). A total of 79 land cover types recorded in 1978 were used to derive this summary and Bunce *et al.* (1981) provide a full breakdown of these at a national level.

Using the definition derived by summing land classes, the area of land, including forest and agricultural uses, is 7.7×10^6 ha (Table 1). This compares with a figure of 8.7×10^6 ha for the LFAs. The difference can be explained in part by urban areas and areas of inland water which are excluded from the land-class definition. However, within the broad definition of approach provided by the land classes, only 4.6×10^6 ha (23 per cent) of Great Britain consists of upland vegetation as determined by the measurement of land-cover types in the sample

TABLE 1. A comparison of two methods of estimating the major components of upland vegetation in Britain. (i) Derived from areal measurements of land-cover types in 256 sample squares (8 from each of the 32 classes). (ii) Derived from the proportion of 1280 200 m² quadrats placed at random within the 32 classes, classified according to TWINSPAN and summarized at a similar level (proportion only — areal measurements not applicable). The top two indicator species are given in each case from the hierarchy

Total area of Britain (excluding water and urban) = 19.8×10^6 ha

Total area of uplands (including forest and agricultural grass) = 7.7×10^6 ha (39%)

Area of upland vegetation (bog, moorland and grassland) = 4.6×10^6 ha (23%)

	Bog	Moorland	Grassland
(i)	1.5×10^6 ha (32%)	1.7×10^6 ha (37%)	1.4×10^6 ha (31%)
(ii)	*(Drosera rotundifolia, Eriophorum angustifolium)* (34%)	*(Calluna vulgaris, Potentilla erecta)* (37%)	*(Agrostis tenuis, Galium saxatile)* (28%)

TABLE 2. Principal characteristics of the 13 upland land classes as summarized by Bunce *et al.* (1981), derived from full numeric data, together with descriptions of the main management regimes associated with the land in each class as defined in the present paper

Characteristics

17	Rounded intermediate slopes, mainly improvable permanent pasture
18	Rounded hills, some steep slopes, varied moorlands
19	Smooth hills, mainly heather moors, often afforested
20	Mid-valley slopes, wide range of vegetation types
21	Upper valley slopes, mainly covered with bogs
22	Margins of high mountains, often afforested
23	High mountain summits with well-drained moorlands
24	Upper, steep mountain slopes, usually bog covered
28	Varied lowland margins with heterogeneous land use
29	Sheltered coasts with varied land use, often crofting
30	Open coasts with low hills dominated by bogs
31	Cold exposed coasts with variable land use and crofting
32	Bleak, undulating surfaces, mainly covered with bogs

Management

17	Mainly inbye with limited outbye but entirely agricultural
18	Mainly outbye with limited inbye but entirely agricultural
19	Mainly inbye but with considerable outbye and both grouse and deer moors
20	Virtually equal proportions of inbye and outbye but entirely agricultural
21	Predominantly deer forest with some inbye and limited grouse moors
22	Outbye is marginally dominant, otherwise inbye, grouse and deer are almost equal
23	Outbye predominates with grouse also important but with some deer forest also
24	Outbye and deer forest almost equally share the land in this class
28	Inbye occupies most of the land, with outbye secondary and limited deer forest
29	Deer forest covers the largest area, with inbye and outbye covering relatively small areas
30	Outbye predominates, with deer forest secondary and limited areas of grouse moor
31	Inbye and outbye occupy the land in almost equal proportions
32	Outbye is the major category with slightly lower proportions of inbye and deer forest

areas. This reduction is due to the presence of better soils and conditions in valleys and coastal plains in the north and also to agricultural improvement where upland vegetation has been converted to productive lowland species by cultivation and fertilization. Thus, in Land Class 17, in the Welsh uplands, improvement has often extended over 300 m, and in Shetland, by contrast, small, intensively managed crofts can be surrounded by otherwise upland vegetation.

The land cover can be broken down into 79 categories as described by Bunce *et al.* (1984), but only 21 of these are upland in nature. They may conveniently be split into three broad groups to reduce the variation further.

(i) Bog, consisting mainly of cover species such as *Eriophorum angustifolium, Trichophorum cespitosum* and *Molinia caerulea* and a moss carpet principally of *Sphagnum* species.
(ii) Moorland, consisting mainly of cover species such as *Calluna vulgaris, Vaccinium myrtillus* and *Nardus stricta.*
(iii) Upland grassland, with species such as *Agrostis tenuis* and *Festuca ovina,* and *Pteridium aquilinum* as an invasive species.

Again it is recognized that there is a problem with definition, and therefore in Table 1 the initial figures are compared with the data from 200 m² quadrats, 5 of which were placed at random in each of the sample 1km². The species data from these quadrats were classified into 75 types using TWINSPAN (Hill 1979) and grouped into the comparable three categories given in Table 1. The figures are consistent in the two estimates, suggesting that they provide a reasonable division of the major categories within semi-natural vegetation.

The cover of individual species rather than assemblages is also an important characteristic. The species cover was recorded in the quadrats and the estimates of cover are given in Table 3. *C. vulgaris* dominates upland vegetation to a major degree, covering 25 per cent of the uplands, with only two other species, *A. tenuis* and *M. caerulea*, forming 10 per cent or more. The residual area consists of many species, often of local importance.

These figures provide a quantitative estimate of the extent of the uplands and the areas within the broad divisions. The procedures described by Mitchell *et al.* (1983) and Bunce (1988) enable these to be compared with other potential uses in order to examine interactions between them.

DEFINITION OF MANAGEMENT REGIMES

The method of recording was based on the experience derived from a detailed survey of management in a representative series. As reported by Barr *et al.* (1986), considerably more detail was recorded on the land-use pattern on that occasion and this will be used in due course to redefine the figures further. More accurate figures could be achieved by assessing individual owners' objectives. These estimates do, however, give a breakdown of the likely proportion of categories present and identify the principal management regimes which are likely to determine ecological change.

It is useful to separate the farmed area into inbye and outbye so as to assess the balance between relatively intensively managed land and free-range grazed areas.

TABLE 3. Major species covering land in the uplands, determined from cover in 1280 200 m² quadrats (40 in each of 32 classes)

	Area (million ha)	Percentage of Great Britain	Percentage of uplands
Calluna vulgaris	1·4	6	25
Agrostis tenuis	0·8	3	14
Molinia caerulea	0·6	3	10
Nardus stricta	0·3	1	6
Pteridium aquilinum	0·3	1	6
Trichophorum caespitosum	0·2	1	3
Festuca ovina	0·2	1	3
Deschampsia flexuosa	0·2	1	3
Anthoxanthum odoratum	0·2	1	3
Vaccinium myrtillus	0·1	1	3

In general this boundary corresponds with the distinction between leys, permanent and improved grassland on the one hand as opposed to other upland categories. There are exceptions to this general rule, e.g. where *Pteridium* or *Juncus* have invaded the inbye areas.

The areas of woodland, mainly coniferous, have been omitted, as have other categories such as water and shorelines. In crofting areas it is more difficult to define where the boundary between the various categories lies. Likewise the overlap of many of the categories with institutional constraints, such as nature reserves as described by Mitchell *et al.* (1983), can elucidate management pattern further. Finally, the dominant use of the land was determined, rather than combined uses.

Summary descriptions of the land classes and the main management regimes associated with them are provided in Table 2. The distribution of the classes and full numerical values are given by Bunce *et al.* (1981). Distinct differences are present between the types of land involved from the predominantly agricultural classes 17, 18, 20 and 28, to the deer forests of 23 and 24 and the grouse moors of 19 and 22. Table 4 presents the breakdown between the major regimes, showing that agriculture is by some considerable margin the main use, which suggests that changes in the Common Agricultural Policy (CAP) and other support are likely to be as critical as ever in the future of ecosystems in the hills. Deer occupy a subsidiary, but still substantial, role and their management is important, particularly in the more remote areas. However, much of this land is under less direct management in contrast to the inbye land. Change caused by, for example, a decline in the price of venison would therefore have a more limited effect. Although grouse moor occupies a lesser proportion, the area concerned, almost 0.5×10^6 ha, is still highly significant in ecological terms because of the value of the heather moors to many species. It would be useful to derive independent figures for comparison from the Red Deer Commission and from other sources for grouse moors, but again definition and coverage are likely to provide problems.

TABLE 4. Proportion of land under different management regimes in uplands of Britain. Definitions given below. Figures are in thousands of hectares

Inbye	Outbye	Grouse	Deer	Other	Total
2141	2698	489	1296	1074	7698
28%	35%	6%	17%	14%	100%

Inbye   = land within the mountain wall that is cultivated by ploughing, reseeding or fertilizing

Outbye = land predominantly in agricultural use but which is not directly managed, i.e. mainly open-range sheep walks

Grouse = land often holding some sheep but which is also managed for grouse

Deer   = land often holding some sheep but which is principally used as deer forest

Other   = land mainly coniferous plantation, but also including other categories such as open water and urban

Although the above analysis does not allow a full comparison between the categories quoted in Table 1, some observations are useful. The total of 4.6×10^6 ha for semi-natural vegetation approximates to the compilation of outbye, deer forest and grouse moor, which gives a figure of 4.5×10^6 ha, suggesting that the figures derived from different definitions partition the total in a similar way. Concerning accuracy, the above estimates are all based on actual data with defined categories giving likely standard errors in the region of ±5–20 per cent, based on previous experience.

THE POTENTIAL FOR CHANGE

It is first necessary to summarize the principal changes in the uplands over recent years. There are two main types of change. First, there are the gradual ecological changes, e.g. in soil chemistry (Miles 1988) and, secondly, there are abrupt changes that are usually caused directly by man's activities, e.g. conversion of land to forestry or agricultural improvement. The former are widely discussed in the symposium, whereas the latter are mainly addressed in the present paper. Barr *et al.* (1986) have reported two main changes in Britain between 1978 and 1984: (i) the loss of upland vegetation through agricultural improvement, usually to grassland, and (ii) the loss of upland vegetation to coniferous forestry. Such results are similar to those reported by Huntings Surveys (1986) for England and Wales and by Nature Conservancy Council (1986) for Cumbria. No comparable data are available for Scotland.

Despite the current difficulties in agriculture, Barr *et al.* (1986) have shown that the trend towards improvement is still continuing — as confirmed by anecdotal evidence in Wales and Scotland. This seems to be in part due to the support system described by Mowle & Bell (1988), but also to the requirement of individual farmers to expand their income to compensate for falling prices. Many of the land-use changes which have been reported for Britain, e.g. the expansion of wheat and oil seed rape (Barr *et al.* 1986), do not apply to the uplands. Whilst many changes are important locally, e.g. the use of Paraquat or the construction of flighting ponds, the above data indicate that relatively few abrupt changes are occurring in the uplands. This is a reflection of the level of economic activity, since the majority of changes are directly due to farmers extending their enterprise. There are, however, indications of indirect changes in management by the extension of wire fences and upland tracks (Barr *et al.* 1986). The former suggest that some intensificiation of use of upland vegetation is taking place through closer control of stock. The increased access through new tracks also implies more access by people using the uplands, which has a variety of indirect effects. If the pattern of recent years continues, those influences are likely to increase. The increase in sheep numbers also suggests that changes may not be adequately reflected in the general land-use categories discussed above. However, there is no guarantee that these sheep are all using the open hill land, since many may be in more intensive management systems in the inbye land closer to lowland sheep regimes. This would be supported by the continued decline in the number of up-land farms. In all these cases further specific studies are required to determine the actual pattern.

A variety of figures have been produced for land which might become surplus to agriculture, varying from 0.7×10^6 ha to 4×10^6 ha by about the year 2000. It has often been assumed that this land could be in the uplands, although Harvey *et al.* (1986) have suggested that this is not necessarily the case and that the predicted pattern depends entirely upon the assumptions used in any model. An alternative that was suggested recently is that, if land were to decline in value in the lowlands, sheep would then become viable, having an indirect effect upon the uplands, because of the latter's dependence upon sheep farming. However, whichever scenario is taken, there seems little doubt that extensification of the present agricultural system in the semi-natural vegetation of the uplands is likely to take place, although the inbye could well see the opposite trend. The figure of the large proportion of agricultural land in the uplands therefore emphasizes the sensitivity of the system to these potentially destabilizing influences.

However, there is an important buffering system: that of the various designated areas, which are designed progressively to modify potential changes, for example the Wildlife and Countryside Act and the recently designated Environmentally Sensitive Areas (Smith 1988). Otherwise there are National Parks, Natural Scenic Areas in Scotland and Areas of Outstanding Natural Beauty, in addition to nature reserves and Nature Conservancy Council sites. Mitchell *et al.* (1983) have shown that some 25 per cent of land potentially available for energy plantations, largely in the uplands, is affected by some degree of statutory constraint. In addition, the attitude of individual landowners, e.g. the National Trust, is often resistant to change. The land held primarily for game pur-poses is also likely to be relatively resistant to the generally perceived financial pressure, as the objectives of management are not primarily economic. These factors suggest that the uplands are likely to be more resistant to change than often expected.

Forestry is, however, different in that extensive documentation is available from the Forestry Commission. Over the last eight years planting has proceeded on average at 2.2×10^4 ha year^{-1}, the majority of which is in the uplands and specifically in Scotland. Assuming that current trends continue to the end of the century, almost 0.3×10^6 ha will have been afforested, which, although small in relation to the 4.6×10^6 ha of semi-natural vegetation, still represents a major rapid change. Mitchell *et al.* (1983) showed that these figures are more susceptible to financial assumptions than to the type of forest system, and this is confirmed by the current extension of planting in north-east Scotland. The Merlewood Land Classification System has been used (Bunce, in press b) to project the potential impact of this afforestation upon upland vegetation. All the major categories, assuming complete afforestation of suitable land after competition with agricul-ture, were affected by between 45 per cent and 60 per cent with the exception of the very wet bogs, which were largely unplantable. Although this suggests a similar impact upon the major divisions of upland vegetation, it is an over-simplification because certain altitude bands are likely to be affected more than others. For example, the high areas of *Calluna* moorland are unplantable and only the lower sections will be taken. Likewise, the variations in extent of the diverse categories will leave very different residual areas. In addition, there are the changes of the balance between ecosystems in an area. Overall, therefore, the effects are likely to be more subtle than the crude figures show, and a full impact study, incorporating the wide array of work currently in progress, needs to be carried out.

The British uplands differ widely from their continental counterparts in their potential for change, because of their difference in both environmental character-istics and socio-economic regime. In the latter aspect the views of various states differ widely in their interpretation of the term 'environment'. Whilst the Scandinavian countries are generally concerned with conserving the natural environment, in both landscape and nature conservation terms, as shown by Gjaerevoll (1988), the French are more directly concerned with the maintenance of a viable rural population. A common factor of all upland areas is, however, the shift of population away from the higher zones towards more equable conditions. In some areas, such as the Spanish Pyrenees, this has led to complete depopulation, but elsewhere the changes are occurring more gradually. This declining human use of upland areas has had major ecological implications but, because of the difficulty of long-term monitoring over large areas, the effects have been inadequately studied.

Whilst the uplands have many designated areas and the protection of the farming population has long been government policy, there seems little doubt that their future is particularly susceptible to sudden political change. The introduction of milk quotas has had many indirect environmental effects that were not foreseen at the time of their introduction. The current pressure on the CAP budget could lead to comparable sudden changes in policy, although, as the agricultural products from the uplands are not currently in surplus, the effects are likely to be indirect. The emphasis on the importance of management in this symposium demonstrates the impact that any such changes are likely to have. The assessment of potential change is therefore inextricably linked to socio-economic factors. Moreover, as measures taken to control production in the uplands will not affect surpluses, the encouragement of afforestation is a separate issue.

The European mountain areas differ markedly in environmental and veg-etation composition and, in the absence of an adequate objective classification, certain general points only are valid. Firstly, in southern upland areas of Spain and France, such as the Pyrenees, the Picos Europa and the Gredos mountains, the tree level is very high with only a limited nival zone. The soils are well able to be readily colonized by trees so that the abundant seed supply means that a de-cline in grazing is followed by rapid scrub development and eventually forest cover. A comparable situation exists in the majority of areas in the Alps, where the tree cover, in comparison with Britain, is still high. In less marked upland areas, such as the Vosges, Ardennes and Massif Central, the tree cover has been highly modified but scrub development is still rapid. By contrast the higher mountain areas of Scandinavia are largely unmanaged for agriculture, so that only the intermediate pastures will be affected by decline. Also, the slower rate of growth and colonization means that the whole process is more gradual. The deforested nature of British uplands, combined with the development of peat since the ice ages, means that the potential for the recovery of forest cover is lower than in the areas mentioned above. The potential for change, as in the western mountains of Ireland, is therefore less.

The recreational use of the uplands is also different. The paper by Bayfield, Watson & Miller (1988) emphasizes the problems of footpath erosion and ski development in Britain. However, the area involved is small in comparison with comparable developments both in the Alps and Pyrenees, where whole valleys and upland areas have been completely modified for recreational purposes. These have led to far greater environmental impacts, epitomized by the disaster in 1987 in one of the Italian valleys of the southern Alps. The more extreme climatic conditions emphasize these impacts but the extent of the developments and the lack of control are major factors separating these areas from Britain.

The extent to which management, especially farming, is important to the conservation, in the widest generally accepted sense, of semi-natural vegetation is debatable and largely a matter of perception. Certainly conservation of wildlife needs to be separated from conservation of amenity (Harvey *et al.* 1986) in order to determine the actual impact of change. Although value judgements will inevitably play an important part in the debate, a variety of basic ecological problems remain unanswered in relation to simple changes in management. The current lack of information enables general emotive statements such as the decline of the uplands to go unchallenged ecologically. The present symposium presents the current state of the art and indicates gaps that exist, but many of the studies need to be integrated so that their relative importance to management can be assessed and so that clear statements of ecological implications can be made.

REFERENCES

Ball, D.F., Radford, G.L. & Williams, W.M. (1983). *A Land Characteristic Data Bank for Great Britain.* Occasional Paper 13. Institute of Terrestrial Ecology, Bangor.

Barr, C.J., Benefield, C.B., Bunce, R.G.H., Ridsdale, H.A. & Whittaker, M. (1986). *Landscape Changes in Britain.* Institute of Terrestrial Ecology, Abbots Ripton.

Bayfield, N.G., Watson, A. & Miller, G.R. (1988). Assessing and managing the effects of recreational use on British hills. *This volume.*

Bunce, R.G.H. (1987). The extent and composition of upland areas in Great Britain. *Agriculture and Conservation in the Hills and Uplands* (Ed. by M. Bell & R.G.H. Bunce), pp. 19–21. Institute of Terrestrial Ecology, Grange-over-Sands.

Bunce, R.G.H. (1988). The impact of afforestation on semi-natural vegetation in Britain. *Wildlife Management in Forests* (Ed. by D. Jardine), pp. 54–59. Institute of Chartered Foresters, Edinburgh.

Bunce, R.G.H. & Heal, O.W. (1984). Landscape evaluation and the impact of changing land-use on the rural environment: the problem and an approach. *Planning and Ecology* (Ed. by R.D. Roberts & T.M. Roberts), pp. 164–188. Chapman & Hall, London.

Bunce, R.G.H., Barr, C.J. & Whittaker, H.A. (1981). *Preliminary Description of the Land Classes for Users of the Merlewood Land Classification System.* Merlewood R. & D. Paper 86. Institute of Terrestrial Ecology, Grange-over-Sands.

Bunce, R.G.H., Tranter, R.B., Thompson, A.M.M., Mitchell, C.P. & Barr, C.J. (1984). Models for predicting changes in rural land use in Great Britain. *Agriculture and the Environment* (Ed. by D. Jenkins), pp. 37–44. Institute of Terrestrial Ecology, Cambridge.

Gjaerevoll, O. (1988). Nature conservation in Norway. *This volume.*

Harvey, D.R., Barr, C.J., Bell, M., Bunce, R.G.H., Edwards, D., Errington, A.J., Jollans, J.L., McClintock, J.H., Thompson, A.M.M. & Tranter, R.B. (1986). *Countryside Implications for England and Wales of Possible Changes in the CAP.* Centre for Agricultural Strategy, Reading.

Hill, M.O. (1979). TWINSPAN — *a FORTRAN Program for Arranging Multivariate Data in an Ordered Two-way Table by Classification of the Individuals and Attributes.* Report of the Section of Ecology and Systematics, Cornell University.

Hill, M.O., Bunce, R.G.H. & Shaw, M.W. (1975). Indicator species analysis — a divisive polythetic method of classification and its application to a survey of native pinewoods in Scotland. *Journal of Ecology,* **63**, 597–613.

Huntings Surveys (1986). *Monitoring Landscape Change. Volume 1, Main Report.* Huntings Surveys, Borehamwood.

MacEwan, M. & Sinclair, G. (1983). *New Life for the Hills.* Council for National Parks, London.

Miles, J. (1988). Vegetation and soil change in the uplands. *This volume.*

Mitchell, C.P., Brandon, O.H., Bunce, R.G.H., Barr, C.J., Tranter, R.B., Downing, P., Pearce, M.L. & Whittaker, H.A. (1983). Land availability for production of wood energy in Britain. *Energy from Biomass* (Ed. by A. Strub, P. Chartier & G. Schleser), pp. 159–163. Applied Science, London.

Moss, D. (1985). An initial classification of 10 km squares in Great Britain from a land characteristic data bank. *Applied Geography,* **5**, 131–150.

Mowle, A. & Bell, M. (1988). Rural policy factors in land-use change. *This volume.*

Nature Conservancy Council (1986). *National Countryside Monitoring Scheme: Project Report for Cumbria.* Nature Conservancy Council, Peterborough.

Sinclair, G. (1983). *The Upland Landscape Study.* Environment Information Services, Dyfed.

Smith, R.S. (1988). Farming and the conservation of traditional meadowland in the Pennine Dales Environmentally Sensitive Area. *This volume.*

Index

Figures in *italic* indicate tables and/or figures

427